Birkhäuser

Trends in the History of Science

Trends in the History of Science is a series devoted to the publication of volumes arising from workshops and conferences in all areas of current research in the history of science, primarily with a focus on the history of mathematics, physics, and their applications. Its aim is to make current developments available to the community as rapidly as possible without compromising quality, and to archive those developments for reference purposes. Proposals for volumes can be submitted using the online book project submission form at our website www.birkhauser-science.com.

More information about this series at http://www.springer.com/series/11668

Maria Teresa Borgato •
Erwin Neuenschwander • Irène Passeron
Editors

Mathematical Correspondences and Critical Editions

Birkhäuser

Editors
Maria Teresa Borgato
Department of Mathematics and Computer Science
University of Ferrara
Ferrara, Italy

Erwin Neuenschwander
Institute of Mathematics
University of Zurich
Zurich, Switzerland

Irène Passeron
Institut de Mathématiques de Jussieu-Paris Rive Gauche
CNRS, Sorbonne Université
Paris, France

ISSN 2297-2951 ISSN 2297-296X (electronic)
Trends in the History of Science
ISBN 978-3-319-73575-7 ISBN 978-3-319-73577-1 (eBook)
https://doi.org/10.1007/978-3-319-73577-1

Library of Congress Control Number: 2018955311

This book is published under the imprint Birkhäuser, www.birkhauser-science.com by the registered company Springer Nature Switzerland AG part of Springer Nature.
The registered company address is: Gewerbestrasse 11, 6330 Cham, Switzerland

Preface

Collected Mathematical Works and Correspondences: A Short History and an Introductory Summary

The publication of collected mathematical works and correspondences has a long and distinguished tradition. Apart from the editions of classical Greek and Roman works in the Renaissance, it started in the sixteenth and seventeenth centuries with the publication of, among others, the works of François Viète (*Opera mathematica*, London 1589, or Leiden 1646), Niccolò Tartaglia (*Opere*, Venice 1592/93 and 1606), Christoph Clavius (*Opera mathematica*, 5 vols, 1612), Simon Stevin (*Œuvres mathématiques*, 6 vols, Leiden 1634), and Galileo Galilei (*Opere*, 2 vols, Bologna 1655/56). Around the same time, the first editions of mathematical correspondences (*Commercium epistolicum* of John Wallis, 1658; and John Collins, 1712, etc.) appeared, and in the eighteenth century, there followed some twenty further editions of collected mathematical papers (Jacob and Johann Bernoulli, Pierre de Fermat, Bernard Lamy, Pierre Louis Moreau de Maupertuis, Jacopo Riccati, Gilles Personne de Roberval, Robert Simson, etc.).

The nineteenth and twentieth centuries saw a highpoint in the publication of multivolume and often nationalistically driven collected mathematical works. The most voluminous of these were the editions of the works of Leonhard Euler (ca. 100 volumes, 1911–), Gottfried Wilhelm Leibniz (ca. 50 volumes, 1923–), Augustin-Louis Cauchy (27 vols, 1882–1974), Johannes Kepler (26 vols, 1937–2017), Christiaan Huygens (22 vols, 1888–1950), Galileo Galilei (20 vols, 1890–1909), Paul Tannery (17 vols, 1912–50), Joseph-Louis Lagrange (14 vols, 1867–92), Pierre-Simon de Laplace (14 vols, 1878–1912), Arthur Cayley (13 vols, 1889–97), René Descartes (12 vols, 1897–1910), Nicolas de Condorcet (12 vols, 1847–49), Carl Friedrich Gauss (12 vols, 1863–1933), etc. The best of these editions include, besides the collected works, also unpublished papers, letters, commentaries, translations, biographies, bibliographies, etc., that are often unavailable anywhere else.

These developments led in the twenty-first century, the age of "Digital Humanities," to large-scale projects such as the D'Alembert edition with its new interface *D'Alembert en toutes lettres*, or platforms like *Circulation of Knowledge and Learned Practices in the 17th-century Dutch Republic*, which currently contains about 20,000 letters that were written by and sent to various scholars who were active in the Netherlands in the seventeenth century. Similar projects are *Early Modern Letters Online,* or even the Stanford project *Mapping the Republic of Letters,* which includes also other media besides printed works and correspondences. For a compilation and description of the approximately one thousand printed editions of collected mathematical works and correspondences published to date, we refer the reader to the "Collected works in Mathematics and Statistics Collection" of the Stanford University Libraries or to the comprehensive bibliography of Steven W. Rockey. The latter appeared in a print version in 1991 and is now available online in an updated version on the website of the Mathematics Library of Cornell University.

The present volume originates from a symposium held at the sixth ESHS Conference in Lisbon in 2014. It presents 16 mostly ongoing projects on editions of collected works and correspondences. The first group of papers deals with the edition of large-scale collected works and correspondences in the past, present, and future. Luigi Pepe gives a short summary of the publications in this field and describes in detail the history and scope of Lagrange's collected works. Eberhard Knobloch deals with the history of Series VII and VIII of the Leibniz edition, whereas Karin Reich and Elena Roussanova offer a critical survey and inventory of the edited works of Carl Friedrich Gauss. Irène Passeron and Alexandre Guilbaud describe the mathematical correspondence of D'Alembert and its digital edition, Sulamith Gehr presents the online edition of the Bernoulli letters, and Philip Beeley contributes a survey of the history of the nine-volume print edition of the correspondence of John Wallis.

A further group of papers considers the renewal of mathematical research in Italy at the time of the Risorgimento, the Italian unification around 1870. Maria Teresa Borgato and Iolanda Nagliati provide an overview of the creation and consolidation of a network of personal relations among Italian mathematicians and leading European scholars in this period. Other papers of this group contain detailed descriptions and evaluations of the correspondences of Giusto Bellavitis, Enrico Betti, Francesco Brioschi, Luigi Cremona, Placido Tardy, etc. The third and last group of papers presents a variety of other projects on European mathematical correspondences from different centuries. Nicolas Rieucau discusses the scientific correspondence of Condorcet, Erwin Neuenschwander describes the major correspondences of B.L. van der Waerden, and Catherine Goldstein and Scott A. Walter present different aspects of the correspondences of Hermite-Lipschitz and Hilbert-Poincaré. For further information about the sixteen contributions, we refer the reader to the following introduction and the abstracts of the authors.

The present volume of course cannot give a comprehensive overview of the vast topic of the publication history of collected works and correspondences over the last five centuries. Nevertheless, we hope that it will prove useful to future editors in accomplishing their task and that it will promote further historical work in important fields of study such as knowledge transfer and communication networks, where scientific, societal, and economic interests all come into play.

Zurich, Switzerland Erwin Neuenschwander

Introduction

Correspondences and editions of collected works: problems, situations, perspectives

Letter writing has always been very important for the spreading of scientific ideas, even in times of a great number of specialized journals.

The correspondences on mathematical issues or those of interest in the history of mathematics involve a vast field of topics, not only those of a scientific nature. They include letters between mathematicians and from mathematicians to politicians, publishers, and men and women of culture. Leibniz, Euler, D'Alembert, Lambert, Lagrange, Laplace, Gauss, Hermite, and Cremona are undoubtedly authors of great interest and their letters are precious documents, but the correspondence of less well-known authors can also make an important contribution to the history of science.

All of these kinds of correspondence constitute an essential component in the reconstruction of biographies, as well as the genesis of scientific ideas, in analyzing relations and debates and, ultimately, in the correct dating and interpretation of various memoirs. Their publication is, therefore, important for the success of critical editions of the works of great mathematicians (Galileo, Newton, Wallis, Huygens, Euler, the Bernoulli family, etc.).

In dealing with our subject, one must also take into account the varying editorial standards and formats for editions carried out in the past, especially in the nineteenth century, the most prolific period for collected works (Galileo, D'Alembert, Lagrange, Laplace, Huygens, Cauchy, Fourier, Weber, Gauss, Riemann, Kronecker, Dirichlet, etc.). They vary greatly in their presentation and structure; generally, they contain only printed works. At times, they are ordered chronologically, or according to discipline or type of publication. Only rarely are the correspondences, whether complete or partial, included in the edition.

Variety in editorial criteria is also to be found in twentieth-century editorial projects, some of which are still ongoing (Galileo *Edizione Nazionale*, Leibniz, Bernoulli, Brioschi, Betti, D'Alembert, . . .) and are gradually being supported by digitalization processes. In fact, the digital editions make mathematical works of the past increasingly available to a wider public and facilitate the research process of scholars by allowing them to easily access and browse rare texts. This poses new problems in addition to those of the traditional printed editions, particularly in the choice of the target audience and corresponding suitable technical tools.

The editors of the present volume invited scholars to reflect on these topics in a symposium within the frame of the 6th International Conference of the European Society of History of Science entitled "Communicating Science and Technology," held in Lisbon, September 4–6, 2014. The topic generated considerable interest and the symposium on "Mathematical Correspondences and Critical Editions" was a great success in terms of participation and debate. Subsequently, a project aimed at collecting these contributions came into being, and other scholars were invited to intervene on the theme, since the publication of collected works and correspondences is of major interest in the field of the history of mathematics.

This volume contains sixteen contributions by various researchers from five different European countries. It offers a fairly broad spectrum, albeit partial, of the research being carried out, as well as the arguments under debate, such as the complementary role of printed and digital editions, integral and partial editions of correspondences, reproduction techniques of manuscripts, pictures and formulas, and tools for identifying dates and correspondents. These problems may involve different approaches according to the period and the subject, in this wide-ranging volume that focuses on correspondences and collected works from the seventeenth to the twentieth century with reference to all mathematical sciences.

Our intention was not to present a simple collection of various projects of editions, but rather to relate correspondences and works and compare the various types of edition, the problems encountered, and the solutions found to solve them. Of particular interest was the way in which the editions of correspondences and works should be linked and prioritized. For example, in the edition of Huygens, letters precede the works; for Lagrange, letters follow the works; and for Favaro's edition of Galileo, letters follow the works in the final volumes of the series, but were collected before and organized within a unique editorial plan. Important editions, however, like those of Laplace and Cauchy, do not contain the correspondence.

All of the contributions are related to editorial projects of correspondences or collected works. In some cases, the papers deal with projects of print editions (Leibniz, Wallis, Lagrange, Gauss) or online or mixed editions (Bernoulli, D'Alembert, Poincaré). In other cases, they refer to a correspondence between two mathematicians, relevant for specific mathematical contents (Germaine–Gauss, Betti–Brioschi, Hermite–Lipschitz), or are aimed at reconstructing a particular period for the history of mathematics (Cremona, Tardy), or a network of relations (D'Alembert, van der Waerden). Other articles discuss policy and methods for dating letters and discovering unknown correspondents (D'Alembert, Condorcet, . . .), or critically examine previous non-satisfactory editions (Lagrange, Gauss).

It is not our aim to create an exhaustive discussion of the best method for producing an edition, which depends on many variables, such as the historical period and range of correspondences, the multiplicity of correspondents and overlapping with other editions, as well as the contents and target audience. We believe, however, that a volume that allows us to compare various situations by presenting a reasonably wide picture may be a publication that arouses considerable interest for many scholars of the history of mathematics.

The very first article of the volume, for instance, poses fundamental questions regarding editions of correspondences: should they be complete or partial? Should they feature only unedited material? Should previously published material be included? In the case of a partial publication of selected letters, on what basis should the criteria be chosen: subject (scientific, political, private etc.), importance, or correspondents? Whatever the choice, a census of all existing documents must be as thorough as possible.

The author of the first article, Philip Beeley, describes the stages that led to the edition of the correspondence of John Wallis, beginning with Christoph J. Scriba's research carried out at Oxford in the early 1960s. Scriba's cultural and methodological background originated from the Leibniz edition carried on by the German Academy of Sciences at Berlin, according to the guidelines of the Leibniz scholar Joseph Ehrenfried Hofmann. Beeley's paper outlines Scriba's profound and systematic investigation into Wallis's manuscripts, letters, and other materials, at Oxford, Cambridge, London, and Vienna, which produced a whole series of card catalogues and a list of Wallis's correspondence of 800 or so letters. From the initial idea of publishing only a significant selection of Wallis's letters, of interest for the history of science in general or the history of mathematics in particular, the project went in a new direction with the discovery of up to over 2000 new letters. After 30 years, this led to a collection of up to ten volumes, the first four of which were published from 2003 to 2014, with the fifth currently being printed. Philip Beeley entered the project when he was a doctoral student at the Technische Universität of Berlin, taking over as the successor of the previous collaborator, Sigmund Probst, at the beginning of November 1996. Together with Scriba, and after Scriba's death, he acted as editor of John Wallis's entire correspondence in chronological order, complete with a critical analysis and introductory essays on the themes discussed in the letters. Beeley explains, in his contribution, the various choices that had to be made during the course of the project, due to the development that took place within the methodology of the historiography of science in the last decades of the previous century, shifting from an internal historiographical approach to a more general survey of the history of ideas.

Eberhard Knobloch's essay introduces us to one of the biggest edition projects ever planned: the Leibniz edition, which cannot possibly be described in a few pages. It includes more than 50 published volumes of the expected 130 and has been a reference point for other edition projects. Eberhard Knobloch provides a detailed examination of its VII series, modified in 1975, and exclusively devoted to the manuscripts concerning mathematics (30 volumes), whereas the scientific, medical, and technical writings were to be published in a new series, the VIII, which was to follow. In 1976, Knobloch was assigned the editorial work on the first two volumes of Series VII, the first to be completed within 10 years, and the second within the following 5 or 7 years, with the help of a research assistant, Walter S. Contro.

Knobloch describes the difficulties and events involved in that edition, which accompanied the increasingly professional and academic growth of the young but already experienced researcher, and which were affected by the period of unification of Germany. One of the first and foremost tasks was the identification and dating of Leibniz's writings during his stay in Paris, which were to fill eight volumes of Series VII. Owing to the fact that most of the handwritten manuscripts are not dated and that dating the manuscripts to a period of exactly half a year is generally not possible, it became clear that a strict chronological ordering of the handwritten manuscripts was impracticable. The solution came in the form of defining thematic groups for the entire Parisian period and developing a chronological order within these groups. In the case of volume VII, 1, the following groups were identified: geometry, number theory, and algebra.

With the assistance of various colleagues, Knobloch was able to publish the first four volumes of Series VII between 1990 and 2008 (the series was then completed in two other volumes). Knobloch was mainly involved in Leibniz's writings on actuarial and financial mathematics, also published in Series IV, Political Writings (IV, 4 (1680–1692), section VII Statistics, Life Insurance, Pensions). Knobloch's subsequent involvement in Series VIII further demonstrated how important it is for a historian of sciences and editor of old scientific texts to have not only the required scientific knowledge but also philological skills, since a lack of either one of these could lead to serious misinterpretations. A project of such dimension also revealed the need for international cooperation among scholars and institutional agreements with other countries (France and Russia in this case), as well as adequate funding. Digitalization of Leibniz's manuscripts was carried out first. Series VIII, leaving aside previously attempted editions of some manuscripts containing serious mistakes, includes up to ninety percent of unedited material. Knobloch points out these errors and presents significant examples (various apparatuses and instruments linked to pneumatics and mechanics) of Leibniz's procedure in the field of Natural Sciences and Technology. The first two volumes of Series VIII appeared in 2009 and 2016.

Sulamith Gehr describes the history of another celebrated edition project, namely, the publication of the letters of the mathematicians of the Bernoulli family. The project started in the 1930s in the form of a classical book edition and continued as an online edition in the past decade.

The Bernoulli family's correspondence includes a vast network of over 400 correspondents, among whom can be found the foremost scientists of the seventeenth and eighteenth centuries, and, as such, presents various problems of classification and organization. Gehr's work reminds us of the important role played by epistolary commerce in the transmission of knowledge and scientific debate in past centuries. In the specific case of the Bernoulli family, there is an important historical precedent: the publication edited by Gabriel Cramer of the *Commercium philosophicum et mathematicum*, which appeared in 1745 and featured the correspondence between Johann I Bernoulli and Gottfried Wilhelm Leibniz, following, as it did, just a few years after the publication of the *Opera omnia* by Johann I Bernoulli in 1742 may be considered as a completion of it. Further plans of epistolary publication followed, namely, the idea of publishing the correspondence between Johann I Bernoulli and the Marquis de L'Hôpital, begun by Johann III Bernoulli but never concluded. In the nineteenth century, the letter exchanges of the Bernoullis with Leibniz, Euler, and further scientists who had been active at the Imperial Russian Academy of Sciences were edited, based on the manuscripts in Hannover and St. Petersburg, by Carl Immanuel Gerhardt and Paul Heinrich Fuss, in different projects. Other partial publications then followed from the manuscripts rediscovered in Gotha and Stockholm by Gustaf Eneström, among others.

So, a comprehensive project finally emerged at the beginning of the twentieth century with Otto Spiess in Basel, part of the wave of new understanding of the history of science: the Bernoulli edition, which included not only letters, but also manuscripts and printed works. Spiess started by organizing the institutional framework and collecting all of the manuscripts in Basel, then compiled an inventory of all known letters sent or received by the Bernoullis and by Jacob Hermann, and subsequently prepared the editorial plan and transcribed the letters. The first volume appeared in 1955, with 162 letters in total, reproduced in the original language and annotated and commented upon in German, and included a rich critical apparatus. After Spiess's death, three other volumes of letters were published in 1988, 1992, and 1993, the first one edited by Pierre Costabel and Jeanne Peiffer, the second by André Weil, and the last one by André Weil with the help of Clifford Truesdell and Fritz Nagel, all following the structural model, methods, and editorial standards set by Spiess. The correspondences of the Bernoullis with Leonhard Euler and Gottfried Wilhelm Leibniz were not included in the Bernoulli edition, because they were destined to be edited within two other important edition projects: the Euler edition and the Leibniz edition. Sulamith Gehr gives a short overview of the state of the Bernoulli correspondence in these further editions.

The contributions of Passeron and Guilbaud refer to the great edition project of the works of D'Alembert, which has involved roughly forty French scholars over several decades. The correspondence constitutes a section of this grandiose project, which is planned in about fifty printed volumes, seven of which have already come out. The edition is organized into five series: the first and third include mathematical works, the second, articles from the *Encyclopédie*, and the fourth, philosophical, historical, and literary writings. The organization of the edition is, therefore, partly thematic and partly chronological. Finally, the critical edition of the complete correspondence constitutes the subject of Series V.

The systematic study of the entire correspondence sent and received by D'Alembert was begun about 20 years ago, by Irène Passeron, in collaboration with A.-M. Chouillet and J.-D. Candaux, and led to the publication of an analytical inventory in 2009. Two thousand three hundred letters have been classified, of which about 500 are unedited, exchanged with over 420 correspondents. In 2015, the first volume of Series V, collecting the letters exchanged between 1741 and 1752, was published. The project aims at the publication of a further ten volumes, with the rest of the letters organized according to chronological periods.

Like his works, D'Alembert's letters also cover a vast domain of knowledge: mathematical sciences, music, literature, and philosophy, and editors have to deal with the twofold problem of research of documents and reconstruction of the vast network of correspondents, as well as the organization of the entire correspondence in order to make it available to scholars. Of particular importance is his correspondence with Gabriel Cramer, Euler, and Lagrange concerning questions of science, whereas matters of philosophy, politics, and morality are to be found in the correspondence with Frederick II and Voltaire. Further correspondences concern personal affairs, and many of his other letters concern academic issues related to his work as an influential member of the Académie royale des sciences in Paris and as secrétaire perpétuel de l'Académie française.

In her paper, Irène Passeron discusses the reconstruction of D'Alembert's network of correspondents and the way in which this research was carried out starting from information collected from D'Alembert's biography and various activities. Conversely, this reconstruction provides not only a deeper insight into his work, but also into scientific and literary debates, as well as the general way of thinking prominent in that century. Thus, all in all, it constitutes an essential contribution to the edition of D'Alembert's Complete Works.

The printed edition of D'Alembert's works is accompanied by a website providing information on its organization and progress and supplies documents such as a bibliography, chronology and studies on D'Alembert, other references and databases on the correspondence, academic reports, and so forth. A parallel project has been developed for the correspondence: *D'Alembert en toutes lettres*, which includes the uploading of the letters as soon as permission has been received for their publication. Alexandre Guilbaud's contribution describes the accomplishment of this project, and so he deals with such issues as the interface that allows for online access to both the description of metadata (place and date of the letter; reference number in the inventory; name of the correspondent; material description of the source; place of conservation; list of edited versions; other manuscript sources, if any; incipit; summary) and, when available, the reproduction of the original exemplars of the letters. The website is, moreover, enhanced by critical information on the letters and the history of the documents presented. Besides a continuously updated dynamic version, the site of the digital edition of D'Alembert's correspondence is intended as a support to the printed edition, allowing us to navigate within the index and, when possible, the text of the letters, equipped with specific research tools.

The next paper contains a contribution by Nicolas Rieucau, which is ideally a continuation of the previous papers, since it concerns the correspondence of the encyclopaedist, and protégé of D'Alembert, Nicolas de Condorcet, who wrote the long eulogy read at the Académie des sciences, on November 12th, 1783 (*Histoire de l'Académie royale des sciences—Année 1783*, Imprimerie royale, Paris, 1786, pp. 76–120). This was later inserted into the first of ten volumes of D'Alembert's selected works (*D'Alembert, sa vie, ses œuvres, sa philosophie*, Paris, Firmin-Didot, 1847).

The nineteenth-century edition of Condorcet's works (1847–1849) is also rather incomplete, particularly as far as the correspondence is concerned, as it reports less than 200 letters, most of which present little of scientific interest. However, the bicentennial anniversary of the French revolution and of Condorcet's death provided an occasion to publish some rare or unedited texts by Condorcet, above all those regarding political arithmetic and the philosophy of history. Moreover, various letters from Condorcet to different interlocutors were published in connection with essays devoted to them.

There can be no question of the importance of correspondences in providing a better understanding of Condorcet's figure as a scientist, almost overshadowed by that of the philosopher as a defender of the core values of the Enlightenment and victim of their degeneration. Thus, around 2010, the project entitled "Inventaire Condorcet" focused on the construction of as complete an index as possible of Condorcet's correspondence, which amounts to more than 2100 letters. Over half of these are scientific in content and unedited, the originals of which still in existence are scattered over 130 different archives worldwide. More than 300 of them, either received or sent, concern integral calculus, calculus of probability, hydraulics, and chemistry. A second group of over 150 are related to disciplines less frequently found in the Condorcet correspondence, such as meteorology, geodesy, mineralogy, optics, geology, botany, and agronomy. The number of correspondents rose to 250. Rieucau's paper gives an account of the goals, difficulties, limitations, and results of the ongoing work. The difficulty of making an inventory of Condorcet's letters was compounded by the lack of a register of correspondences and the multiplicity of his correspondents (given his position as permanent secretary to the Académie des sciences), but above all, by the dispersion of his letters after his condemnation and death. Many of them were sold on the market of autographed letters. Besides the difficulties involved in researching lists, catalogues of sales, archives, and libraries, there were also problems of identification, classification, and dating of the letters, for which methods of investigation had to be developed on the basis of various data, not only of an intrinsic nature but also deduced from original documents (like watermark and other characteristics of paper support, origin, location, and so on).

The paper by Luigi Pepe gives us a critical view of past editions, above all, the editions of collected works of the nineteenth century, which was the most prolific for this type of publication. The aim was to collect and make more accessible the works of great mathematicians, published in many volumes and academic journals, and finance them by public funds for the glory of the nation: the works of Lagrange, Laplace, Cauchy, Fourier, Arago, Galilei, Huygens, Gauss, and others. In particular, Pepe provides an in-depth analysis of the structure and criteria used for the publication of Lagrange's works, edited by Serret, Darboux, and Lalanne (1842–1917), taking up and developing the critique by George Sarton. Academic memoirs were grouped according to the journal in which they were published, and the annotations were modernized to fit contemporary mathematical writings; in the case of more than one edition, the latest was reprinted, with no historical-critical commentary. If such an edition could be useful from a mathematical point of view, it is, however, completely useless from the point of view of the history of science. Lagrange's correspondence with D'Alembert, Condorcet, Laplace, Euler, and other scientists, which takes up the last two volumes of the series (XIII–XVI), was, however, critically edited by the historian Lalanne, but many other letters were published later and still others remain unedited. Furthermore, the *Œuvres de Lagrange* contain only a minimal part of the manuscripts by Lagrange conserved in the library of the Institut de France set out in sixteen volumes. Several of these manuscripts were published at a later date, as were other unedited ones belonging to other archives. Pepe concludes with some suggestions for integration of the *Œuvres* with other volumes containing newly found documents and hopes that a modern project similar to that being carried out in France for D'Alembert may be devoted to Lagrange and that at least a site with references to all printed materials after the publication of the *Œuvres* could be set up.

Remaining within the context of editions of works and correspondences initiated in the nineteenth century, there follows the study by Karin Reich and Elena Roussanova, which deals with the works of Gauss, a series of 12 volumes (14 tomes) published between 1863 and 1933. This edition, which was to be the complete edition of the works by Gauss, does, in fact, collect almost all the works published by Gauss, as well as posthumous writings, or rather manuscripts, letters, and documents extracted from the archive of the State and University Library of Göttingen, with the addition of other material or comments on the part of the editor, material for a scientific biography, and writings by other authors. This is not the only case in which we find comments inserted by the editor (see, for example, the works of Fourier); however, the preponderancy of unsuitable material imposed upon Gauss's text is surprising, as pointed out by the authors, who also criticize the lack of precise references concerning the location of the unedited material reproduced.

Reich and Roussanova reconstruct the stages of the edition project, originally entrusted to a pupil of Gauss, Ernst Christian Julius Schering, who gained fame and inspiration from it, and under whose direction seven volumes were published between 1861 and 1873. In these, organized according to theme, were collected all the printed works and some manuscripts. Not only did Schering select from among Gauss's manuscripts those which he deemed worthy of publication, he also excluded some tables and charts from the reproduction of published works. There followed a second edition of the first five volumes (1870–1877) with the addition of a significant number of unedited writings. It was not until 20 years later that the edition was once again taken up with a new series, under the direction of Felix Klein with the collaboration of Martin Brendel and Ludwig Schlesinger. Klein was obliged to leave the direction of the work in 1922, due to ill health. He was succeeded by Max Born and then, in 1928, by Richard Courant. Between 1907 and 1933, the following items were published: an anastatic reproduction of Volume 6, a new edition of Volume 7 with the addition of many unedited works on astronomical matters, and Volumes 8–12, devoted to other unedited works by Gauss. The publisher also changed as the series progressed: from the Royal Society of Sciences in Göttingen to the publishing house Perthes in Gotha, then Teubner in Leipzig, and, finally, Springer in Berlin. Editorial plans constantly underwent changes under Klein's direction, as may be seen from the reports he periodically presented to the Royal Society of Sciences in Göttingen. Reich and Roussanova's analysis highlights his plans both for a scientific biography, to be set out in various chapters of Volume 11 and entrusted to experts in the field, and a general index, which was never carried out.

It would not seem possible at present to re-propose a new edition of the works of Gauss that follows modern criteria, especially considering the fact that some documents may no longer be traceable. So, in spite of their defects, recourse to anastatic copies of old editions took place (Georg Olms: Hildesheim, New York, 1973 and 1981), fortunately available nowadays in a digital version. As the authors suggest, the necessity of a complete table of contents of Gauss's works, as well as a keyword index for all available volumes, can now be solved by an online database, where bibliographic details should also be supplemented and improved upon.

The paper by Andrea Del Centina and Alessandra Fiocca focuses on a female figure, practically isolated in the overwhelmingly male-dominated panorama of the works and correspondences dealt with in this volume. Sophie Germain attempted, not always successfully, to communicate with some of the leading mathematicians of her day and to take part in mathematical research at the highest level. Because she was a woman, she was not allowed access to adequate university studies and was excluded from the academic career she so deserved, which would have enabled her to participate in the ongoing scientific debate. Recognition of her contributions to the theory of numbers and the theory of elasticity has been given further drive by the finding and analysis of her documents and manuscripts, which belonged to Guglielmo Libri. In this paper, the authors reconstruct the complicated phases of a progressive rediscovery of the correspondence between Sophie Germain and Gauss, as well as the mathematical notes attached to the letters, the unfinished publication project by Baldassarre Boncompagni and Angelo Genocchi, and the correspondence with Guglielmo Libri. The chapter finishes with some references to the publications of unedited letters and mathematical notes on the part of the authors, who have contributed to a reevaluation of Sophie Germain's life and the part she played in the field of the theory of numbers. All of the active and passive correspondence of Sophie Germain that has been published is classified in the appendix.

A set of contributions in this volume deals with the correspondences of the great protagonists of the Italian Risorgimento: Brioschi, Cremona, Betti, Tardy, etc. It was a period of extraordinary scientific, cultural, social, political, economic, and technological reawakening, which placed Italy at the same level as the most advanced countries in Europe. There are many correspondences that testify to this immense effort and success, which occurred over a relatively short period of time. The protagonists of this transformation corresponded with one another and with scientists from other parts of Europe. The fact that there is no single protagonist to whom others may be referred has meant that editorial choices have also been fortuitous and that correspondences linked to one another have come to light at different times with no comprehensive methodological plan. Similar correspondences are still being published nowadays. This situation can only be changed by means of an extensive national plan involving many researchers over a prolonged period of time in an attempt to recreate the transmission and evolution of ideas through diverse and interconnected epistolary networks.

The question arises as to whether it would be possible to organize, within the scope of a single project, the entirety of such intertwined correspondences among the mathematicians active during the Italian Risorgimento, by a group of researchers who have devoted studies to them and even edited some of them. This would allow us close comparison of their contents, which are often found to be repeated.

Italy, on the other hand, has a solid tradition of publishing collected works of great mathematicians. Owing to the policy of nationalism in the first half of the twentieth century, Italy celebrated its glorious past with a national edition of the works of Galileo Galilei, edited by Antonio Favaro, nine volumes of which are devoted solely to the correspondences. Similarly, the protagonists of the Italian school of mathematics, which, at that time, was at the top of contemporary European research, had editions of their collected works printed (Betti, Brioschi, Cremona, . . .). The Italian Mathematical Union initiated a series of collected works of eminent mathematicians (above all, selected works), which still continues today, but only recently, and only occasionally, have we found correspondences inserted into these publications. This would explain why, in Italy, many scholars of the history of mathematics have directed their efforts toward editions of correspondences of the great protagonists of the Risorgimento, which constitute the completion of previous editions of works.

This volume on this theme collects papers by the following authors: Cinzia Cerroni, on the vast source of only partially published letters of Placido Tardy's correspondence; Ana Millán Gasca, Giorgio Israel, and Luigi Regogliosi, on the publication of the collection of 1122 letters received by Luigi Cremona from foreign correspondents; Maria Teresa Borgato and Iolanda Nagliati, on Francesco Brioschi's correspondence with Enrico Betti and Tardy, as well as all their correspondences with other foreign scientists; and Paolo Freguglia, Giuseppina Fenaroli, and Giuseppe Canepa, on the correspondence of Giusto Bellavitis. The collections come from the Polytechnic University of Milan, the Scuola Normale of Pisa, the Department of Mathematics of Rome, the Mazzini Institute and the University Library of Genoa, the Veneto Institute in Venice, and the Historical Archives of Göttingen University.

The main difficulty surrounding these editions lies in the selection of the material to be published, since it is linked to other correspondences in a wide network of intertwined relationships. The themes under debate are mainly of a scientific nature, but there are others that deal with politics, administration, culture, state education, university and higher education, academies, and so forth.

Cinzia Cerroni deals with important collections preserved in the archives of Genoa, and in particular, the correspondences of Placido Tardy and Luigi Cremona; Tardy's letters are preserved at the Genoa University Library and Cremona's letters at the Mazzini Institute of Genoa. The University Library of Genoa hosts an important archive, donated by the historian of mathematics Gino Loria, containing 784 letters sent by prestigious Italian and foreign mathematicians to Placido Tardy. Tardy was at the center of a wide network of correspondents, a fact that allows us to investigate "the connections between the development of Italian mathematics in the second half of the nineteenth century and the main political issues of Italian history." This correspondence has been partially published (letters sent by Beltrami, Bellavitis, Betti, Cremona), in some cases completed with the letters sent by Tardy contained in other Italian archives. The Mazzini Institute in Genoa possesses another important archive, donated by Cremona's daughter, mainly consisting of Cremona's correspondence, part of which has also been published.

In her paper, Cerroni describes these funds and provides insight into the Cremona–Tardy, Betti–Tardy, and Cremona–Guccia correspondences, their substantialness, and, with abundant quotations, the main issues contained in those letters: the foundation of the journal *Annali di Matematica pura ed applicata*, the foundation of the Circolo Matematico of Palermo, the discussion on non-Euclidean geometry, Riemann's theory and Abelian functions, references to Giuseppe Garibaldi and the Italian wars of independence, the educational reforms, and university policy.

Luigi Cremona's correspondence is also preserved in different places. The principal sources in Italy are at the Mazzini Institute in Genoa, where the 6000 documents consist mainly of correspondences with Italian scientists and politicians or state officials, as well as with 34 foreign mathematicians, and at the Department of Mathematics in Rome, which houses letters addressed to Cremona from 176 mathematicians, most of whom are foreigners, and from representatives of three scientific societies. The latter archive source, presented in the paper by Ana Millán Gasca, Giorgio Israel, and Luigi Regoliosi, was the subject of a recent edition overseen by Giorgio Israel. Among those correspondents can be found Carl Wilhelm Borchardt, Alfred Clebsch, Eugène Prouhet, Olry Terquem, Maximilian Curtze, Rudolf Sturm, Heinrich Schröter, Arthur Cayley, Thomas Hirst, George Salmon, Rudolf Sturm, Elwin Bruno Christoffel, Wilhelm Fiedler, Johann Nicolaus Bischoff, Theodor Reye, Carl Friedrich Geiser, Ludwig Schläfli, Emil Weyr, and translators such as Eugène Dewulf.

The paper presents the distribution not only of the letters over time, but also of the correspondents according to their nation or geographical area, and discusses some aspects of the edition. It is not a complete collection, since more than half of the letters have been lost; the remaining 1122 are published in alphabetical order of the correspondents, with a chronological index, critical apparatus, and bibliography. A team of nearly 20 researchers from six European countries have contributed to the edition. Research on the letters sent by Cremona has not gone forward, given that the high number of correspondents from various countries would have greatly delayed the publication without any hope of completing it within a reasonable period of time.

The extensive network of correspondents and the multiplicity of the languages used in the letters and their countries of origin, especially those in Europe, provide a vivid picture of the mathematical community in the second half of the nineteenth century, which was actively involved not only in cultural, social, and political issues, but also in a process of modernization. This allows us to reconstruct the political thinking of the day, as well as its scientific interests and cultural goals. Mathematics itself was not merely a vehicle for scientific progress, but encompassed cultural and social issues as well.

Maria Teresa Borgato and Iolanda Nagliati focus their paper on the figure of Francesco Brioschi, who was instrumental in the scientific, political, and administrative development during the unification of Italy. He played many roles: editor of the *Annali di Matematica Pura ed Applicata*, founder of the Polytechnic Institute of Milan, for many years senator and general secretary of the Department of Education, influential member of many ministerial commissions regarding the railways, fluvial hydraulics, and finance, and president of scientific societies and academies. This paper, however, mainly concentrates on the scientific themes discussed in his correspondences with Enrico Betti and Placido Tardy: the theory of invariants of binary forms, the resolution of fifth degree algebraic equations by elliptic functions, and the theory of fractional integrals.

Starting from references contained in the letters, the authors reconstruct the contributions made to these mathematical theories. Furthermore, they present the current picture of the epistolary relationships of these three mathematicians with their foreign correspondents, which allows for a reconstruction of the frequent journeys abroad undertaken by Italian scholars to further their studies, as well as the journeys of foreign scholars to Italy. Of particular importance for their number and contents are Brioschi's correspondences with Felix Klein and Charles Hermite. The documents studied for this paper come from the historical archives of the Polytechnic Institute in Milan, the Institute Library of Genoa, the Scuola Normale in Pisa, and the University of Göttingen.

The contribution made by Paolo Freguglia, Giuseppina Fenaroli, and Giuseppe Canepa explores another area of the variegated world of Italian mathematics during the Risorgimento, in particular that of the Veneto centering around the University of Padua. The region's previous political history (first as the Republic of Venice, and then under Habsburg dominion) gave rise to its cultural diversity, which is also reflected in the different interests represented by research studies. Giusto Bellavitis was an eminent mathematician of the "Studio Padovano," the University of Padua, and founder of the calculus of equipollences, originating from the study of geometric foundations of complex numbers, which is related to the works of Moebius, Hamilton, and Grassmann. He also represents a sort of link between the previous generation of Italian mathematicians of the universities of Turin and Pavia (Antonio Bordoni, Gabrio Piola, Felice Chiò, and Ottaviano Fabrizio Mossotti) and the new group of researchers whose outlook was more international (Enrico Betti, Francesco Brioschi, and Felice Casorati).

The present contribution provides a general survey of Bellavitis's letters, preserved in Venice, Genoa, Rome, and Piacenza. Among his best known correspondents are Luigi Cremona, Placido Tardy, Domenico Chelini, and Angelo Genocchi. Part of this correspondence has been published; however, most of his letters (about 1270 letters and minutes, donated to the Istituto Veneto in Venice) remain unpublished. Besides private subjects, the topics in the letters are related to academic questions (the role of the Società dei XL), social and political situations, opinions on scientific papers and their mutual exchange, and various mathematical and scientific items, in particular the calculus of equipollences.

Catherine Goldstein's paper also focuses on the second half of the nineteenth century and centers on the French mathematician Charles Hermite, who was one of the most important of the century. In the case of Hermite, the same problem of publishing the entire correspondence arises, since he wrote thousands of letters to dozens of correspondents, on different subjects: personal, political, academic, and mathematical. Even if the letters received were lost during a fire, important collections of letters preserved in various archives and libraries survive, some of which have been published, for example, those sent or exchanged with Thomas Stieltjes, Paul Du Bois-Reymond, Andrei Markoff, Gösta Mittag-Leffler, Ernesto Cesàro, Angelo Genocchi, and Georg Cantor. Other selected letters sent to or received from James Joseph Sylvester or certain Italian mathematicians have been published. The problems to be dealt with and the choices to be made in the case of a complete edition are presented, starting from the correspondence between Hermite and Rudolf Lipschitz. After comparing the similarities and differences in the scientific and academic formation of the two protagonists, as well as providing a general picture of each one's published and unedited letters, there follows an analysis of the Hermite–Lipschitz correspondence, which is mostly preserved in the Lipschitz collection in Bonn, and consists of 148 letters and 9 postcards sent by Hermite, as well as 70 drafts of letters from Lipschitz. Two letters from Lipschitz to Hermite can be found in the Archives of the French Academy of Sciences. The letters were written in the last quarter of the century, 1877–1900, at the end of the two mathematicians' careers, with a peak around the year 1884.

The contents concern issues of publishing and dissemination of mathematics, mathematical research, academic and scientific policies, university teaching, as well as personal matters and political opinions. Starting from a detailed analysis of some of the letters, Goldstein is able to give a general outline of the themes broached throughout the correspondence, many of which may also be present in just one letter, and which also provide further insight not only into personal and scientific relationships among the scientists of that period, but also the role of correspondence in the scientific community. In the second part of the paper, the author considers the influence this correspondence had on the scientific output of Hermite and Lipschitz; in particular, this correspondence also testifies to the resumption of Franco-German relationships following the Franco-Prussian War of 1870.

In the conclusion, the possibility of the digital edition of this correspondence is discussed. By taking some existing projects as examples, the creation of an open platform is hypothesized, to which new documents and references can be added in real time, with a selective display that allows users not only to access the text of the letters, but also to search for specific concepts and references, and with a "homoiconic" structure in which "links should be treated as data, as well as the texts of the letters, capable of receiving themselves links and commentaries."

Correspondences require recourse to external elements if they are to be faultlessly interpreted and edited, yet at the same time, they themselves give precious information not only on the lives and characters of the correspondents, but also on external events and the general historical and cultural situation of the time in which they lived. Furthermore, they help us to reconstruct the role played by the correspondents within the scientific community, as well as the type of research they carried out and their reciprocal influence. Scott Walter's paper leads us to the beginning of the twentieth century, when David Hilbert and Henri Poincaré were at the height of their renown and influence within the international scientific community. Holding opposing positions concerning the foundations of mathematics and the relationships between mathematics and the physical world, the study of the correspondence between the two brings to light new details in the Hilbert–Poincaré relationship and in Poincaré's approach to questions of theoretical physics.

The Hilbert–Poincaré correspondence, transcribed here, is made up of seven letters written in the period between November 1908 and March 1909. It concerns Hilbert's invitation to Poincaré to hold a cycle of lectures at the Society of Mathematics of Göttingen (supported by the Paul Wolfskehl foundation). Topics discussed are the planning of themes to be treated, like the reduction of Abelian integrals, applications of Fredholm's method, the theory of tides and Fredholm's equation, Hertzian waves and Fredholm's equation, and the notion of transfinite cardinal numbers. Hilbert had asked that the themes of theoretical physics and mathematical logic be added, but in the end, Poincaré chose to add a conference on the theory of relativity instead.

This set of letters forms part of Poincaré's large correspondence of over 2000 letters, exchanged with over 290 interlocutors; it has been indexed and put online (in images or transcriptions) on the site Henri Poincaré Papers of the University of Nantes, together with manuscripts and publications by Poincaré, as well as sources relative to his work. A Sphinx search engine enables the user to find the documents.

Erwin Neuenschwander's contribution brings us well into the twentieth century. It describes the extremely rich collection of van der Waerden's papers housed at the Library of the Eidgenössische Technische Hochschule (ETH) in Zurich, including around 15,000 letters, stretching from 1943 until his death in 1996. Most of van der Waerden's correspondence has been catalogued and made available to the public. Neuenschwander was van der Waerden's last assistant and longtime coworker at the Research Center for the History of Science at the Institute of Mathematics at the University of Zurich. He is therefore able to provide a detailed reconstruction of van der Waerden's activity during his Zurich years, enriched by personal memories. After a brief biography of van der Waerden, which states van der Waerden's position with regard to the Nazi regime, Neuenschwander gives a brief overview of his ongoing research of van der Waerden's Zurich years, accompanied by a select edition of his correspondence. In the present paper, he provides a detailed list and discussion of the ninety most extended correspondences, which contain at least 25 letters. Among these are those with Hans Freudenthal, Edward S. Kennedy, Otto Neugebauer, and Clifford A. Truesdell, which comprise more than a hundred letters and which are discussed in more detail in the paper. The correspondence with Hans Freudenthal contains, in particular, information about van der Waerden's unpublished textbook *Introduction to Topology and Riemann Surfaces* and van der Waerden's views about synthetic a priori knowledge and its role in natural science. With Edward S. Kennedy, a specialist in medieval Islamic astronomy, van der Waerden discussed ideas about the transmission of Babylonian and Hellenistic astronomical notions. Otto Neugebauer was one of van der Waerden's oldest friends, with whom van der Waerden discussed his research in the history of astronomy in ancient cultures.

Van der Waerden also had important exchanges of letters with Walter Burkert, Richard A. Parker, David Edwin Pingree, William Kendrick Pritchett, Abraham Sachs, Derek Thomas Whiteside, and Clifford A. Truesdell. The correspondence with the last of these contains a great deal of information about the journal *Archive for History of Exact Sciences*, of which van der Waerden was one of the coeditors. A particular focus is also devoted to van der Waerden's polemic with David Pingree, about the transmission of astronomical theories between the Near East and India.

We hope that this volume, which not only deals with the edition of collected works and the publication of correspondences of mathematicians, but also with the significance of correspondences within the context of editions of complete works, will provide interesting reading and that it may be of help and serve as a stimulus to historians of mathematics in their research.

A substantial commitment on the part of publishing houses involved in the field of science, combined with the financial aid of public bodies or institutes, is, however, of paramount importance to sustain this fundamental activity and avoid fragmentation of publication aimed at supporting specific historiographical theses.

Ferrara, Italy Maria Teresa Borgato
Paris, France Irène Passeron

Contents

Historiographical Change and Editorial Practice: The Origins of the Edition of the Correspondence of John Wallis (1616–1703)

Philip Beeley

Abstract

The edition of the correspondence of John Wallis (1616–1703) has had a relatively long and varied history, being conceived originally by Christoph J. Scriba as a small companion volume of those letters of the great Oxford mathematician that had not already appeared elsewhere, in publications such as the *Correspondence of Isaac Newton* or the *Œuvres complètes de Christiaan Huygens*. In this chapter, the author charts the evolution of the Wallis Edition from those initial plans through to the major critical edition that is now in progress. Drawing on Scriba's own letters and papers, he argues that the history of the Wallis Edition mirrors changes that have taken place in the nature and outlook of the history of mathematics itself; changes on which Scriba reflected intensely during his lifetime and to which he contributed both through his writings and his historiographical practice.

1 Introduction

For the young German historian of mathematics, Christoph J. Scriba (1929–2013), the two years he spent in Oxford in the early 1960s were something like an intellectual home-coming. By the time he arrived at the university, he had already developed a strong leaning towards pre-Newtonian mathematics in the British Isles, and while he was there he quickly became an authority on a group of figures that had largely been overlooked by contemporary scholarship such as Thomas Brancker (1633–1676), Nicolaus Mercator (1620–1687), John Pell (1611–1685), and John Collins (1625–1683). But it was John Wallis (1616–1703), Savilian professor of

P. Beeley (✉)
History Faculty, University of Oxford, Oxford, UK
e-mail: philip.beeley@history.ox.ac.uk

M. T. Borgato et al. (eds.), *Mathematical Correspondences and Critical Editions*, Trends in the History of Science, https://doi.org/10.1007/978-3-319-73577-1_1

geometry in the University of Oxford from 1649 until his death in 1703, who grabbed Scriba's attention most. With only very little in the way of secondary resources available, he set about creating them himself. From painstakingly exact archival work in Oxford, Cambridge, and London emerged a series of groundbreaking investigations as well as a stunning array of catalogues for his personal use—of Wallis's books, his correspondence (ordered both chronologically and according to holding libraries and archives), and his manuscripts. While these catalogues were mainly intended to form the basis for his *Habilitationsschrift*, already at this time he devised his first ideas for an edition of Wallis's letters. What Scriba first had in mind was quite a modest endeavour, namely to publish those letters which could be deemed significant for an understanding of the history of mathematics. And not even all possible letters which conceivably fitted into that category were to be considered. Scriba felt it was unnecessary to republish any letters of Wallis which had already appeared in major editions such as those of Christiaan Huygens (1629–1695), G. W. Leibniz (1646–1716), Isaac Newton (1643–1727), and Henry Oldenburg (1618–1677). The main purpose of an edition was according to his view at that time to make correspondence accessible to the broader research community. It was more a question of reconstructing the emergence of scientific ideas than creating resources for biographical studies or understanding the intellectual dynamics of the Republic of Letters.

For various reasons nothing came of these early plans to publish the correspondence and it was nearly 30 years later before a successful project application was submitted and work could begin in earnest. However, by that time the history of mathematics itself had undergone a transformation requiring an edition quite different from that originally conceived. The story of the Wallis edition mirrors historiographical developments on which Scriba not only reflected in his own writings but also to which he decisively contributed. It is with this interplay of historiography and editorial practice that this chapter is primarily concerned.

2 The Early Years

Having written his doctoral thesis on James Gregory (1638–1675) under the supervision of Egon Ullrich (1902–1957) at the University of Giessen,[1] a heavily bomb-scarred city in central Hesse, Scriba left the moral and material deprivations of post-war Germany behind him and, with the help of a two-year Fulbright scholarship, pursued the next stage of his academic career in the United States. He adapted quickly to the new environment. A year spent teaching and researching at the University of Kentucky was followed by an appointment as assistant professor of mathematics at the University of Massachusetts at Amherst. A year later, in

[1] Scriba's thesis was published under the title *James Gregorys frühe Schriften zur Infinitesimalrechnung*, Giessen: Selbstverlag des Mathematischen Seminars 1957 (= Mitteilungen aus dem Mathematischen Seminar Giessen No. 55).

1959, he moved across the border into Canada, where he became lecturer and subsequently assistant professor of mathematics at the University of Toronto. As Scriba's carefully prepared and regularly updated list of lectures and courses documents, he was during his years in North America continuing to pursue his historical interests while carrying out the core duties of a mathematics professor. His first major publication in the history of mathematics came out while he was still in Toronto.[2] The focus of his investigations was, as it would continue to be up to the end of his life, the history of mathematical ideas in the seventeenth century, especially the work of James Gregory, John Wallis, John Pell, and Isaac Newton. However, his first major study after completing his doctorate touched only fleetingly on early modern science. While still in Toronto, Scriba produced a course book for a graduate course in education on the development of mathematical thought in which he investigated the history of the concept of number.[3] He later published the text in book form in Germany.[4]

The figure who in many ways guided Scriba's research in the history of mathematics, both thematically and, at the outset at least, methodologically, was the great Leibniz scholar Joseph Ehrenfried Hofmann (1900–1973). Having served as director of the Leibniz edition of the German Academy of Sciences during the time of the Second World War, Hofmann subsequently took up a teaching position at the grammar school in the Bavarian town of Günzburg, the Dossenberger Gymnasium, while lecturing regularly as honorary professor at the University of Tübingen. Pre-eminent among post-war Germany's historians of mathematics, Hofmann initiated, in 1954, the renowned annual colloquium on the history of mathematics at the Mathematisches Forschungsinstitut in Oberwolfach in the Black Forest. Scriba described himself as a pupil of Hofmann's,[5] and with good reason. Shortly after accepting Ullrich's invitation to work on a doctoral thesis, in early 1955, Scriba accompanied him to Oberwolfach where he attended the second colloquium. Hofmann, who was a good friend of Ullrich's and who had been invited a number of times to give talks in Giessen, suggested to him in discussion that the promising young scholar he had brought along with him might make a detailed study

[2]Christoph J. Scriba, 'Zur Lösung des 2. Debeauneschen Problems durch Descartes. Ein Ausschnitt aus der Frühgeschichte der inversen Tangentenaufgaben', in: *Archive for History of Exact Sciences* 2 (1961), pp. 406–19.

[3]Christoph J. Scriba (with the assistance of Dormer Ellis), *The concept of number. A chapter in the history of mathematics, with applications of interest to teachers*, University of Toronto: Ontario College of Education 1961. Shortly afterwards, Scriba produced another course book for the Toronto secondary school teachers' course in mathematics, 1961–2: *A brief historical survey of elementary algebra and geometry*. In contrast to the former, this course book has remained unpublished.

[4]Christoph J. Scriba (with the assistance of Dormer Ellis), *The Concept of Number. A chapter in the history of mathematics, with applications of interest to teachers*, Mannheim: Bibliographisches Institut 1968 (BI Hochschultaschenbücher No. 825/825a).

[5]Christoph J. Scriba to Harold Hartley, 18 October 1962; Archive of the Wallis Edition: "I am a pupil of Prof. J. E. Hofmann, the German historian of mathematics".

of the early writings of the Scottish mathematician James Gregory the topic of his thesis. Scriba took up the proposal and soon set to work, ordering and obtaining Photostat copies of both the *Vera circuli et hyperbolae quadratura* (1667) and the *Geometriae pars universalis* (1668) for this purpose. Tragically, Ullrich died shortly before his doctoral student was able to defend his thesis, in June 1957. His place on the examining committee was taken by Hofmann.

Sometime later, Hofmann drew Scriba's attention to the significance of John Wallis's surviving letters and writings for the history of mathematics and suggested that he might make a study of them. Nor did he simply make this proposal. It was Hofmann, too, who secured the support of the German Research Foundation (Deutsche Forschungsgemeinschaft, DFG) for Scriba to spend two years in Oxford carrying out investigations on Wallis's papers. In fact, Hofmann's intention went further. Through the writing of a *Habilitationsschrift* based on these investigations Scriba was to obtain the venia legendi necessary for a professorial post at a German university. Promoting Scriba's academic work was part of a clear strategy on Hofmann's part to put the history of mathematics in post-war Germany on a strong institutional footing.[6]

Soon after his arrival in Oxford, in 1962, Scriba began immersing himself in the intellectual life of the university. Not only did he work long hours in the Bodleian Library, meticulously going through all the relevant printed and manuscript sources, but also he occasionally attended seminars and above all developed what would become lifelong friendships with Adolf Prag (1906–2004) as well as with A. Rupert Hall (1920–2009) and Marie Boas Hall (1919–2009), who lived close by at Tackley. A refugee from Nazi-Germany, Prag was a highly respected historian of mathematics who could claim the honour of having published the only noteworthy study of Wallis's mathematics up to that time.[7] He assisted D. T. Whiteside (1932–2008) in his monumental edition of Newton's mathematical papers and was a regular attendee of the Oberwolfach colloquia. The Halls had already produced an impressive body of work on Newton and Robert Boyle (1627–1691) respectively and had recently jointly embarked upon the massive task of editing the complete correspondence of the influential early secretary of the Royal Society, Henry Oldenburg.[8]

To say that Scriba's approach to his investigations on Wallis was systematic would be a gross understatement. Not only did he list every book and manuscript he

[6]It is no accident that the two major centres of the history of mathematics in West Germany, at the Technische Universität Berlin and the Universität Hamburg, were institutes in which Scriba successively played a decisive role as statutory professor.

[7]Adolf Prag, 'John Wallis (1616–1703). Zur Ideengeschichte der Mathematik im 17. Jahrhundert', in: *Quellen und Studien zur Geschichte der Mathematik, Astronomie und Physik*. Abt. B, vol. 1, 1931, pp. 381–412.

[8]*The Correspondence of Henry Oldenburg*, ed. A. Rupert Hall and Marie Boas Hall, 13 vols, Madison, University of Wisconsin Press; London: Mansel; London: Taylor & Francis, 1965–86.

found, in the case of books noting the day or days on which they were consulted, but he also produced a whole series of card catalogues, the two most impressive of which, a chronological catalogue of Wallis letters and a catalogue of Wallis manuscripts recording in note form the contents of each, he constructed out of improvised materials: he recycled the insides of envelopes used for sending him letters, some white, some brown, some grey, cutting them down to a uniform size (Fig. 1). These makeshift cards, each carefully inscribed, were not housed in a varnished wooden box such as could then be found in libraries or offices up and down the country, but instead in re-used cartons, one for coffee filters here, another for a presentation table lighter there. He also produced a catalogue reconstructing the Savilian Library of mathematical books as it would have been available to Wallis and successive contemporary astronomy professors such as Seth Ward (1617–1689) and Edward Bernard (1638–1697) (Fig. 2). The appearance does not diminish the content or the value in any way. These painstakingly produced catalogues and records of scholarly practice are not only accurate and reliable. They also contain a wealth of information which has only very rarely been superseded. Historical research instruments born of the shortages of post-war Europe and prepared by a careful mind averse to wastefulness, they are still in use today.

In populating the various catalogues with detailed information on books, manuscripts, or letters, Scriba was aided considerably by his mastery of shorthand. Relevant facts could be set out in a compact manner, saving time, space, and ink. Many a card is replete with acquired knowledge which, however, can only be unlocked by the skilled interpreter. Here, too, we have an example of Scriba's foresight and industry. Aware that he might need a financial back-up to fund his future studies, he had taken courses in shorthand and typing while still at school. These skills stood him in good stead. When Hofmann made his far-reaching proposal back in 1955, that Scriba work on Gregory's early mathematical writings, the prospective young doctoral student had no immediate means of financing such an endeavour. It was only because Ullrich was able to offer him a part-time position as secretary to the mathematics department in Giessen that Scriba accepted the proposal. Shorthand also played another important role in the beginning of Scriba's academic career. After attending the colloquium in Oberwolfach in 1955, Scriba became a member of Hofmann's select circle of scientific assistants, who would seek out rare books, check dates, and collate other factual information for their intellectual master. Hofmann was at the time engaged in producing the first volume of mathematical correspondence for the third series of the Academy Edition of the complete letters and papers of Leibniz, the *Sämtliche Schriften und Briefe*. In return, he would send Scriba proof-sheets for correction along with a steady stream of scholarly letters, mainly written in shorthand, and often supplemented by explanatory coloured drawings. Indeed, Scriba claimed that it was these epistolary

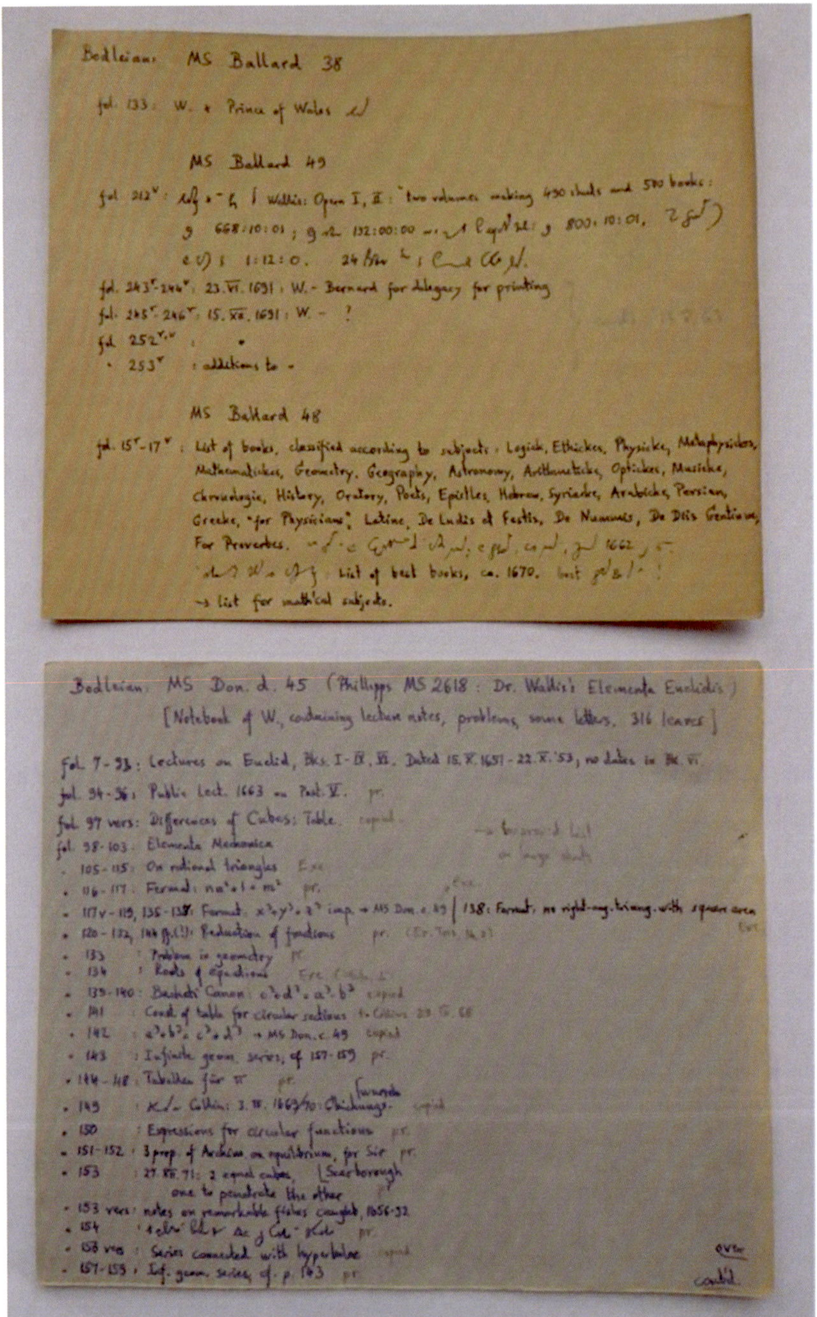

Fig. 1 File cards produced by C. J. Scriba during his stay in Oxford, containing detailed information on Wallis letters and papers in the manuscript holdings of the Bodleian Library. Typically, some of this information is written in shorthand

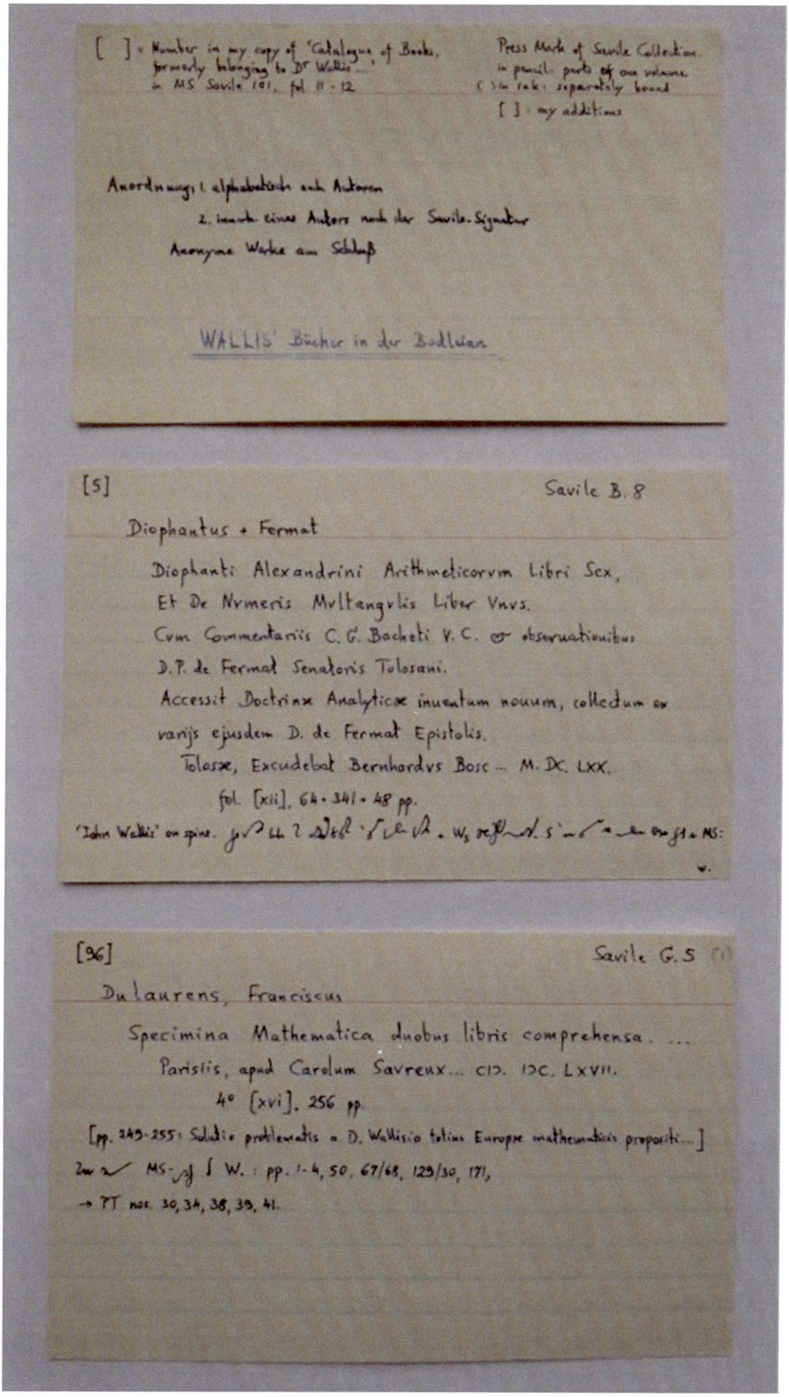

Fig. 2 File cards forming part of C. J. Scriba's reconstruction of the Savilian Library of mathematical books. Again, some of the information is written in shorthand

exchanges with Hofmann which gradually introduced him to serious academic research in the history of mathematics.[9]

3 Scriba's 'Check-List' of Wallis Correspondence

Christoph Scriba's search for Wallis letters was extensive. Apart from the Bodleian Library and the college libraries at Corpus Christi and Christ Church in Oxford, he investigated the book and manuscript holdings of the British Library (then still housed in the British Museum), as well as the University Library in Cambridge and the Library of the Royal Society. He also sent enquiries to the National Register of Archives in London and the Austrian National Library in Vienna. These efforts soon bore fruit. By the autumn of 1963, he had succeeded in compiling a list with details of around eight hundred Wallis letters, either written by or sent to the Savilian professor of geometry. Scriba called this compilation his 'check-list' of Wallis's correspondence and its creation provided the pretext for his acquaintanceship with the pinnacle of Great Britain's scientific establishment. Alistair Crombie (1915–1696), Oxford's first professor of the history of science, had already given him the necessary letter of introduction to the Bodleian Library. But the library of the Royal Society was a different matter entirely. Through early contacts with Tom Whiteside in Cambridge, he was aware that Sir Harold Hartley (1878–1972), the "grand old man of British science" and prominent Fellow of the Royal Society,[10] provided the keys to that institution and possibly also to the extensive private collection of the Earl of Macclesfield at Shirburn Castle in Oxfordshire.[11] Whiteside offered to write to Hartley for a letter of introduction to the Royal Society, but Scriba also wrote to him on his own accord. Fortunately, he found in Hartley someone keen to

[9]Christoph J. Scriba, 'Geschichte der Mathematik im Spiegel der Zeit. Zugleich eine Würdigung des Schaffens von Joseph E. Hofmann', in: *Joseph Ehrenfried Hofmann zum 70. Geburtstag*, Giessen: Selbstverlag des Mathematischen Seminars 1971 (= Mitteilungen aus dem Mathematischen Seminar Giessen No. 90), 2–24, pp. 2–3: "Nun, bald wurde auch ich einer jener Hilfsknechte, die aus seltenen Büchern Daten und Fakten ausgruben und nach Ichenhausen sandten. Dazu traten Korrekturen, die ich mitlesen durfte, und dazu kam jener Strom von Briefen, meist in Stenographie verfaßt, häufiger durch farbige Zeichnungen ergänzt, der mich allmählich in die mathematikgeschichtliche Forschung einführte." After the war, Hofmann and his wife Josepha had moved back to the family home in Ichenhausen in Bavaria.

[10]Preface to *The Mathematical Papers of Isaac Newton*, ed. D. T. Whiteside, Vol. VI (1684–1691), Cambridge: Cambridge University Press 1974, p. viii.

[11]Scriba wrote to D. T. Whiteside for possible assistance in gaining access to the Macclesfield Collection at Shirburn Castle once work with the author to publish the Wallis edition was well underway. In typical style he replied by email on 12 May 1998; Archive of the Wallis Edition: "Can Beeley gain access to Shirburn Castle Library? There is nothing lost by writing to him [sc. The Earl of Macclesfield] (but include a stamped addressed envelope if you expect a reply)". A reply was indeed received, but informing us that access could not be granted, because of restorative work taking place on the manuscripts. Only since their acquisition by Cambridge University Library have they become accessible—fortunately in time for the first volume of the *Correspondence of John Wallis*.

promote his work in the history of mathematics and who also shared his interest in Wallis. But formalities had to be fulfilled first. In his opening letter to Scriba, dated 17 October 1962, he requested a meeting with the young German scholar, for personal acquaintance was presupposed for a letter of introduction.[12] As a token of his interest, he enclosed the copy of an article on the early president of the Royal Society and skilled mathematician William Brouncker (1620–1684). This was a meeting of like-minded men.[13] Between Hartley and Scriba a close, friendly relationship developed and over the years they regularly exchanging greetings at birthdays or sent copies of recent publications to each other.

While being a distinguished chemist in his own right and a high-ranking official in the Central Electricity Generating Board, Hartley also served for many years editor of the history of science journal, *Notes and Records of the Royal Society*. This provided another avenue for Scriba to advance his career. In a letter to Hartley, dated 24 November 1963, he floated the idea that his 'check-list' might "despite its shortcomings" be published in the journal.[14]

But what exactly were these shortcomings? During his investigations into Wallis's correspondence, Scriba had decided in general to exclude all or most of the letters concerning subjects which he described as being "completely alien" to his field of research.[15] These were

i. The letters on the Doctrine of the Trinity and other theological questions, some of which were printed during Wallis's lifetime.
ii. The letters Wallis exchanged with government officials relating to his deciphering of intercepted political correspondence, and the deciphered letters themselves.
iii. Letters written or received by Wallis on affairs relating to the University of Oxford in consequence of his forty-five year tenure of the post of Keeper of the University Archives.

---

[12]Harold Hartley to Christoph J. Scriba, 17 October 1962; Archive of the Wallis Edition: 'My friend Derek Whiteside has told me that you are anxious to read regularly at the Royal Society Library. Whiteside has told me about the subject of your thesis which you will see from the enclosed paper is of particular interest to me. I should, however, like to see you first so that I can say that you are known personally to me.' The article he enclosed was one he had written jointly with J. F. Scott: 'William, Viscount Brouncker, F.R.S. (1620–1684)', in: *Notes and Records of the Royal Society* 15 (1960), pp. 159–65.

[13]Hartley's interest in the history of science is reflected in his later publication *Studies in the History of Chemistry*, Oxford: Clarendon Press 1971.

[14]Christoph J. Scriba to Harold Hartley, 24 November 1963; Archive of the Wallis Edition: "Do you think it would be desirable to publish this list, in spite of its short-comings, in a form similar to the index of Boyle's correspondence given by R. E. W. Madison in *Notes and Records of the Royal Society* 13?"

[15]See Christoph J. Scriba, 'A Tentative Index of the Correspondence of John Wallis, F.R.S.', in: *Notes and Records of the Royal Society of London* 22 (1967), pp. 58–93, p. 58.

iv. Letters written or received by Wallis that are known to have existed, but which are currently missing.

While Scriba therefore recognized that the eight hundred or so letters on his list were not exhaustive and that, as he wrote to Hartley, it was not unlikely that further letters might "come to light if a systematic search is made in other British and continental libraries",[16] he felt that those he had succeeded in locating constituted the bulk of Wallis's scientific legacy at least as far as his correspondence was concerned. Indeed, he made it clear that he had no intention of making such a search, since, as he writes, "there is no hint that important new material on Wallis's mathematics, in which I am mostly interested will be found".[17]

Hartley evidently accepted the rationale of Scriba's exclusion principles and therefore encouraged the young German scholar to prepare his 'check-list' for publication. It took, however, almost another three years before Scriba finally sent in his manuscript, and a further twelve months after then, before it finally appeared in print in the *Notes and Records*.[18]

Part of the reason why Hartley was keen to support Scriba's work on the 'check-list' went beyond the latter's scholarly endeavours. In 1967, the Historical Manuscripts Commission and the Royal Society established a joint committee to examine the existing arrangements in the United Kingdom for "locating, preserving and making available records and documents (including personal correspondence) of value for the history of science and technology".[19] Hartley was the first chairman of this committee with its forward-looking brief. In his letter to Scriba of 18 January 1967, written, as he points out, the day after he had chaired the first meeting, he describes the particular value of Scriba's contribution in the light of the Committee's remit as being precisely that it not only identifies Wallis's extant letters but also gives their precise location (Fig. 3). Almost anticipating modern-day efforts to create central data-bases of early-modern scholarly correspondence,[20] Hartley explains part of the Committee's aim as being to harvest the epistolary records created through investigations into the papers of those individuals who have played a role in the development of modern science: "Part of our programme is to approach scholars like yourself who have studied the work of individual scientists to ask them to let us have the results of their researches as to the location of documents".[21]

[16]Christoph J. Scriba to Harold Hartley, 24 November 1963; Archive of the Wallis Edition.

[17]Ibid.

[18]See note 13. The date of publication is given as September 1967.

[19]The agenda, minutes, correspondence, and other papers relating to this commission are deposited under the reference 'HMC 6' in The National Archives, Kew.

[20]See for example the online resources of Early Modern Letters Online (EMLO).

[21]Harold Hartley to Christoph J. Scriba, 18 January 1967; Archive of the Wallis Edition.

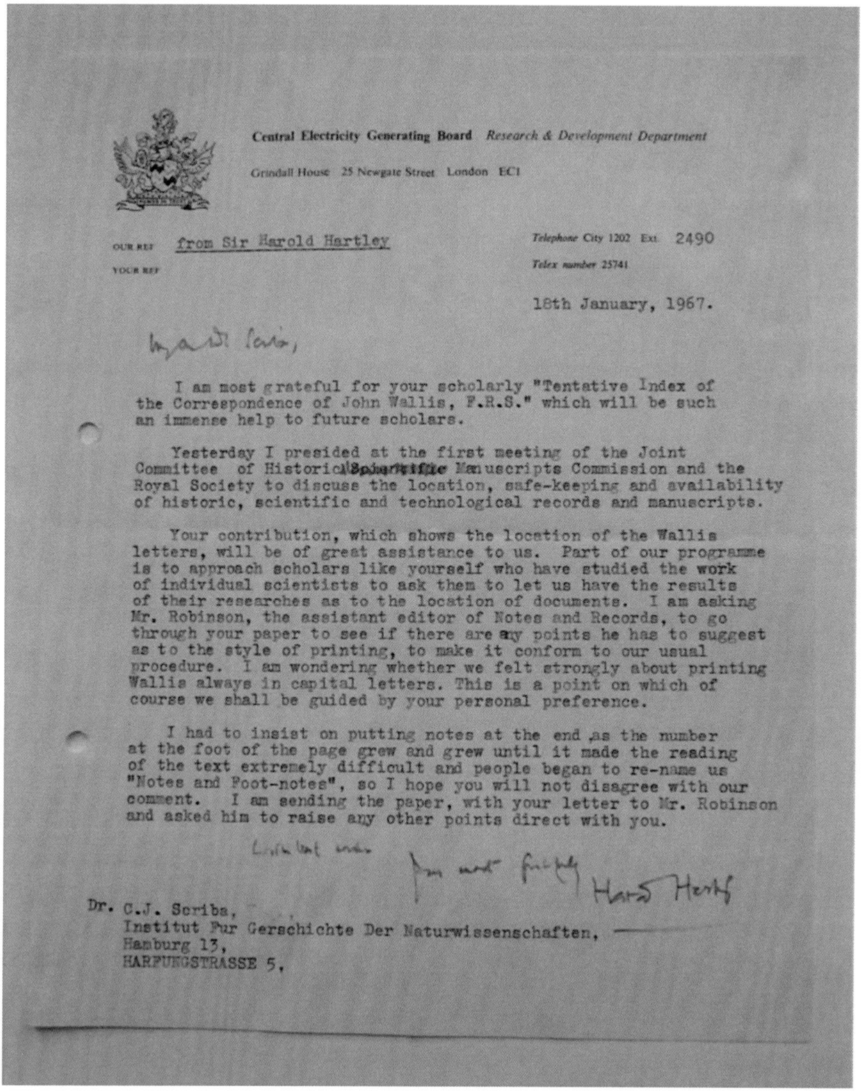

Central Electricity Generating Board *Research & Development Department*
Grindall House 25 Newgate Street London EC1

OUR REF from Sir Harold Hartley
YOUR REF

Telephone City 1202 Ext 2490
Telex number 25741

18th January, 1967.

I am most grateful for your scholarly "Tentative Index of the Correspondence of John Wallis, F.R.S." which will be such an immense help to future scholars.

Yesterday I presided at the first meeting of the Joint Committee of Historical Manuscripts Commission and the Royal Society to discuss the location, safe-keeping and availability of historic, scientific and technological records and manuscripts.

Your contribution, which shows the location of the Wallis letters, will be of great assistance to us. Part of our programme is to approach scholars like yourself who have studied the work of individual scientists to ask them to let us have the results of their researches as to the location of documents. I am asking Mr. Robinson, the assistant editor of Notes and Records, to go through your paper to see if there are any points he has to suggest as to the style of printing, to make it conform to our usual procedure. I am wondering whether we felt strongly about printing Wallis always in capital letters. This is a point on which of course we shall be guided by your personal preference.

I had to insist on putting notes at the end, as the number at the foot of the page grew and grew until it made the reading of the text extremely difficult and people began to re-name us "Notes and Foot-notes", so I hope you will not disagree with our comment. I am sending the paper, with your letter to Mr. Robinson and asked him to raise any other points direct with you.

Dr. C.J. Scriba,
Institut Fur Gerschichte Der Naturwissenschaften,
Hamburg 13,
HARFUNGSTRASSE 5.

Fig. 3 Letter from Sir Harold Hartley to C. J. Scriba, dated 18 January 1967, in which he outlines the programme of the Joint Committee of the Historical Manuscripts Commission and the Royal Society for collecting data on the manuscript letters and papers of scientists

4 First Plans for an Edition

Shortly before the end of his two year sojourn in Oxford, in February 1964, Scriba fashioned his first plan for an edition of Wallis's correspondence. This was to be a fairly modest affair: a small collection of selected letters, previously unpublished and considered to be of interest for the history of science in general or the history of mathematics in particular. Which letters he had in mind is not apparent from surviving papers, and it is possible that he never got as far as drawing up a possible list. Indeed, the enterprise seems to have been hurriedly conceived at the last moment, for it is first mentioned in the context of a conversation Scriba had with Alistair Crombie in Oxford less than a week before his planned departure for Germany. On his return, he was to take up a teaching post in the Institute for the History of Science at the University of Hamburg which had recently been established by the astronomer and science historian Bernhard Sticker (1906–1977).[22] Crombie recommended that Scriba write immediately to Daniel Davin (1913–1990), a senior figure at Oxford University Press (OUP), in order to arrange a meeting to discuss the proposal.[23] Under normal circumstances, Crombie would have sent a letter of introduction to Davin beforehand, but in view of the shortage of time this formality was omitted (Fig. 4).

A meeting was arranged a few days later at the offices of OUP in Walton Street, but with the science editor A. M. Wood, not with Davin.[24] During their discussion, it was agreed that Scriba would, early in the following year, send Wood a specimen Wallis letter, edited and provisionally typeset, together with an outline of the proposed publication. Although this demand might not have seemed particularly onerous, Scriba probably underestimated the amount of time and effort involved in setting up house again in Germany, in getting acclimatized in the new institute, and above all in completing work on his *Habilitationsschrift*.

Ten months later, with the agreed deadline rapidly approaching, Scriba set pen to paper. Writing to Wood on 28 December 1664, he had no alternative than to explain to his prospective editor that "due to diverse circumstances" he had been unable to advance as fast with his work on the Wallis edition as he had hoped.[25]

---

[22]Sticker had been directing and at the same time building up the Institute, which came to be known under its German acronym IGN (Institut für Geschichte der Naturwissenschaften) since 1960. Part of the funding arrangement for the institute was that it would train young scholars so that they could take up professorial chairs in the history of science at other German universities and in this way enable the growth of the discipline. It was agreed by Hofmann and Sticker that Scriba after the end of his DFG-funded research in Oxford would take on teaching responsibilities at the IGN in Hamburg. See Christoph J. Scriba, 'Bernhard Sticker (2 August 1906-30 August 1977)', in: *Studia Leibnitiana* 9 (1977), pp. 159–67; Andreas Kleinert, 'Christoph J. Scriba (1929–2013)' in: *Sudhoffs Archiv* 97 (2013), pp. 136–42.

[23]In his letter to Davin, dated 25 February 1964, Archive of the Wallis Edition, Scriba points out that he, his wife, Inge, together with their three-year old son, Friedemann, had intended to stay in Oxford for another month, but had then to change their plans.

[24]Davin, a Fellow of Balliol College, was Assistant Secretary to the Delegates of the University Press at the time. It was no doubt felt more appropriate that Scriba should meet with Michael Wood, the commissioning editor for science and medical publications. On the roles of Davin and Wood at OUP during the 1960s, see *The History of Oxford University Press*, ed. William Roger Louis, vol. III (1896–1970), Oxford: Oxford University Press 2013, p. 108; Keith Ovenden, *A Fighting Withdrawal: the life of Dan Davin, writer, soldier, publisher*, Oxford: Oxford University Press 1996.

[25]Christoph J. Scriba to A. M. Wood, 28 December 1964; Archive of the Wallis Edition.

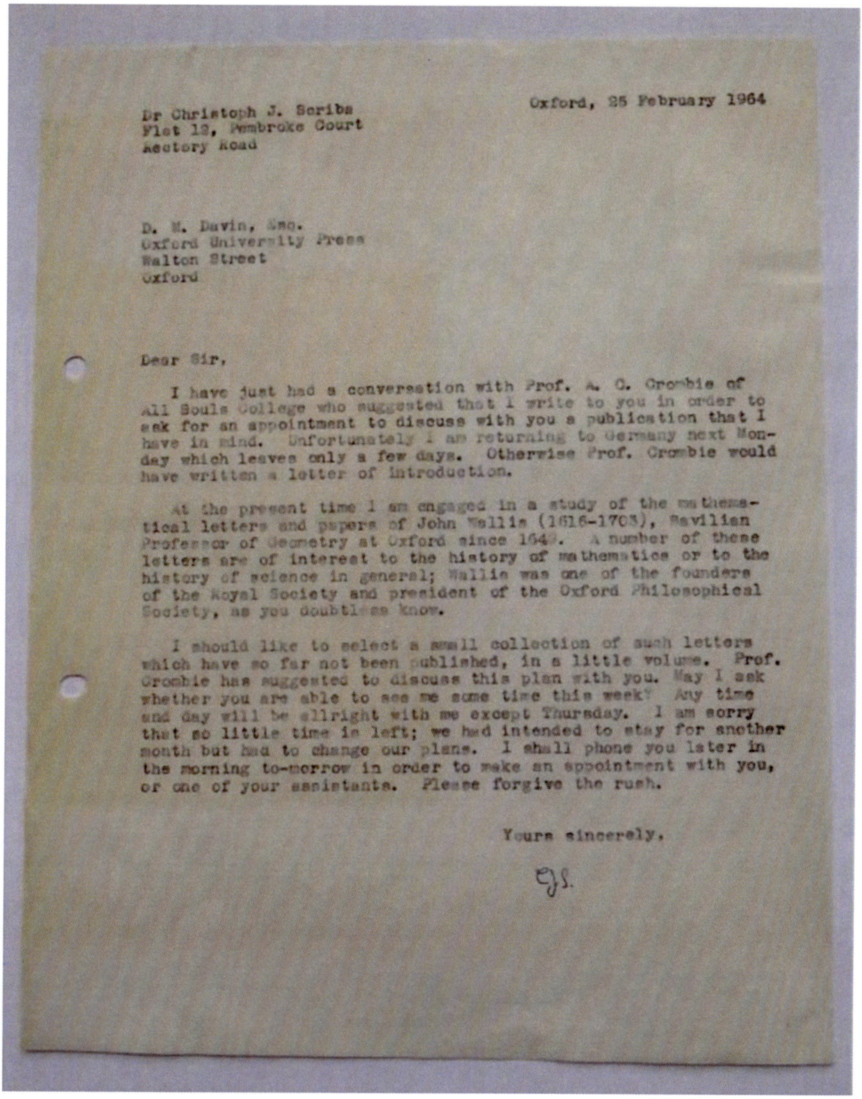

Dr Christoph J. Scriba
Flat 12, Pembroke Court
Rectory Road

Oxford, 25 February 1964

D. M. Davin, Esq.
Oxford University Press
Walton Street
Oxford

Dear Sir,

I have just had a conversation with Prof. A. C. Crombie of All Souls College who suggested that I write to you in order to ask for an appointment to discuss with you a publication that I have in mind. Unfortunately I am returning to Germany next Monday which leaves only a few days. Otherwise Prof. Crombie would have written a letter of introduction.

At the present time I am engaged in a study of the mathematical letters and papers of John Wallis (1616-1703), Savilian Professor of Geometry at Oxford since 1649. A number of these letters are of interest to the history of mathematics or to the history of science in general; Wallis was one of the founders of the Royal Society and president of the Oxford Philosophical Society, as you doubtless know.

I should like to select a small collection of such letters which have so far not been published, in a little volume. Prof. Crombie has suggested to discuss this plan with you. May I ask whether you are able to see me some time this week? Any time and day will be allright with me except Thursday. I am sorry that so little time is left; we had intended to stay for another month but had to change our plans. I shall phone you later in the morning to-morrow in order to make an appointment with you, or one of your assistants. Please forgive the rush.

Yours sincerely,

CJS.

Fig. 4 Typescript copy of letter from C. J. Scriba to Daniel Davin, dated 25 February 1964, setting out his plan for a small edition of Wallis's letters

Conservatively, he estimated that another year might pass before he would be able to submit the material Wood required. For an academic editor, of course, delays of that nature are common place. Wood duly responded to Scriba's request for a postponement with corresponding kindness and understanding. "We will wait

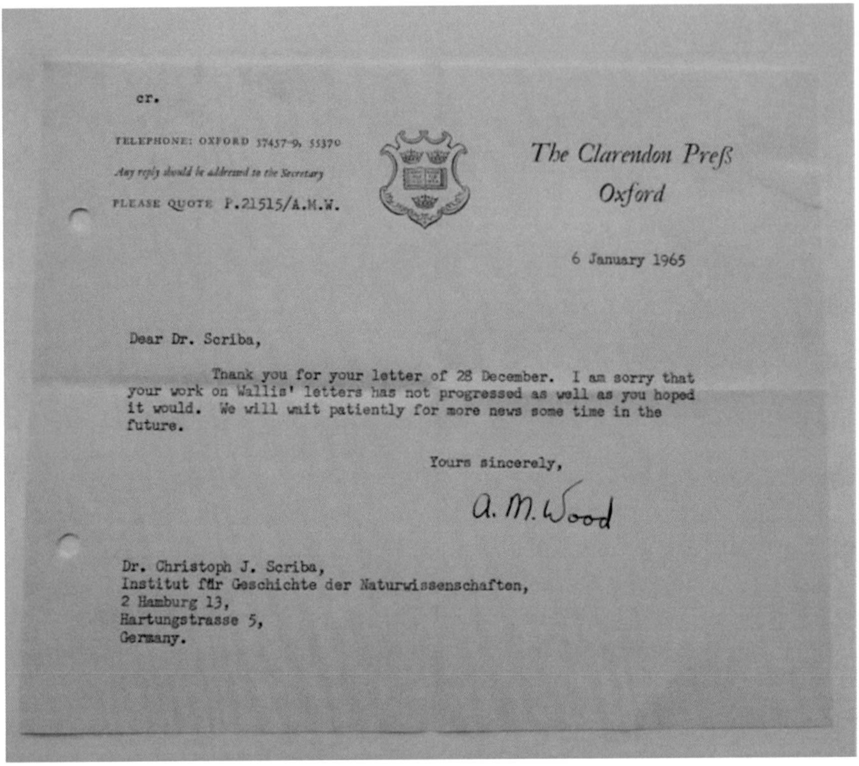

cr.

TELEPHONE: OXFORD 57457-9, 55370
Any reply should be addressed to the Secretary
PLEASE QUOTE P.21515/A.M.W.

The Clarendon Preß
Oxford

6 January 1965

Dear Dr. Scriba,

Thank you for your letter of 28 December. I am sorry that your work on Wallis' letters has not progressed as well as you hoped it would. We will wait patiently for more news some time in the future.

Yours sincerely,

A. M. Wood

Dr. Christoph J. Scriba,
Institut für Geschichte der Naturwissenschaften,
2 Hamburg 13,
Hartungstrasse 5,
Germany.

Fig. 5 Letter from A. M. Wood to C. J. Scriba, dated 6 January 1965, in response to news of the delay in the realization of the planned Wallis edition

patiently for more news sometime in the future," he wrote at the beginning of the following year.[26] (Fig. 5)

5 The Second and Third Plans for an Edition

Patience is certainly what was required, for it was a quarter of a century later that Scriba next contacted OUP about his proposal to edit the correspondence of John Wallis. Writing to Martin Gilchrist, senior mathematics editor, in December 1989, he refers to his earlier exchanges with Michael Wood and indicates that he is now not merely intending to return to his earlier plan, but has actually made a start in carrying it out: "I should add that we are just seriously beginning to type and edit the

[26] A. M. Wood to Christoph J. Scriba, 6 January 1965; Archive of the Wallis Edition.

letters".[27] With the editorial work about to get underway, Scriba sought Gilchrist's advice on the most suitable text processing system from OUP's point of view.

By the time Scriba wrote to Gilchrist, he had been ordinary professor for the history of science at the University of Hamburg for fourteen years, having previously occupied the chair for history of exact sciences at the Technische Universität Berlin. His teaching and administrative commitments at the time were enormous, not to mention his involvement in scientific bodies such as the DFG or the Leopoldina, his editorial roles for various academic journals, conference organization, and so on. Apart from the evidence of his letter little seems to have happened by way of re-starting the project at the end of that momentous political year in Germany, although it is possible that Scriba made some attempts at entering the texts of letters using his preferred text processing system.[28]

Nothing substantial altered this prolonged state of dormancy until around five years later, when, with retirement approaching, Scriba drew up a first application to the DFG for funding for a three-year project to produce an edition of Wallis's correspondence. This application, submitted on 30 June 1993, was phrased largely in the terms of the plan he had sketched out in Oxford some thirty years earlier, although there was now no talk of a "selection of letters". Instead, the proposal formulated by Scriba envisaged the publication of around 350 scientific letters of the 800 or so he had listed in his 'Tentative Index', published in 1967. The presentation of the letters was to follow the model adopted by the editors of the Oldenburg correspondence. There was to be no critical apparatus, but persons mentioned in the letters were to be identified and factual problems were to be explained concisely, referring where necessary to relevant primary and secondary literature. Explicitly not to be included in the edition were those letters which had been published in other collected editions, namely the *Correspondence of Isaac Newton*, the *Correspondence of Henry Oldenburg*, the *Sämtliche Schriften and Briefe* of G. W. Leibniz, and the *Œuvres complètes* of Christiaan Huygens. Essentially, therefore, the edition was conceived as serving as a supplement to those great works of scholarship, as Scriba made clear in the section of the application outlining the project's overall aims. Importantly, all the scientific correspondence, whether texts were published or not, was to be listed:

> The aim of the edition is to make the scientific correspondence of Wallis completely accessible for the first time. Since, despite his major importance for the scientific development in England in the second half of the seventeenth century, Wallis does not count among the really great mathematicians or natural scientists, and since also in particular his correspondence with Huygens, Leibniz and Newton is already available in their editions, only previously unpublished letters and the odd letter which has appeared in some remote publication will be included. Of course, *all* known letters (as well as those known to have

[27]Christoph J. Scriba to Martin Gilchrist, 27 December 1989; Archive of the Wallis Edition.

[28]In his letter to Gilchrist, Scriba expresses a preference for Word Perfect over LaTeX which he knew to be OUP's choice, suggesting it provides "more convenient editorial features for normal text". His main reason, however, was the practicality of the former: "also it is easier to learn (when one has frequently changing student assistants who do the typing) than TeX or LaTeX".

> existed, but which up to now have not been found) will be listed in chronological sequence. Corresponding to the model of the Leibniz edition, which in this respect proceeds in exemplary fashion, for every letter details of all extant manuscripts and print manifestations will be given, but the text itself only in those cases indicated.[29]

Another aspect of what Scriba intended only emerged when the DFG questioned the considerable sum, foreseen within the financial specification of the application, to cover the costs of checking grammatically, and stylistically, the English language introduction to the whole edition and the commentaries to the individual letters. Perhaps understandably, it was not clear to the DFG why English should be required at all, but Scriba explained that the reason was precisely in order that the proposed publication could be used alongside the correspondence editions of Newton and Oldenburg. But there was another reason, too. In accordance with the practice of those editions, all letters in the Wallis edition that were written in languages other than English were to be translated:

> Since in the intended edition of the correspondence of John Wallis none of those letters will be re-edited which have already been included in the published correspondences of Isaac Newton and Henry Oldenburg, it is planned that in its outer form it will be like those two editions. This requires in particular that English translations are provided for letters written in Latin, since nowadays unfortunately many readers have difficulties with Latin or cannot understand it at all.[30]

Quite simply, Scriba's argument was that the Wallis edition should have precisely the same format as those editions it was intended to complement.

The collaborator for this project had already been chosen when the application went in. During a stay in Hamburg, in November 1991, Siegmund Probst, a graduate student of Imre Tóth (1921–2010) from Regensburg, had discussed with Scriba his plan to write a doctoral thesis on the long-lasting intellectual war which took

[29]Christoph J. Scriba, 'Antrag auf eine Sachbeihilfe', 30 June 1993; Archive of the Wallis Edition: "Ziel der Edition ist es, die wissenschaftliche Korrespondenz von Wallis erstmals vollständig zugänglich zu machen. Da dieser trotz seiner großen Bedeutung für die Wissenschaftsentwicklung in England in der 2. Hälfte des 17. Jahrhunderts nicht zu den ganz großen zu rechnen ist, da ferner insbesondere seine Korrespondenz mit Huygens, Leibniz und Newton in deren Ausgaben bereits vorliegt, sollen hier nur die bisher unveröffentlichten und einzelne, versteckt publizierte Briefe aufgenommen warden. Selbstverständlich werden *alle* bekannten Briefe (auch erschlossene, aber bisher nicht aufgefundene) in chronologischer Anordnung verzeichnet. Nach dem Vorbild der Leibniz-Ausgabe, die hierin mustergültig verfährt, sollen für sämtliche Briefe die Hinweise auf erhaltene Handschriften und Drucke aufgenommen werden, doch der Text selbst nur in den genannten Fällen."

[30]Christoph J. Scriba to Sylvester Rostosky, 16 October 1993; Archive of the Wallis Edition: "[. . .] da in der vorgesehenen Ausgabe der Korrespondenz von John Wallis auf die Wiedergabe all jener Briefe verzichtet warden soll, die bereits in den publizierten Korrespondenzen von Isaac Newton und Henry Oldenburg enthalten sind, ist beabsichtigt, die äußere Form diesen beiden Ausgaben möglichst anzugleichen. Das bedingt insbesondere, daß die lateinischen Briefe auch in englischer Übersetzung gebracht warden, weil sich leider heute viele Leser mit dem Latein schwer tun oder es gar nicht mehr verstehen."

place between John Wallis and the philosopher Thomas Hobbes (1588–1679) over a large part of the second half of the seventeenth century. Soon convinced of his ability, Scriba helped secure funding for Probst's subsequent investigations. At one of their meetings, he told Probst of his intention to produce an edition of Wallis's correspondence and asked him to work as his assistant in Hamburg after completing his thesis. Meanwhile, Scriba submitted his application to the DFG to fund work on the edition with the start-date set to coincide with Probst achieving the necessary qualification to take up his position. Although much of the planning had taken place three years earlier, work on the edition of the correspondence did not begin in earnest until May 1994.

During the following months, Probst began to create a filing system with all the letters to be edited ordered chronologically, while student assistants began the laborious task of entering the texts electronically using the text editing system LaTeX. Initial steps were undertaken to obtain copies of all the manuscript and printed manifestations of each letter, and a reference library of all literature that would be needed for everyday editorial work was initiated. However, despite these promising beginnings, work soon came to a practical standstill, when Probst, at the beginning of 1995, left Hamburg to take up a permanent position as editor of Leibniz's mathematical writings at the Niedersächsische Landesbibliothek (now the Gottfried-Wilhelm-Leibniz Bibliothek) in Hanover. Difficulty in finding a suitable replacement led to an interregnum of almost 2 years. The present author, a former doctoral student of Hans Poser and Eberhard Knobloch at the Technische Universität Berlin, took over as Probst's successor at the beginning of November 1996.

As work got underway again, fundamental questions of presentational style, critical apparatus, and inclusiveness needed to be addressed. Considerable time was expended in developing a catalogue of editorial principles. Since all three editors, Scriba, Probst, and the author, had close professional ties to the Leibniz edition and had learned to appreciate the elegance of it critical apparatus, with variant readings nicely contained at the foot of the page, leaving the text uncluttered by editorial interventions, it was decided to adopt this model in the Wallis edition, too, but in a somewhat simplified version: only exceptionally would variant readings of sources other than the lead manuscript be recorded. This was a concession to the demand for efficiency, given the financial limitations of the edition both in regards to project time and personnel resources. Otherwise things were to be very similar. There was to be a complete list of manifestations in which a letter or its enclosure had been handed down, footnotes were to explicate unclear references in the text (books, persons, events, and so on), but were not to be interpretive in character, and the orthography of the original was, with minor exceptions, to be retained. Furthermore, in contrast to what Scriba had applied for, there were to be no translations of letters written in languages other than English. Indeed, this was one of the provisions made by the DFG in awarding the grant to finance work on the edition.

Progress was slower than originally anticipated and the disruption caused by bringing in a new editorial assistant did not help matters either. There were also considerable delays in obtaining copies (photocopies, photographs, microfilms, or later digital images) of letters from innumerable libraries and archives around the world. It did not help matters that in some cases the shelf-marks that Scriba had carefully recorded during his investigations in the 1960s had been superseded by more modern referencing systems.

On some occasions, the libraries approached were able to inform the editors of other Wallis letters which either had been overlooked earlier or which had since come to light. Notices about the ongoing work on the edition combined with a call for assistance in finding letters appeared in journals such as *Isis* and *Historia Mathematica*. Encouraged by the responses and the willingness of colleagues to pass on information about Wallis letters they had come across during their investigations, it was decided to widen the net further and contact European and North American libraries directly. While a major search of Italian institutions revealed no other letters than those which were known already, enquiries to repositories in Denmark, France, Germany, the Netherlands, Sweden, and the United States, proved more successful. Most fruitful of all, however, was a repository that Scriba quite consciously—and in retrospect probably wisely—had avoided consulting in any depth during his stay in Oxford: the University Archives, housed in the Tower of the Five Orders, the magnificent entrance to the Bodleian Library, where Wallis had served as Custos archivorum for much of his professional life. In the Lower Archive Room, where most of the cabinets with their sturdy cupboards and drawers can be traced back to Wallis's time, vast quantities of his letters were found.

6 The Fourth Plan for an Edition

In the spring of 1997, with the original funding through the DFG coming to a close, and against the backdrop of an ever increasing number of Wallis letters on file, a new application was submitted to the DFG, proposing for reasons based on the changing requirements of the scholarly community, an extension to the original project. Part of the reason for an extension, Scriba argued, was that whereas originally he had started out from there being around eight hundred extant letters, a conservative estimate now put this sum at around eleven hundred. (By the time the present author had finished sifting through the material in the University Archives in 2012 this figure had reached two thousand.) Consequently, the number of letters to be edited would need to be corrected upwards from the originally conceived three hundred and fifty to at least six hundred "and probably rather more", as he set out:

> For two reasons these figures now unfortunately need to be corrected heavily upwards. On the one hand we have in the meantime knowledge of more than 1100 letters written by or to Wallis. On the other hand—and this is of even more consequence—through recent editorial work it has become apparent that the texts of a large number of the letters edited in the fourteen volumes of the Oldenburg Correspondence between 1965 and 1986 are incorrect or so full of mistakes that they need to be edited again. (This is true in particular in the many

cases where both a draft and the letter sent and/or a copy thereof exist, since that edition only considered one manifestation.) Moreover, those letters cannot be separated from other epistolary communications which in Wallis's lifetime were published (often only partially) as articles in the *Philosophical Transactions*; for there are here many points of intersection of content.[31]

It is no doubt a reflection of the high regard in which Scriba was held in the DFG, where for seven years he had served as subject referee for the history of science, that a two year extension was subsequently granted. Remarkably, the editors' argument that the relatively small number of Wallis letters which had appeared in the editions of Christiaan Huygens, Newton, and Leibniz should be re-edited for the Wallis edition was accepted. However, despite Scriba's persuasive argument, a line was drawn at the three hundred or so Wallis letters that had appeared in the publication of the Oldenburg correspondence.[32] In true scholarly fashion, those letters were entered by the editors in their spare time.

Over the course of his five years working on the Wallis edition in Hamburg, the author was successful in convincing Scriba that the needs of contemporary scholarly audiences would only be met when editions of the correspondence of figures such as Wallis did not introduce artificial limits as to what can be counted as scientific. His argument was strengthened as a result of a meeting with Elizabeth Johnston, Gilchrist's successor as commissioning mathematics and medical science editor at OUP, in 1999, when the proposal to publish the correspondence of John Wallis was discussed at length.[33] From the University Press's point of view there was not the slightest doubt that the edition should be of Savilian professor's complete correspondence. On this understanding, though with some hesitation regarding

[31]Christoph J. Scriba, Antrag auf eine Sachbeihilfe, 17 April 1997; Archive of the Wallis Edition: "Aus zwei Gründen müssen diese Zahlen jetzt leider stark nach oben korrigiert warden. Einerseits haben wir inzwischen Kenntnis von über 1100 Briefen von bzw. an Wallis. Andererseits—und das ist viel gravierender—hat sich bei der bisherigen Bearbeitung herausgestellt, daß die Wiedergabe der großen Zahl der in den 14 Bänden der Oldenburg-Correspondenz zwischen 1965 and 1986 edierten Briefe oft ungenau oder gar derart fehlerhaft ist, so daß eine Neuedition erforderlich wird. (Dies gilt insbesondere für die Vielzahl der Fälle, wo sowohl ein Konzept wie eine Abfertigung und/oder eine Abschrift existiert, da die dortige Edition häufig nur eine Fassung berücksichtigte.) Auch sind jene Briefe nicht von der eigentlichen Korrespondenz zu trennen, die zu Lebzeiten von Wallis (oft auszugsweise) als Aufsätze in den *Philosophical Transactions* publiziert wurden; denn hier gibt es viele inhaltliche Berührungspunkte." Inadvertently, Scriba spoke in his application of fourteen rather than of thirteen volumes of the Oldenburg Correspondence.

[32]Guido Lammers to Christoph J. Scriba, 10 October 1997; Archive of the Wallis Edition: "Die Briefe aus der Oldenburg-Korrespondenz können auf keinen Fall in die beantragte Edition eingeschlossen warden."

[33]See Christoph J. Scriba to Elizabeth Johnston, 29 April 2000; Archive of the Wallis Edition: "Our aim is to present the complete correspondence chronologically, since this approach allows the letters best to reflect the day-to-day activity of the author."

future funding, the delegates of OUP accepted the proposal to publish the edition.[34] The news was also greeted with enthusiasm from Scriba's friends and colleagues.[35]

When Scriba responded to the DFG's misgivings over his application for an extension, including their rejection of a re-edition of the Oldenburg letters, he set out the reasons why now in his view a complete edition of Wallis's correspondence was scientifically desirable, even if the restrictions imposed by that funding body, not least with a view to what had originally been applied for and approved, were to be observed. Remarkably, he now turned the original argument—that the Wallis edition should complement that of Newton or Oldenburg—on its head:

> Purely from the point of view of the history of science only a complete edition of the Wallis correspondence would be a satisfactory solution. For it is precisely the interconnections of scientific, theological, political and university-related correspondence which are so revealing both when it comes to assessing the role of a certain figure and more generally of the state of science in seventeenth-century England. Nonetheless, in consideration of the difficult question of financing I decided to undertake a partial edition, as described in the original application and in that for its extension. The fact that later users of this edition will need at the same time to have access to a series of other editions of the letters of important contemporaries if they want to study Wallis's mathematical and scientific correspondence in context, will represent a considerable barrier to its use and to the scientific evaluation of its content. This disadvantage has been noted in my discussions with colleagues both at home and abroad.[36]

7 Historiographical Change

In many ways the transformation of the Wallis edition from its first conception in the 1960s until the publication of the first volume in 2003, mirrors the transformation in the conception of the history of mathematics itself. Scriba's original plan for

[34]Elizabeth Johnston to Christoph J. Scriba, 10 March 1999; Archive of the Wallis Edition: "Your plans for the publication of the Wallis Correspondence were presented to the Delegates at their recent meeting and I am pleased to tell you that your plan met with approval."

[35]Thus Marie Boas Hall writes in her letter to Scriba of 1 November 1999; Archive of the Wallis Edition: "I am very pleased to learn that you are working on the Wallis correspondence, a most worthy task and I fear involving much labour."

[36]Christoph J. Scriba to Guido Lammers, 30 October 1997; Archive of the Wallis Edition: "Zunächst eine Vorbemerkung: Allein unter wissenschaftshistorischen Aspekten gesehen, wäre eine vollständige Edition der Wallis-Korrespondenz die einzig zufriedenstellende Lösung. Denn es ist gerade das Ineinandergreifen der wissenschaftlichen, theologischen, politischen und Universitätskorrespondenz, das für die Beurteilung der Persönlichkeit wie der Situation der Wissenschaft im England des 17. Jahrhunderts so aufschlußreich ist. Dennoch hatte ich mich mit Rücksicht auf die schwierige Frage der Finanzierung zu einer Teiledition entschlossen, wie im Erstantrag und im Fortsetzungsantrag beschrieben. Die Tatsache, daß spätere Benutzer dieser Ausgabe dann zur gleichen Zeit Zugang zu einer Reihe von weiteren Briefeditionen bedeutender Zeitgenossen haben müssen, wollen sie die mathematisch-naturwissenschaftliche Wallis-Korrespondenz im Zusammenhang studieren, bedeutet eine erhebliche Erschwerung der späteren Benutzung und wissenschaftlichen Auswertung. Das wurde in Unterhaltungen mit Kollegen aus dem In- und Ausland auch schon moniert."

a selection of Wallis's unpublished scientific correspondence was perhaps an idea born of the moment, but the intention of publishing only those letters which were considered to have been of importance to the growth of early modern science was of longer duration. Indeed, it formed the basis of his first funding application to the DFG. Of course, among historians of science active in the mid-twentieth century the view was widespread that a fairly clear distinction could be made between scientific and non-scientific letters. Such correspondence as pertained to theological questions or legal matters would, for example, clearly be excluded from the former category. But equally if not more important is the intellectual environment in which Scriba began to make his mark in the history of mathematics. For, when Scriba writes that it was Hofmann who introduced him to the history of mathematics he refers to an approach to the subject with which the name of the great Leibniz scholar is very much associated, namely internal history or, to use the more precise German expression, Problemgeschichte. Scriba followed this approach for a while, but then came to realize its limitations. Indeed, he felt that Hofmann in his professional career took this particular approach, in which mathematical problems themselves constitute the focal point of interest, to perfection. So much so that in his view there was little prospect of it being able to progress any further.[37]

After his return to Germany following his time in North America and England, Scriba organized the yearly colloquium on the history of mathematics at Oberwolfach together with Hofmann according to his teacher's preferred internal history approach. Only slowly was room made for other historiographical methods such as that of the history of ideas (Ideengeschichte), for which Scriba, in the 1970s, argued increasingly strongly. He saw the rationale for an approach rooted in the history of ideas as being the conceptual interrelationships between mathematics and other disciplines such as philosophy or the physical sciences particularly when it came to the development of new ideas. Among the historical examples he cites are the development of the binary system in Leibniz or contemporary advances in the concept of space.[38]

Scriba's history of ideas approach, while recognizing that the inspiration for creative mathematical work often comes from outside the discipline, was still firmly rooted within the mid-twentieth century tradition of the history of science—as could be seen most readily through a comparison with contemporary publications of Adolf Prag or D. T. Whiteside. Although he witnessed the emergence of the

[37]Scriba, 'Geschichte der Mathematik im Spiegel der Zeit', p. 7: "Erstens glaube ich, daß die problemgeschichtliche Behandlung der mathematischen Entwicklung, wie sie Herr Hofmann in zuvor unerreichter Meisterschaft entwickelt hat bei Untersuchungen zur Geschichte der Mathematik bis ins 19. Jahrhundert hinein, in dieser Form nicht viel weiter fortgesetzt werden kann." Scriba also discusses this question in a number of other publications. See 'Über Aufgaben und Probleme mathematikhistorischer Forschung', in: *Beiträge zur Methodik der Wissenschaftsgeschichte*, ed. Walter Baron, Wiesbaden: Franz Steiner Verlag 1967, pp. 54–80, and 'Geschichtsschreibung der Mathematik', in: *Gießener Universitätsblätter* 2 (1970), pp. 44–51.

[38]Scriba, 'Geschichte der Mathematik im Spiegel der Zeit', pp. 8–9. On this topic see also his article 'Die Rolle der Geschichte der Mathematik in der Ausbildung von Schülern und Lehrern', in: *Jahresbericht der Deutschen Mathematiker-Vereinigung* 85 (1983), pp. 113–28.

social history of science, this was not an approach that he welcomed or embraced in any meaningful sense. Nonetheless, it is a reflection of his intellectual openness that during the many years he was chiefly responsible for organizing the history of mathematics colloquia at Oberwolfach following Hofmann's death in 1973, he readily accommodated the different approaches to the subject.[39] It was this historiographical change that led to a major study being carried out on developments in the history of mathematics in the 1990s. Supported by the International Commission on the History of Mathematics (ICHM), Christoph Scriba and Joseph Dauben assumed overall editorial responsibility for the resulting publication.[40]

8 The Correspondence of John Wallis in Print

The extension of the range of letters to be included in the edition of the *Correspondence of John Wallis* met the changing demands not only of contemporary historians of mathematics but also corresponded to the wishes of the publisher. However this extension quickly revealed the insufficiencies of present day funding regimes to deal with such projects. Long-term employment restrictions at German universities were the source of additional problems. After five years at the Institute for History of Science at Hamburg, the author left to take up a permanent position working on the Leibniz edition at the University of Münster. This move was necessary, because the statutory limit for employment on a non-permanent project basis had been reached. His successor, Uwe Mayer, was employed on the remainder of the DFG grant until the end of 2003; soon thereafter he took up a permanent position working on Leibniz's mathematical papers for the Leibniz edition in Hanover. With DFG funding all but exhausted, work on the Wallis edition was continued by Scriba in retirement and the present author in his spare time until the latter succeeded in obtaining funding to move the whole edition to Oxford in September 2007.[41]

[39]See Christoph J. Scriba, 'Die Tagungen zur Geschichte der Mathematik im Mathematischen Forschungsinstitut Oberwolfach/Schwarzwald', in: *Nachrichtenblatt der Deutschen Gesellschaft für Geschichte der Medizin, Naturwissenschaft und Technik e.V* 26 (1965), pp. 63–7. In this report, Scriba reveals that at one of the meetings he had given a talk on unpublished mathematical correspondence and manuscripts, no doubt those of Wallis (p. 65). See also his article 'Dreißig Jahre Tagungen zur Geschichte der Mathematik im Mathematischen Forschungsinstitut Oberwolfach', in: *Berichte zur Wissenschaftsgeschichte* 8 (1985), pp. 47–9.

[40]*Writing the History of Mathematics: its historical development*, ed. Joseph W. Dauben and Christoph J. Scriba, Basel, Boston, Berlin: Birkhäuser Verlag 2002. On the background to this study see especially pp. xxv–xxvii.

[41]Numerous hurdles had to be overcome to bring about these working arrangements up to 2007, for the Department of Mathematics at the University of Hamburg, of which the Institute for History of Science was a constituent part, at first insisted that the author could not even continue to work on the project without remuneration. Access to the project rooms was also forbidden due to potential problems of legal precedence in establishing a case for permanent employment. Fortunately, Scriba and the university administration were able to find a way around these issues so that the author was able to work in the project rooms whenever he was in Hamburg.

In contrast to the "800 or so letters" listed in Scriba's 'Tentative Index', the Wallis edition now has over two thousand letters on file, making it one of the major scientific correspondences of the seventeenth century. Up to now, four volumes, edited jointly by Christoph Scriba and the author, have appeared, while volume five is currently in preparation:

Volume I (1641–1659) Oxford: Oxford University Press 2003
Volume II (1660–September 1668) Oxford: Oxford University Press 2005
Volume III (October 1668–1671) Oxford: Oxford University Press 2012
Volume IV (1672–April 1675) Oxford: Oxford University Press 2014
Volume V (May 1675–1678) Oxford: Oxford University Press (in preparation)

Each volume of *The Correspondence of John Wallis* comprises around two hundred and fifty items of correspondence, so that the edition when complete will consist in eight volumes. It is proposed that a ninth volume will contain additions to the previous ones—essentially letters which since publication have come to light—along with summaries of all the letters written in languages other than English. A persistent complaint in reviews of the volumes that have appeared up to now is that they do not contain translations of letters written in Latin, French, or Dutch, the absence of translations being non-standard in English language publications of the present day. Providing summaries will go some way to meet this complaint.

A simple comparison between the 'Tentative Index' and the latest published volume shows how large a component of Wallis's epistolary exchanges fell outside of the narrow classification of scientific correspondence employed by Scriba earlier in his career.

Year	Tentative Index	Volume IV
1672	30	67
1673	23	73
1674	22	78
1675 (to end of April)	4	33
Total	79	251

However, some of this discrepancy cannot simply be explained by Scriba's decision when preparing the 'Tentative Index' to exclude certain categories of letters as being alien to his interests. In a number of cases it is due to the nature of early modern correspondence in general and of mathematical correspondence in particular. Scientific discourse in the seventeenth century often involved numerous authors across the Republic of Letters. Indeed, the intersecting networks of correspondence criss-crossing Europe not only served to promote but also to engender such participation. Sometimes an individual's contribution to a topic would be conveyed by word of mouth or by the letters of other scholars. In such cases not to include the correspondence of third parties might and often would lead to serious gaps in the historical account. An example of this phenomenon is provided in volume IV of the *Correspondence of John Wallis* by the debate over the method of tangents in which a whole range of mathematicians participated, including René François de Sluse (1622–1685), Jan Hudde (1628–1704), Christiaan Huygens, Isaac Barrow (1630–1677), James Gregory, Isaac Newton, and Wallis himself. Important parts of the Savilian professor's contribution to that debate do not figure in his extant correspondence at all, but are reported in the letters of others such as Gregory and the mathematical intelligencer John Collins. Including the letters of third parties can help to remove such gaps in the transmission of ideas as well as those that have come about through the inevitable loss of parts of an author's correspondence over time.

But there are other reasons, too, why it is often necessary to go beyond an individual's direct personal correspondence. In order to lend a certain kind of objectivity to scientific discourse, letters were often addressed to intermediaries on the understanding that they would nonetheless reach their intended destination. Thus the scientific amateur Francis Jessop (1638–1691) discussed Wallis's hypothesis of tides and the cycloidal line which in his view was traced by the Earth's centre in its annual rotation around the sun in letters sent to his friend, the natural philosopher Martin Lister (1639–1712) in York. Lister, a respected member of the Royal Society, sent these letters to the Society's secretary, Henry Oldenburg, who subsequently forwarded them to his friend in Oxford, John Wallis. Sometimes the original letters were sent on as enclosures, sometimes salient passages were copied by Lister or Oldenburg into their own letters. Wallis's replies took the same route in reverse order. To capture this debate, it was necessary in volume IV to include a number of letters Wallis neither wrote himself nor which were directly addressed to him. It would not have been practicable to consign this kind of information to footnotes or to the introductory comments to individual letters. Not only would such footnotes or comments be overly complex, there would also be a strong likelihood of editorial interpretation. Furthermore, the inclusive approach adopted by the Wallis edition avoids the need to refer users to unpublished and possibly inaccessible material. Indeed, a comparison with similar correspondence editions shows that the necessity for such an approach has long been recognized in the scholarly community.

The Correspondence of Isaac Newton, ed. H. W. Turnbull
Volume I (1661–1675), Cambridge 1959
136 letters; 64 not by or addressed to Newton

The Correspondence of Henry Oldenburg, ed. A. Rupert Hall and Marie Boas Hall
Volume IX (1672–1673), Madison and London 1973
296 letters; 13 not by or addressed to Oldenburg

The Correspondence of John Wallis, ed. Philip Beeley and Christoph J. Scriba
Volume IV (1672–April 1675), Oxford 2014
251 letters; 30 not by or addressed to Wallis

While some of the need to print letters exchanged between third parties will inevitably disappear as more editions of early modern mathematical correspondence are published, it is unlikely that it can ever be completely avoided precisely when it is a question of recreating the transmission and evolution of ideas across discrete epistolary networks. And precisely this was one of Christoph Scriba's concerns when he was advocating his approach to the history of mathematics based on the history of ideas in the 1970s. His argument, set out in his communications with the DFG for the need to facilitate research into Wallis's mathematical legacy through his letters continues to hold true.

9 Conclusion

As this chapter has shown, the edition of the correspondence of John Wallis had an extraordinarily long period of gestation. Between the time when Christoph Scriba first floated the idea of publishing a selection of letters and the appearance of the first volume of the *Correspondence of John Wallis* almost forty years had passed. In some respects this long delay reflects the considerable effort that goes into producing an edition of letters, beginning with the need for an exhaustive search for all extant correspondence. Scriba's 'check-list' made no claim to contain details of all surviving letters, even though he was convinced that all correspondence significant for the development of science in seventeenth-century England had in fact been gathered. But the main reason for the delay in Scriba's case was his teaching and administrative duties in Berlin and Hamburg and his heavy commitments to scientific bodies such as the DFG and the Leopoldina. It is no coincidence that he revived the plan for an edition only at the point when retirement and a corresponding reduction in such professional commitments was in sight—although it has to be said that the so-called historiography project took up most of his research efforts at that time.

The modest edition for which Scriba sought and subsequently received funding from the DFG, a small volume to complement the editions of Newton, Huygens, and Leibniz, would, had it come about, soon have outlived its purpose. The changes in the historiography of mathematics which Scriba and others were documenting in *Writing the History of Mathematics* were beginning to have their impact on research work being carried out on the history of mathematical thought in seventeenth-century England. An arbitrary decision as to what could be counted as scientific correspondence was no longer historically tenable, for it was now recognized that that the whole figure had to be in view if we were to understand that figure's contribution to the growth of scientific thought. This was in many ways a development that Scriba's argument for a history of ideas based approach anticipated. It is a credit to the DFG that it was willing to provide funding for the Wallis edition through the changes in its conception that he and the author introduced along such lines. However, the problem of funding large-scale editorial projects aimed at serving the scholarly needs of historical research remains to this day.

Notes on Series VII and VIII of the Leibniz-Edition

Eberhard Knobloch

Abstract

The article describes the difficult establishment of two of the eight series of the so-called Academy Edition of Leibniz's *Complete Writings and Letters*. In 1976, Series VII, *Mathematical writings*, was realized by means of a collaboration between Knobloch in Berlin and Contro in Hannover. Juridical, staff, and technical problems had to be solved before the editorial work could begin. Series VIII, *Scientific, Medical, and Technical writings*, was realized in 2001, this time as an official project of the Berlin-Brandenburg Academy of Sciences and Humanities. The international collaboration of co-workers from Berlin, Russia, and France was carried out through electronic means, especially through a digitalization of everything concerning the Leibnizian manuscripts.

1 Introduction

On September 28th, 2009, the Berlin newspaper "*Der Tagesspiegel*" published the following advertising text for a new book by Hazel Rosenstrauch:

> Alexander von Humboldt's personality is unimaginable without his wife, Caroline von Dacheröden, mother of his five children. A partner who was his equal in inquisitiveness, literacy, appreciation of art and active humanitarianism.[1]

[1] *Der Tagesspiegel* of the 28.9.2009, p. 29 on: H. Rosenstrauch: *Wahlverwandt und ebenbürtig, Caroline und Wilhelm von Humboldt*. Frankfurt a. M. 2009.

E. Knobloch (✉)
Berlin-Brandenburg Academy of Sciences and Humanities, Berlin University of Technology, Berlin, Germany
e-mail: eberhard.knobloch@tu-berlin.de

M. T. Borgato et al. (eds.), *Mathematical Correspondences and Critical Editions*, Trends in the History of Science, https://doi.org/10.1007/978-3-319-73577-1_2

This aspect of Alexander von Humboldt's character was hitherto unidentified. Have we perhaps, to this day, fostered a false image of Humboldt?

The same newspaper released a further statement concerning the Prussian Chamberlain on November 27th, 2011:

> When the naturalist died in Berlin in 1859, he had debts amounting to exactly 1211 thalers and 4 silver pennies, which, at that time, was equivalent to the 5 year income of a carpenter. His longstanding benefactor, King Friedrich Wilhelm the VI, settled the debt.[2]

All that was known of this Prussian King until now was, in fact, that he did not exist. What do these two examples tell us? That this world, in accordance with the Leibnizian doctrine, is indeed only the best possible world of all worlds, not the actual best: that it is imperfect; that its perfection lies in the fact that it can be perfected. If, then, there are, at no point in time, ideal conditions for any given intention, one will have to abide by some old, pragmatic words of wisdom: if one wants something, one finds a way; if one does not want something, one finds a reason.

In accordance with this saying, I would like, in the following, to explain why and how we came to establish Series VII and VIII of the Leibniz-Edition despite all of the difficulties we were confronted with. Let me say that, in the first instance, it concerned a personal desire, and in the second, a commission.

2 Series VII: The Desire

In order to understand the impetus for Series VII of the Leibniz-Edition from 1976, one has to picture the historical circumstances of the year 1975. At that time, neither Series III, *Mathematical natural science and technical exchange of letters*, nor Series VII, *Mathematical natural scientific and technical writing*, existed in the form originally planned for them. In 1939, Dietrich Mahnke had completed around 40% of Volume III, 1 when he passed away as a result of a traffic accident. His work was continued by the new head of the Leibniz-Edition in Berlin, Joseph Ehrenfried Hofmann.[3] When he then died 34 years later in 1973, also as a result of a traffic accident, the work on the volume was practically complete. But the painstaking and unrewarding final editing of the publication in 1976 was taken on by Heinz-Jürgen Heß. Work on Series VII had not gotten past the stage of preparatory work, that is, the transcriptions completed by Conrad Müller in Hannover.

This situation seemed highly unsatisfactory to me. I was in personal contact with Hofmann as a result of the work on my PhD. It was Hofmann who had originally suggested the theme for my dissertation: *The Mathematical Studies of G.W. Leibniz*

[2] *Der Tagesspiegel* of the 27.11.2011, p. S7.

[3] M. Folkerts: "Die Leibniz-Edition zwischen Wissenschaft und Politik" in: H. Hecht et al (ed.): Kosmos und Zahl. Beiträge zur Mathematik- und Astronomiegeschichte, zu Alexander von Humboldt und Leibniz (= Boethius 58), Stuttgart 2008, p. 23–45, here p. 30–36.

on Combinatorics. The dissertation was published as a monograph in 1973.[4] Two further volumes were in print; the text the dissertation was based on and the bilingual edition of *Dialogue on an introduction to Arithmetic and Algebra*.[5] A fourth volume containing Leibniz's main works on elimination theory and determinant theory was a work in progress. It was to be published in the year 1980.[6]

Owing to my education as a mathematician and classical philologist and the experiences that I had had with editing Leibniz's handwritten manuscripts, I considered myself sufficiently well-prepared to amend the aforementioned situation. The first step was to obtain the rights to edit the volumes of Series VII. The rights to this were in the possession of the Leibniz Research Institute in Münster, founded in 1956, and headed by director Heinrich Schepers.[7]

As a result of an abounding plethora of handwritten manuscripts, which had, to a considerable degree, been the reason why the series had not been tackled successfully up until that point, Schepers and I agreed, in 1975, to recreate Series VII, which was now "only" to contain the roughly 30 volumes of mathematical writings. The natural scientific, medical and technical writings were to be published in a new series, VIII, which was to follow.

In order to complete the work on the original handwritten manuscripts on-site in Hannover, and by virtue of a contract yet to be concluded, I was granted the rights to work on the first two volumes of Series VII under the condition that a research post be created in Hannover.

For this reason, it was necessary to hold talks with the director of the Lower Saxony State Library Hannover, Wilhelm Totok. The workplace of the as-yet-unspecified researcher was, as with the other positions in the Leibniz-Archive, to be situated in the library itself. Totok agreed with me that the supervision of this employee would be handed over to the director of the library, and the actual technical supervision to me. I would sincerely like to thank Wilhelm Totok for his cooperative attitude. In the years to follow, he made sure, through petitioning the ministry responsible in Lower Saxony, that the research post that was originally paid for through external funding was itself established as a fixed position in the Leibniz Archive.

---

[4]E. Knobloch: *Die mathematischen Studien von G. W. Leibniz zur Kombinatorik. Auf Grund fast ausschließlich handschriftlicher Aufzeichnungen dargelegt und kommentiert* (= *Studia Leibnitiana, Supplementa* 11), Wiesbaden 1973.

[5]E. Knobloch: *Die mathematischen Studien von G. W. Leibniz zur Kombinatorik. Textband, im Anschluss an den gleichnamigen Abhandlungsband zum ersten Mal nach den Originalhandschriften herausgegeben* (*Studia Leibnitiana, Supplementa* 16), Wiesbaden 1976; G. W. Leibniz: *Ein Dialog zur Einführung in die Arithmetik und Algebra, nach der Originalhandschrift herausgegeben, übersetzt und kommentiert*, Stuttgart-Bad Canstatt 1976.

[6]E. Knobloch: *Der Beginn der Determinantentheorie, Leibnizens nachgelassene Studien zum Determinantenkalkül, Textband* (=*arbor scientiarum: Beiträge zur Wissenschaftsgeschichte*, series B: *Texte*, vol. II), Hildesheim 1980.

[7]E. Knobloch: "Die Kunst, Leibniz herauszugeben", in: *Spektrum der Wissenschaft* 9 (2011), p. 48–57.

The next step, by means of an application to the Volkswagen Foundation, was to finance the first 4 years of this research post. The application, which was signed and supported by Christoph J. Scriba, was granted in 1976. A fifth year was then paid for through lottery funding before the position was finally established in the archive itself.

In the meantime, Heinrich Schepers had drafted a two and half page contract for us (pending agreement from the Minister of Science and Research of the State of North Rhine-Westphalia). The essence of this contract was to give me the rights to edit the volumes of Series VII, 1 and VII, 2, for a maximum of 19 years. Of the twelve paragraphs, I will list the five most important ones here:

> 1. The Leibniz-Research Centre confers to Mr. Knobloch the editorial work on the first two volumes of Series VII, the mathematical, natural scientific, technical writings of the complete works and letters of G.W. Leibniz. The publishing rights have been granted by the German Academy of the Sciences in Berlin. This transfer of rights is for a fixed-term of 10 years for the first volume and a further five years for the second volume. An extension of this term is possible if the work is not completed following the termination of the term and it can be foreseen that the work will be completed within a further 2 years.
>
> [...]
>
> 3. In order to reach an agreement on the content and method of the work on the two volumes, Mr. Knobloch is to present a list of the scripts to be included in both volumes within one year of employing his research assistant in Hannover (p. 11). He is to report back every two years on the development of his editorial work.
>
> [...]
>
> 6. Both volumes are to be published with two covers. The general cover is to state: "Edited by the Academy of Sciences in the GDR," and the specific cover is to state "Edited by the Leibniz-Research Center of the University of Münster."
>
> [...]
>
> 9. The director of the Leibniz-Research Center reserves the right to add a preface to both volumes, and will confirm, as the official representative for the editorial rights, the final manuscript and that it is ready for print.
>
> [...]
>
> 11. Mr. Knobloch is to have access to all material and the catalogues of the Leibniz-Edition for review in Münster or Hannover. He, for his part—supported by a research assistant who is to be employed in Hannover—is to continually enter the results of his work into the catalogue.
>
> The contract will come into effect with the approval of the Minister for Science and Research NW.

I would sincerely like to thank Heinrich Schepers for the trust he placed in me at that time. Certainly, owing to the political developments of the time, quite a different yet positive departure from some of these statements came to pass. Yet, the main purpose was achieved: 19 years following commencement of the work on Series VII, the first two volumes lay printed in book form before us. I will return to this later.

In the year 1976, a series of important events unfolded. On the same day that Schepers signed our contract, I completed my professorship (*Habilitationsverfahren*) at the Technical University of Berlin in the subject *History of Mathematics and the Exact Natural Sciences*. The dean who signed the Professorial Certificate was the Leibniz researcher Hans Poser. I considered this timely concurrence a good omen for the tasks that lay before me. Indeed, following on from this, on September 1st, 1976, the appointed research assistant, Dr. Walter S. Contro, began his work in Hannover.

A little later, on October 23rd, 1976, Heinrich Schepers formulated a much-anticipated letter, in which he informed me that the minister had agreed to our contract:

> Dear Mr. Knobloch,
>
> It gives me great pleasure to inform you that the Minister of Science and Research of the State of North Rhine-Westphalia has no issue with our Agreement of the 8th/12th of January 1976—which is thus legally binding [. . .]. I have briefly made acquaintance with Mr. Contro by telephone. The sample pages you requested for the creation of the EDP-text should be with you in the next few days.
>
> Yours sincerely, Heinrich Schepers

It was now necessary, in accordance with paragraph 3 of the Münster contract, to gain an overview of Leibniz's mathematical writings from the Parisian period. I presented my results at the symposium *Leibniz à Paris (1672–1676)*, which took place on November 14th to 18th, 1976, in Chantilly, near Paris: *Overview of the unpublished work on mathematics by Leibniz (1672–1676)*. The talks I held were then published 2 years later.[8]

A conservative calculation resulted in estimating that there would be around eight volumes of the Leibniz-Edition from the Parisian period. But this meant that half a year's worth of his mathematical studies would fill one volume. Owing to the fact that most of the handwritten manuscripts are not dated and that dating the manuscripts to a period of exactly half a year is generally not possible, it became clear that a strict chronological systematization of the handwritten manuscripts was impracticable. The solution came in the form of defining thematic groups from the entire Parisian period and developing a chronological order within those groups. In the case of Volume VII, 1 this concerned the groups of geometry, number theory and algebra. The principle of creating a complete edition for the Leibniz-edition meant that it was not possible, for the first two volumes, to refer back to the transcriptions by Conrad Müller, which had been stored in the Leibniz Archive. Leibniz's studies on *calculus*, which the mathematician Conrad Müller had primarily concentrated on, did not refer to the first period of the mathematical investigations in Paris—a further difficulty in producing the first volume.

In the latter stages of 1981, Contro's research position was finally integrated as part of the archive, having been funded externally for five long years. On December

---

[8]E. Knobloch: "Übersicht über die unveröffentlichten mathematischen Arbeiten von Leibniz (1672–1676) mit einem Anhang über die ersten Ansätze zur algebraischen Indexbezeichnung während der Pariser Zeit", in: *Leibniz à Paris (1672–1676), Symposion de la G. W. Leibniz-Gesellschaft (Hannover) et du Centre National de la Recherche Scientifique (Paris) à Chantilly (France) du 14 au 18 novembre 1976*, T. I: *Les sciences (=Studia Leibnitiana, Supplementa* 17), Wiesbaden 1978, p. 3–43.

1st, 1988, a second research position was added for Series VII and filled by Dr. Nora Gädeke. The series was—even if on a minimal level—secured institutionally. From this point onwards, I was working with two research assistants simultaneously. The first volume was in print. The production itself, however, was considerably protracted and difficult, owing to the fact that the Altenburger printing house printed it in lead typeface.

In the meantime, the Academy Program for projects in research and editing of the Academies of Science in the Federal Republic of Germany had been created in 1985. Owing to the political situation of a divided Germany, the Academy of Sciences in Göttigen took over the supervision of both Leibniz editorial posts in Münster and Hannover. This had several important consequences for Series VII: the series was ultimately and permanently handed over to Lower Saxony, i.e., to the Leibniz-Archive in Hannover. The two covers of Volumes VII, 1 and VII, 2 do not reference the Leibniz Research Center of the University of Münster, as was planned in the contract (§ 6) in the year 1976, but instead, the Leibniz-Archive of the Lower Saxony State Library in Hannover. And indeed, Schepers, the director of the Münster Research Center, did not write the preface, but rather it was written by the director of the Leibniz-Archive in Hannover, Albert Heinekamp. The preface begins with the words:

> The fact that the editing of Leibniz's mathematical writings can begin with this volume is, above all else, due to the great effort and merit of Eberhard Knobloch.[9]

The volume that appeared in the midst of the turmoil of Germany's reunification was, in fact, still published by the Academy of Sciences of the GDR, as is stated on the first cover page; on the second, it says: "Under the supervision of the Academy of Sciences in Göttingen." The exceedingly arduous task of making a start had been accomplished. The contributors are named on p. IV: E. Knobloch, Walter S. Contro. When Nora Gädeke switched jobs on the April 1st, 1995, from Series VII to Series I, Siegmund Probst was employed as the second full-time assistant for Series VII. It is accurate to say that I had a consistently untroubled and good working relationship with all three assistants over many years, and in the case of Walter Contro, over a period of 32 years. I would like to express my heartfelt thanks to all three scientists.

The further volumes appeared after considerably shorter time intervals than the period from 1976 to 1990 had allowed: Volume VII, 2 was published in 1996 (Contributors: Knobloch, Contro, assisted by Gädeke), VII, 3 in 2003 (Contributors: Probst, Knobloch, Gädeke), VII, 4 in 2008 (Contributors Contro, Knobloch), VII, 5 also in 2008 (Contributors: Uwe Mayer, Probst, Heike Sefrin-Weis), and VII, 6 in 2012 (Contributors: Mayer, Probst). After 32 years, with the publication of the fourth volume, I handed over control of Series VII, as I simply had far too many other responsibilities. Of those 32 years, I spent 23 working on behalf of the Academy of Sciences in Göttingen. In the meantime, I had, most importantly,

---

[9] A VII, 1, p. XIX.

established Series VIII. Yet, the new research assistants on Series VII, with whom I was no longer able to work, were not entirely unknown to me: I was an academic referee for Uwe Mayer's dissertation in Halle on Ehrenfried Walther von Tschirnhaus. I had also gotten to know Heike Sefrin-Weis in connection with my work as an academic referee.

3 Series VIII: The Commission

The first unsuccessful attempt to establish Series VIII can be traced back to the year 1987. That year, on October 10th, the founding ceremony of the Berlin Academy of Sciences took place in the Plenary Hall of the Reichstag in Berlin (West).[10] President Horst Albach commented in his address:

> What is exceptional does not, of course, lie in the intention to publish these editions, since the publication of Leibniz's writings on natural science and technology is in line with the very purpose of this Academy.[11]

One founding member of the Academy was Jürgen Mittelstrass. In 1988, he invited me to speak in front of the plenary assembly of Academy members on the publication of Series VIII, more particularly, on Leibniz and the publication of his hitherto unpublished scientific estate. The contents of the speech were published 1 year later.[12] I spoke on the following six points: the structure and status of the Academy edition; an overview of the natural science and technical manuscripts; the existing working materials for the edition; practical problems with the edition; the soloist of the history of science and technology as the publisher of Leibniz; an expert on the general: the challenge of the Leibniz-Edition.

The Academy's plenary agreed to undertake the assignment of publishing Series VIII.

Then, however, something unexpected occurred: The new senate in Berlin comprising the SPD and Alternative List Parties decided to dissolve the Academy on December 31st, 1990. And thus it happened. The first attempt to establish Series VIII had failed for political reasons.

Six years went by before a new attempt was made in 1996. In that year, the Saxony Academy of Sciences (SAW) in Leipzig celebrated their 150th year anniversary, which led them back to the idea of Leibniz. And thus, in their anniversary year, the SAW organized an international symposium from April 9th to

[10] E. Lack: "Festakt aus Anlaß der Gründung der Akademie der Wissenschaften zu Berlin am 10. Oktober 1987", in: *Akademie der Wissenschaften zu Berlin, The Academy of Sciences and Technology in Berlin* Jahrbuch/Yearbook 1987, Berlin – New York 1988, p. 129 f.

[11] H. Albach: "Die Akademie der Wissenschaften zu Berlin – Ein Experte fürs Allgemeine", in: *Akademie der Wissenschaften zu Berlin, The Academy of Sciences and Technology in Berlin* Jahrbuch/Yearbook 1987, Berlin – New York 1988, p. 135–145, here p. 141.

[12] E. Knobloch: "Leibniz und die Herausgabe seines wissenschaftlichen Nachlasses", in: *Akademie der Wissenschaften zu Berlin, The Academy of Sciences and Technology in Berlin* Jahrbuch/Yearbook 1988, Berlin – New York 1989, p. 475–483.

11th in Leipzig to celebrate the 350th birthday of G.W. Leibniz. As a corresponding member of this academy, I was on the preparatory committee for the symposium. The various speeches given at the symposium were published 3 years later.[13]

One of the speakers was Heinrich Schepers, who spoke about the Leibniz-Edition. His words were of considerable significance for the future conception and realization of Series VIII. Among other things, he said:

> But, above all, it can scarcely be acceptable that, for the work on 30 full volumes of series VII of the mathematical texts [...] two full-time jobs were created, and for the six to eight volumes in the new Series VIII of material separated into natural scientific, medical and technical texts, not even one. Under precisely these circumstances, it will be necessary to recruit external collaborators who are prepared to take on the toil of editing. The difficulties that arise from working in different places and at different institutions should be resolvable in this age of information. There is growing potential to work in an interconnected fashion on the same data and with the same catalogues.[14]

Schepers was right, yet to this day, in spite of the technical possibilities available, they are not being used sufficiently amongst the four Leibniz workplaces. The latest development since 2012 does, nevertheless, give rise to hope of a change in the situation. The president of the Berlin-Brandenburg Academy of Sciences, Dieter Simon, was in the audience. Later events from 1996 are certainly directly linked to this very fact.

In the course of 1996, I was elected a full member of the Berlin-Brandenburg Academy of Sciences. On August 29th, 1996, Dieter Simon called me and commissioned me to establish Series VIII and to breathe life back into our German-French collaboration. I would like to thank him sincerely for the trust he placed in me. I was, at that moment, determined not to disappoint him, even though this meant an arduous period of preparatory work lasting several years. On January 16th, 1998, he called me again and appealed to me to include Russia. It appears that we were both still mindful of Schepers' words.

Thus, it was clear from the beginning that I would have to create a decentralized system of work. The first move was to establish an *international collaboration.* In line with this, September 30th, 1997, at the Académie des sciences in Paris, a conversation took place between Vice Presidents Guy Ourisson of the Académie des sciences and Manfred Bierwisch of the Berlin-Brandenburg Academy of Sciences and Humanities (BBAW). The French group was represented by, among others, Claude Debru, a philosopher and historian of science. The German group was represented by the secretary of the Humanities Class of the BBAW, Jürgen Trabant, as well as myself. Debru and I were paired together and entrusted with the task of organizing the German-French collaboration. No institutional agreement was reached beyond this, a drawback whose effects are still felt today.

---

[13] K. Nowak/H. Poser (ed.): *Wissenschaft und Weltgestaltung, Internationales Symposion zum 350. Geburtstag von Gottfried Wilhelm Leibniz vom 9. bis 11. April 1996 in Leipzig,* Hildesheim – Zürich – New York 1999.

[14] H. Schepers: "Zur Geschichte und Situation der Akademie-Ausgabe von Gottfried Wilhelm Leibniz", in: ibid., p. 291–298, here p. 296.

Things developed differently in Russia. Following my first visit on July 4th, 1998, to the St Petersburg branch of the Moscow Institute for the History of Science and Technology, there was a meeting at presidential level on November 20th in Moscow between Yurii Sergeevich Osipov, of the Russian Academy of Sciences, and Dieter Simon, of BBAW. Also taking part on behalf of the German group were the mathematicians Martin Grötschel and Hans Schilar, an administrator of BBAW, as well as myself. The contract of cooperation signed by both presidents is still valid to the present day. My future colleague and assistant Vladimir Kirsanov became the contact for the Russian group and I myself for the German group.

A second requirement was the creation of the necessary infrastructure for such an international collaboration. The handwritten manuscripts had to be digitalized in order to make them accessible abroad. This required the consent of the director of the library, Wolfgang Dittrich, and his successor Georg Ruppelt. Both directors have strongly supported my intent from the outset. I would like to express my sincere gratitude for this. The advantage of such an approach, from the point of view of the authorized parties as well, was obvious: the precious, original manuscripts only had to be handled and examined in special cases.

In Hannover, on August 26th, 1999, Peter Cassiers, Simone Rieger and I produced some 75 trial digitalizations of Leibniz's handwritten manuscripts on medicine, optics and mechanics. Both information technology experts advised Dittrich on drafting the application for the German Research Foundation, whose program *Retrospective Digitalization of Library Collections* would potentially authorize the necessary financial means for the digitalization of all of the approximately four thousand handwritten pages that were to be edited within the scope of Series VIII. The GRF's letter of approval arrived on November 28th, 2000. The digitalization work was carried out within the following one and a *half* years. The concept for the database was developed by Simone Rieger and Peter Cassiers.

The three images on every page and the transparent photographs for the watermark, which are available in three resolutions, can be accessed worldwide without a password at the website: http://ritter.bbaw.de

The example on the following page (Fig. 1) relates to mechanics and Leibniz's work on Galileo's solution to the problem known as *Aristotle's Wheel.*

The third step entailed finding suitable personnel abroad and raising finances for equipping them with computers and payment, as well as necessary travelling costs. On January 19th, 2000, the German Research Foundation granted me the required funding for 5 years. This enabled me to conclude relevant contracts in Moscow and St. Petersburg. The BBAW then assumed this financial responsibility. In the case of Russia, the existence of an official, contractual provision proved helpful. On May 1st, 2000, Vladimir Kirsanov and the Latinist Olga Fedorova, who had been drafted by Kirsanov, started work. On June 1st, 2000, Alena Kuznetsova and the Latinist Ekaterina Basargina, who was brought in later by Kuznetsova, began work in St. Petersburg. The deaths of Kuznetsova (September 25, 2005) and Kirsanov (May 12, 2007) brought this developing and fruitful collaboration to a standstill for a considerable period of time. Thanks to Sergei Demidov from the Institute for the History of Science and Technology, it was then possible, in December 2009, to find

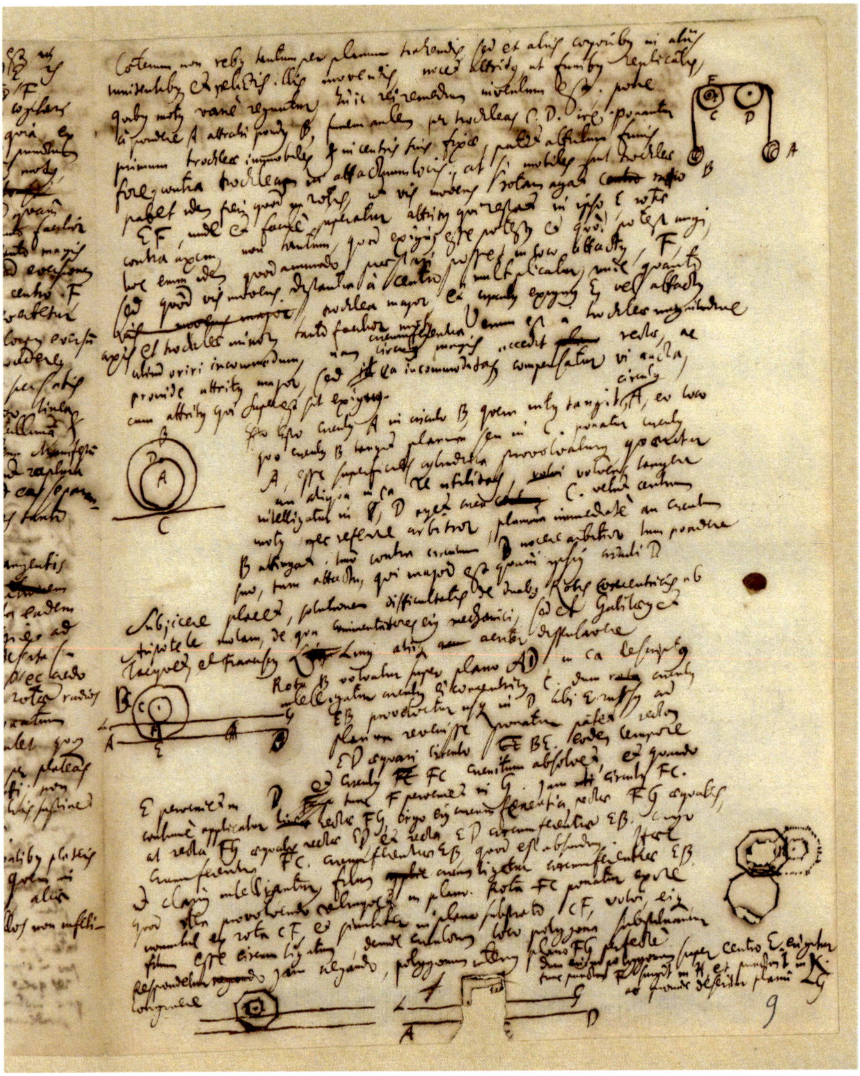

Fig. 1 Leibniz studies the problem known as *Aristotle's Wheel*. Leibniz manuscript LH XXXVII, 5 folio 9 obverse

a new colleague, at least in Moscow: Dimitri Bayuk. He once again called upon the linguistic expertise of Olga Fedorova. In France, contrary to expectation, the search proved more difficult than in Russia. Claude Debru and I spoke to a variety of institutes, for instance, the School for Advanced Studies in the Social Sciences and the French National Centre for Scientific Research, having established that the Academy of Science had neither the means nor the personnel to allow for a relevant collaboration. A first attempt did not prove successful. Since October 30th, 2005,

Anne-Lise Rey, from the University of Lille, and I had been prepared to collaborate, but due to other commitments, we could only do so to a very limited extent.

There still remained the institutional assurance of the BBAW itself, that is to say, the creation of a small post, dedicated full-time to editing Series VIII and supporting foreign employees. In a plenary session on March 16th, 2000, members of the Academy of the BBAW voted on the above. Of the 97 full members present, 96 voted in favor, with one abstention. It was the best result of the eight positions being voted upon.

The new post was created in Berlin on January 1st, 2001, initially without scientific or research infrastructure, in other words, without a reference library, catalogues or electronic access to the handwritten manuscripts. However, on June 28th, 1999, The Hermann and Elise (née Heckmann) Wentzel Foundation of the BBAW granted me funds to produce copies of the relevant microfilms and re-enlargements of the images. There was one research position available, which was filled by the physicist and philosopher Hartmut Hecht, as well as a part-time secretarial position, which was converted into a 12-hour-a-week contract for Simone Rieger. She left the post after just 1 year, the first of numerous changes in personnel that were very harmful to the project. Only years later, and thanks to the efforts of Jürgen Mittelstraß, was it possible to organize a second full-time post, which, since November 1st, 2006, has been filled by the chemist and theologian Sebastian Stork, who is already the third successor to Ms. Rieger.

In a plenary session of the BBAW on February 16th, 2001, I gave a presentation on the situation at that time. In it, I gave my view on six issues: On the structural content of Series VIII: transdisciplinary and transclassical; On financing; On Desire and Ability Part 1: International collaboration; On Desire and Ability Part 2: The Moment of Truth; On the Level of Aspiration: Content-related Insights; On the electronic work environment. The seventh aspect—digitalization and the internet edition—was covered by Simone Rieger as the new colleague.

We had an important sponsor during this difficult initial phase of the project who magnanimously gave us his support: The Landschaftliche Brandkasse Versicherungsgruppe Hannover (VGH). How did this support come to pass? In 2000, the company celebrated its 250 year anniversary. It felt bound to the fundamental concept of community solidarity, as represented by Leibniz when he, in his day, supported the establishment of insurance companies to cover fire and water damage. For this reason, a huge oversized portrait of Leibniz had hung in the headquarters of the company for a long time and was the catalyst for the chairman's portrait gallery.

In the late 1990s, Günter Schmidt, the then chairman, commissioned me to put together, for the first time, Leibniz's main writings on actuarial and financial mathematics, and in two languages at that, in other words, with a translation supported by J.-Matthias Graf von der Schulenburg. The volume appeared, as agreed, in time for the anniversary celebrations and to the great satisfaction of the commissioner.[15] Twenty four of the fifty two pieces were published for the very

[15]G. W. Leibniz: *Hauptschriften zur Versicherungs- und Finanzmathematik*. ed. by E. Knobloch and J.-M. Graf von der Schulenburg with commentaries by E. Knobloch, I. Schneider, E. Neuburger, W. Karten and K. Luig, Berlin 2000.

first time. As a result of this, on August 29th, 2001, the newly founded VGH Foundation allocated me a considerable amount of money spread over 4 years, a *contingency reserve*, as it were, which, over many years, helped to pay assistants, as well as Peter Cassiers in Berlin, who looked after the server and also published the digitalized copies on the net. This was possible because the third-party funding was not a budgetary fund that had to be paid out within a financial year—which was an important prerequisite for an economical and targeted use of the resources.

In addition, this volume had a welcome side effect for the Leibniz-Edition. Most of the handwritten manuscripts identified by me should, for chronological reasons, have already been published in 1986 in Volume 3 (1677–1689) of Series IV, *Political Writings*. For this reason, I made my materials and transcripts available to the Leibniz Editorial Centre in Potsdam, so that these manuscripts could be incorporated as a supplement into the work in progress on Volume 4. As a consequence, 213 pages, or 25.5% of Volume IV, 4 (1680–1692) or Section VII, *Statistics, Life Insurance, Pensions*, are based on the anniversary volume financed by the VGH. The VGH generously agreed to this course of action in the end, with Volume IV, only appearing in 2001, and thus after the anniversary publication. The leaflet produced in 2011 by the Potsdam office even portrays the facsimile of a handwritten note by Leibniz, which was edited and reproduced in the VGH volume for the first time.[16]

Series VII also benefits from work on Series VIII. Since a division of the handwritten manuscripts into two series cannot, due to the content, be achieved easily, several hundred handwritten pages were digitalized with the signature LH XXXV, which will presumably be published in Series VII.

Apart from the digitalized manuscripts, the new office in Berlin had no infrastructure worth mentioning and, in particular, no access to the catalogues that were available to the other offices. Meanwhile, in Potsdam, they began work producing an ACCESS-database from the Ritter catalogue of Leibniz's handwritten manuscripts. Thus, in June 2002, I applied to the Alfried Krupp von Bohlen und Halbach Foundation for a significant amount of money to enable the work to be completed. The application was granted. The Gottfried Wilhelm Leibniz Library in Hannover received a copy of this database. The development of the Leibniz-Edition, which was to be initiated in spring 2012, raised hopes that this centralized working tool would be accessible in a suitable form to all four workplaces. Today this is the case.

A characteristic feature of Series VIII is that certainly more than ninety percent of the relevant handwritten manuscripts have never been edited before. Yet, there had been a few previous attempts to edit these manuscripts. However, the results show a dramatic discrepancy between ability and desire. This is well illustrated in the following two examples.

In 1906, the physicist and electrical engineer Ernst Gerland published 134 works on physics, mechanics and engineering.[17] He dramatically underestimated

[16]Ibid., p. 121.

[17]*Leibnizens nachgelassene Schriften physikalischen, mechanischen und technischen Inhalts*, ed. by E. Gerland. Leipzig 1906 (Reprint Hildesheim 1995).

the difficulty of editing Latin texts from the seventeenth and eighteenth centuries: his texts abound with serious reading errors, and on numerous occasions, he was unable to decipher the text at all, as can be seen in Fig. 2. For example, Page 90, depicted here, contains 26 reading errors. The entire text was transcribed error free and *in toto* by our office in Berlin. To Gerland's credit, it can be said that he was a scientist and engineer, but certainly not a philologist.

Even worse is the unsuccessful publication of the first version of Leibniz's *Specimen dynamicum* by the classical philologist Glenn Most.[18] Owing to the fact that I have already commented in detail on this in a previous review,[19] a few comments here should suffice.

This publication contains numerous distortions and incorrect readings, such as those shown in Fig. 3: on line 304, "in" is missing, line 306 says "communi" instead of the correct word "compositis," line 307 has "fit" instead of "sit," line 308 says "distatur" instead of "sistitur," line 310 has "per naturam data" instead of "qua natura utitur," etc.

The "translation" betrays the fact that the meaning of the text was often not even remotely understood. For example, a "method of indeed astounding effectiveness, which has no problem dealing with irrational quantities" (referring to differential calculus)—becomes a method "which, certainly, does not take account of the irrational of an incredible capacity." Clearly, Glenn Most also dramatically underestimated the task of editing a physics text composed in Latin.

When viewing Series VIII, one has to take into consideration that it is more demanding than its predecessor, since it not only contains the physical texts, but also generally all natural scientific, medical and technical Leibniz texts. Leibniz's slogan was *Theoria cum praxi.* On March 2nd, 1691, he wrote the following to Huygens:

> I prefer a Leeuwenhoek, who tells me what he sees, to a Cartesian, who tells me what he thinks. Nevertheless, it is necessary to unite one's thoughts with one's observations.[20]

I would like to present, using four non-trivial examples, how Leibniz thus proceeded in the fields of natural science and technology.

3.1 Pneumatics

The first example concerns a bundle of handwritten manuscripts Hartmut Hecht worked on. In 1673, Leibniz elucidated the famous barometric experiments of Torricelli and the experiments on the vacuum by Otto von Guericke: "Celeberrimus

[18]G. W. Leibniz: *Specimen dynamicum*, ed. and trans. by H. G. Dosch, G. W. Most and E. Rudolph, Latin-German, Hamburg 1982.

[19]Review by E. Knobloch: "G. W. Leibniz: *Specimen dynamicum,* ed. and trans. by H. G. Dosch, G. W. Most and E. Rudolph. Latin-German. Hamburg 1982", in: *Annals of Science* 40 (1983), p. 501–504.

[20]A III, 5, p. 62 f.

90 Physikalischer Teil.

Solutio: Efficitur hoc: si omnes superficies refringentes sunt sphaericae concentricae, et faciant radios convergentes.

Demonstratio. Est objectum *abc*. Superficies refringens sphaerica objectiva *def*, cuius centrum *g*, puncti *b* radius perpendicularis refractionis expers *beg*. continuetur ultra *g*. Radius *bd* refractus in *d* ad perpendicularem in medium densius ex rariore versus *h*. incidat in *h* in aliam superficiem sphaericam *hik*, superficiei *def* concentricam, per quam rursus in medium rarius egrediatur. Ne igitur divergat radius *bdh* à irrefracto *beg* continuato, patet superficiem *hik* debere concavitatem obvertere medio densiori. Ita radius *bdh* secabit radium *beg* in *l*. Eodem modo radius *bf* refractus ad *f* in medium densius ad *k* ex densiore refringetur ad *l*. Idemque dicendum est de omnibus punctis superficiei *def* distantibus à puncto *b*, quantum ab eo distat punctum *d*. Id est, qui continentur circumferentia circuli in superficie sphaerica, cuius diameter est *df*. Idem dicendum de radiis *ad*, *an* et omnibus aliis in plano non designabilibus, qui continentur circumferentia circuli in superficie sphaerica, cuius diameter *dn*. Colligentur enim omnes in puncto *O*.

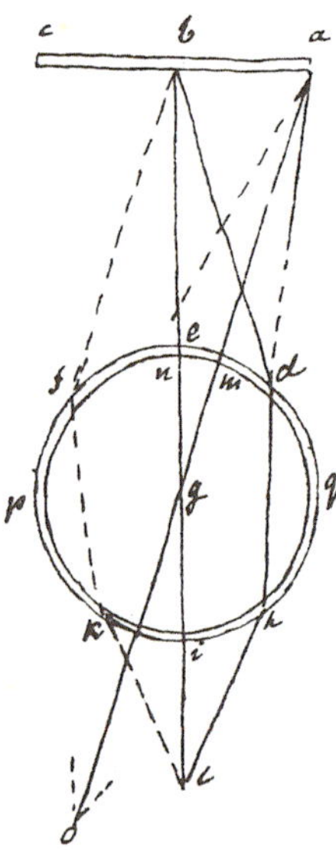

Fig. 58.

Observandum est, nihil referre sive superficies *def* et *hik*. sint portiones eiusdem sphaerae sive sphaerarum concentricarum. posse item vel adhiberi vel corpus cylindricum *def*, *kih*, contentum superficiebus sphaericis *def*, *hik* et planis *dh*, *fk* vel sphaeram integram *defpkih*.

Cum Hyperbola et Ellipsis colligant omnes radios ex puncto in axe optico et vicinissimos tanto . . .,[1]) licet tam minus accurata, quando ipsa Hyperbola et Ellipsis obtusior. Hinc fieri ut figura Optica quasi perfecta . . .[2]) vel Ellipsibus Hyperbolisque sibi oppositis, quasi mechanice quadam construendi . . .,[3]) ut huic hoc illi aliud objecti punctum sit in axe optico, ita totum simul perfectè, quantum possibile est, detegetur: imprimis si illae variae projectiones inter se . . .,[4]) ut si in unum speculum concavum incidant ubi . . .[5]) ob auctam magnitudinem. Aut si in convexum ubi . . .[6]) ob arctitatem . . .[7]) poterunt inde projici in amplificans speculum tubumque. Amplificantur in puncto.

41. [1 Blatt 4°, halb beschrieben.]

Hugenius et Newtonus statuunt, imperfectionem vitrorum opticorum maximè oriri à diversa refrangibilitate radiorum et ideò magis opus esse

1) Unleserlich, wohl in uno puncto.
2) Unleserlich, wohl non circulis utitur. 3) Unleserlich, vielleicht arte.
4) Unleserlich, wohl consentiunt. 5) Unleserlich, wohl distinctiores fiunt.
6) Ebenso vielleicht minores. 7) Ebenso vielleicht videri.

Fig. 2 Gerland's attempt to edit Leibniz's physical manuscripts. E. Gerland (ed.), *Leibnizens nachgelassene Schriften physikalischen, mechanischen und technischen Inhalts*. Leipzig 1906, p. 90

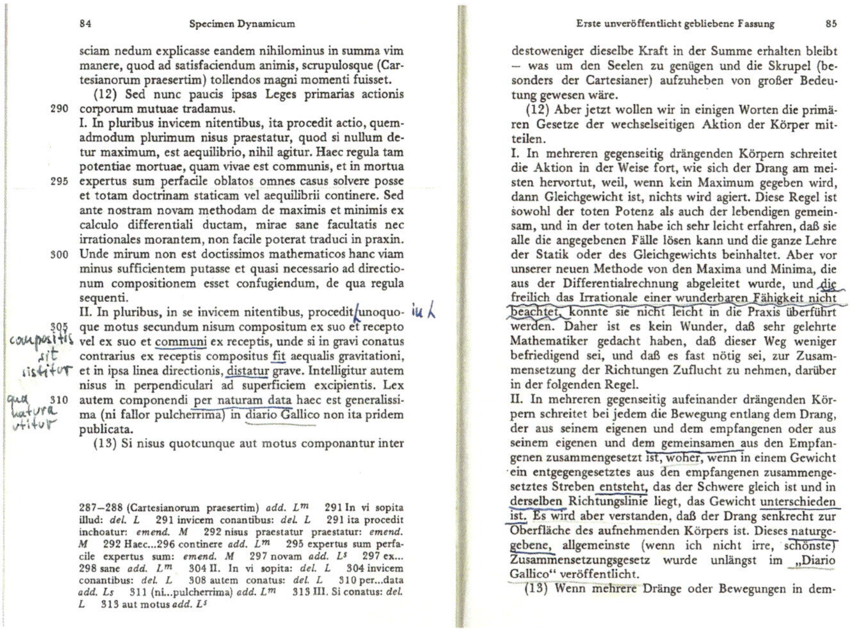

84 Specimen Dynamicum

sciam nedum explicasse eandem nihilominus in summa vim manere, quod ad satisfaciendum animis, scrupulosque (Cartesianorum praesertim) tollendos magni momenti fuisset.

(12) Sed nunc paucis ipsas Leges primarias actionis corporum mutuae tradamus.

I. In pluribus invicem nitentibus, ita procedit actio, quemadmodum plurimum nisus praestatur, quod si nullum detur maximum, est aequilibrio, nihil agitur. Haec regula tam potentiae mortuae, quam vivae est communis, et in mortua expertus sum perfacile oblatos omnes casus solvere posse et totam doctrinam staticam vel aequilibrii continere. Sed ante nostram novam methodam de maximis et minimis ex calculo differentiali ductam, mirae sane facultatis nec irrationales morantem, non facile poterat traduci in praxin. Unde mirum non est doctissimos mathematicos hanc viam minus sufficientem putasse et quasi necessario ad directionum compositionem esset confugiendum, de qua regula sequenti.

II. In pluribus, in se invicem nitentibus, procedit unoquoque motus secundum nisum compositum ex suo et recepto vel ex suo et communi ex receptis, unde si in gravi conatus contrarius ex receptis compositus fit aequalis gravitationi, et in ipsa linea directionis, distatur grave. Intelligitur autem nisus in perpendiculari ad superficiem excipientis. Lex autem componendi per naturam data haec est generalissima (ni fallor pulcherrima) in diario Gallico non ita pridem publicata.

(13) Si nisus quotcunque aut motus componantur inter

287–288 (Cartesianorum praesertim) *add.* L^m 291 In vi sopita illud: *del.* *L* 291 invicem conantibus: *del.* *L* 291 ita procedit inchoatur: *emend.* *M* 292 nisus praestatur praestatur: *emend.* *M* 292 Haec...296 continere *add.* L^m 295 expertus sum perfacile expertus sum: *emend.* *M* 297 novam *add.* L^s 297 ex... 298 sane *add.* L^m 304 II. In vi sopita: *del.* *L* 304 invicem conantibus: *del.* *L* 308 autem conatus: *del.* *L* 310 per...data *add.* *Ls* 311 (ni...pulcherrima) *add.* L^m 313 III. Si conatus: *del.* *L* 313 aut motus *add.* L^s

Erste unveröffentlicht gebliebene Fassung 85

destoweniger dieselbe Kraft in der Summe erhalten bleibt – was um den Seelen zu genügen und die Skrupel (besonders der Cartesianer) aufzuheben von großer Bedeutung gewesen wäre.

(12) Aber jetzt wollen wir in einigen Worten die primären Gesetze der wechselseitigen Aktion der Körper mitteilen.

I. In mehreren gegenseitig drängenden Körpern schreitet die Aktion in der Weise fort, wie sich der Drang am meisten hervortut, weil, wenn kein Maximum gegeben wird, dann Gleichgewicht ist, nichts wird agiert. Diese Regel ist sowohl der toten Potenz als auch der lebendigen gemeinsam, und in der toten habe ich sehr leicht erfahren, daß sie alle die angegebenen Fälle lösen kann und die ganze Lehre der Statik oder des Gleichgewichts beinhaltet. Aber vor unserer neuen Methode von den Maxima und Minima, die aus der Differentialrechnung abgeleitet wurde, und die freilich das Irrationale einer wunderbaren Fähigkeit nicht beachtet, konnte sie nicht leicht in die Praxis überführt werden. Daher ist es kein Wunder, daß sehr gelehrte Mathematiker gedacht haben, daß dieser Weg weniger befriedigend sei, und daß es fast nötig sei, zur Zusammensetzung der Richtungen Zuflucht zu nehmen, darüber in der folgenden Regel.

II. In mehreren gegenseitig aufeinander drängenden Körpern schreitet bei jedem die Bewegung entlang dem Drang, der aus seinem eigenen und dem empfangenen oder aus seinem eigenen und dem gemeinsamen aus den Empfangenen zusammengesetzt ist, woher, wenn in einem Gewicht ein entgegengesetztes aus den empfangenen zusammengesetztes Streben entsteht, das der Schwere gleich ist und in derselben Richtungslinie liegt, das Gewicht unterschieden ist. Es wird aber verstanden, daß der Drang senkrecht zur Oberfläche des aufnehmenden Körpers ist. Dieses naturgegebene, allgemeinste (wenn ich nicht irre, schönste) Zusammensetzungsgesetz wurde unlängst im „Diario Gallico" veröffentlicht.

(13) Wenn mehrere Dränge oder Bewegungen in dem-

Fig. 3 Most's attempt to edit Leibniz's *Specimen dynamicum*. H. G. Dosch, G. W. Most, E. Rudolph (eds. and translators), *G. W. Leibniz: Specimen dynamicum*, Latin-German. Hamburg 1982, p. 84f

Gerickius noster," "Our celeberrimous Guericke," as Leibniz says. Guericke had published his experiments on the vacuum in his 1672 monograph. Leibniz sought an explanation for what he had observed as the behavior of water, quicksilver, and air in a glass tube.

The handwritten manuscripts afford us a typified insight into Leibniz's manner of thought and writing.[21] The text is altered, crossed out and adapted innumerable times. The critical apparatus developed by Mr. Hecht reconstructs this creative process with the help of guidelines on formalisms, which Simone Rieger and Peter Cassiers have further developed with me.

3.2 The Mariotte-Leibnizian Pendulum

We assume that two pendulums connected to each other, bh, bc, are balanced in a vertical or, respectively, horizontal position. We seek to know the movement of the coupled pendulum system when the pendulum bh deflects, while the horizontal plane bac is fixed and the system as a whole is then left to itself (Fig. 4).

If with term φ, the angle of deflection of the pendulum bh, with l as its length, with λ_1, λ_2 as the length of the balances, ψ as the angle of deflection of the anterior

[21] A VIII, 1, p. 243–543.

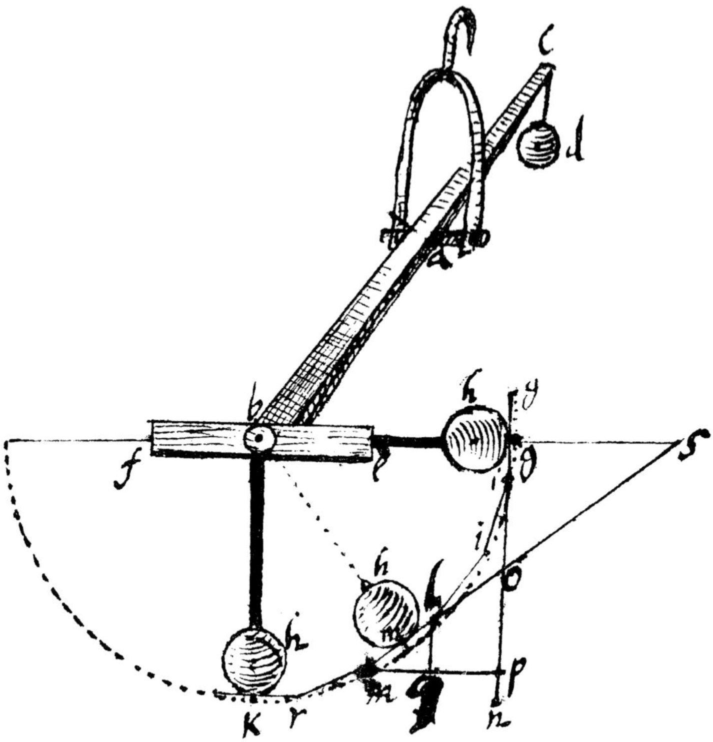

Fig. 4 The Mariotte-Leibnizian pendulum. K. Holzemer/I. Szabó, "Über das Mariotte-Leibnizsche Pendelproblem", in: Humanismus und Technik 22 (1978), p. 29

balance from λ_2, then one obtains a coupled system of non-linear differential equations of a second order in these quantities. Closed solutions to the system cannot be given. Approximate solutions are ascertained through the Runge-Kutta-method. For Leibniz and Mariotte, the problem was irresolvable.[22] The model test shows: with a small φ, the balance exercises an oscillatory motion. With a bigger φ, the pendulum bob is raised upwards.

3.3 Brake Mechanism for Vertical Apparatus (Rotation Speed Control of March 1686)

The controlled breaking of the cross-axle for the prevention of harmful overstraining of transmission components through a self-activating controller is to be resolved in the following manner: the principle of automatisation consists of the fact that, with

[22] K. Holzemer, I Szabó: "Über das Mariotte-Leibnizsche Pendelproblem", in: *Humanismus und Technik* 22 (1978), p. 23–36.

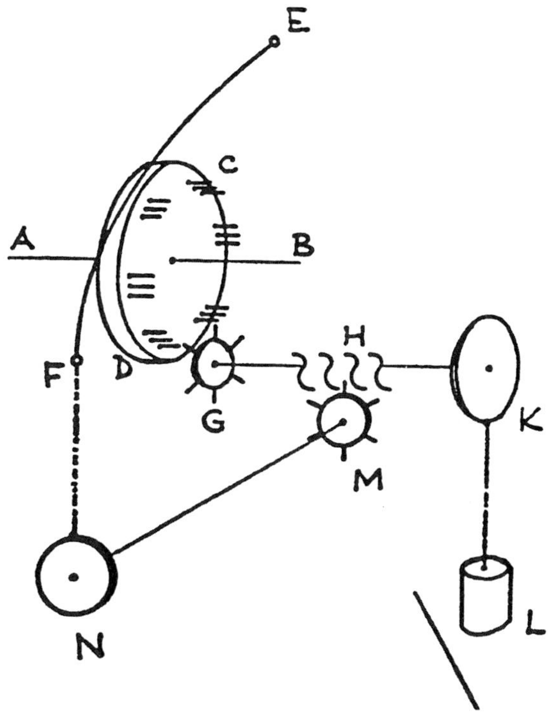

Fig. 5 Brake mechanism. J. Gottschalk: "Proposals for engineering improvements in mining in the Harz mountains", in: K. Popp, E. Stein (eds.), *Gottfried Wilhelm Leibniz, The work of the great universal scholar as philosopher, mathematician, physicist, engineer*. Hannover 2000, p. 130

an increasing velocity of circulation of the main driving shaft, the velocity of fall of a weight is negative in the attached rule ordering, and owing to this, the weight is pulled upwards. As a result, a braking mechanism slows down the velocity of circulation of the main driving shaft until the velocity of fall of a weight again predominates and loosens the brake (Fig. 5).

The usual evaluation ensues according to the technical criteria of trial, testing and further development. A model of this device has been in existence since 1990. It was exhibited in the atrium of the TU Berlin, along with other technical models, on the occasion of the VIIth International Leibniz-Congress from September 10th to 14th, 2001.

3.4 Horizontal Wind Apparatus

Vertical wind apparatuses were connected with pump leverages that could only be moved with difficulty. In 1684, Leibniz developed plans for a wind machine, a horizontal wind apparatus that was to propel Archimedean screw pumps. The handwritten script LH XXXVIII, sheet 313r, contains drafts and construction designs for this, including initial ideas on calculating the forces that act on the horizontal revolving blades. Leibniz has a horizontal revolving rotor with four

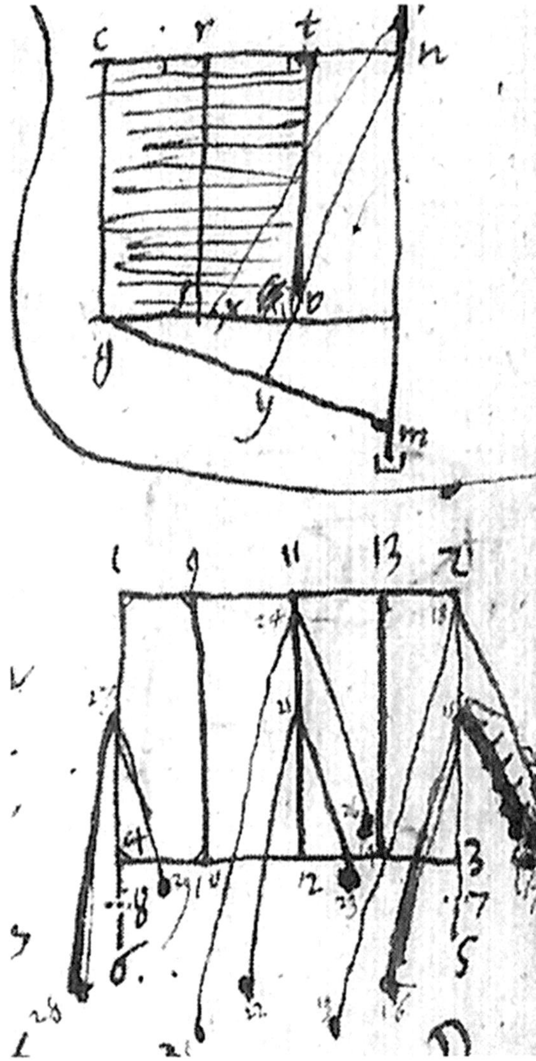

Fig. 6 Horizontal wind apparatus. J. Gottschalk: "Proposals for engineering improvements in mining in the Harz mountains", in: K. Popp, E. Stein (eds.), *Gottfried Wilhelm Leibniz, The work of the great universal scholar as philosopher, mathematician, physicist, engineer*. Hannover 2000, p. 110f

blades in mind, which is covered and surrounded by eight fixed, large-scale conveying shields (Fig. 6).

The wind machine was to keep an encased, presumably three-geared and 45 *schuch* (*feet*) long Archimedean screw pump in continual revolution.[23]

[23] J. Gottschalk; "Proposals for engineering improvements in mining in the Harz mountain", in: K. Popp, E. Stein (eds.), *Gottfried Wilhelm Leibniz, The work of the great universal scholar as philosopher, mathematician, physicist, engineer*. Hannover 2000, p. 109–137, here p. 111.

Heinrich Schepers had made a further remark at the Symposion in Leibzig in 1996, which was of vital importance to Series VIII:

> Even if it is not possible to alter the fact that the Academy Edition is to be published down to the last volume as a book, this cannot mean that the editors should forego using different mediums suited to making the arduous and completed work accessible to research.[24]

The internet is a medium of this type. The very conception of Series VIII entailed that it had to rely on an electronic work environment to a greater degree than had been typical until then. At the same time, as the post in Berlin was created, Martin Grötschel founded a workgroup at the BBAW on December 19th, 2000, called *electronic publishing* (TELOTA). In the report to the board of the BBAW on January 22nd, 2001, he stated:

> Modern methods of information technology [. . .] aid collaboration between internationally dispersed working groups in an effective way and enable excellent platforms designed for communication and presentation [. . .] Every scientific project develops a strategy that is oriented towards subject-specific needs in electronic [. . .] publication and presentation.

With the explicit support of President Dieter Simon, it was, therefore, decided that we would publish Series VIII in book form and as an online edition, hence not merely as an additional PDF placed on the internet. The TELOTA-team of the BBAW, in particular, Markus Schnöpf, became an important contact at the BBAW. The online edition of Volume VIII, 1 is based on the digitalized handwritten manuscripts and features animated drawings, which allows the user insight into the process of the development of the drawings that would not otherwise be possible with the printed version alone. The reconstruction of the genesis of the text is made possible by using different colored *layers*. The texts can be viewed at: http://leibnizviii.bbaw.de

This project was presented,[25] amongst other venues, at the 7th International Leibniz Congress, which took place from September 10th to 14th, 2001, in Berlin. The project caught Ulrich Schneider's eye, who, at the time, was the acting director of the Herzog August Library Wolfenbüttel. On May 28th, 2002, he informed us that, within the framework of a digitalization of selected archives of the library, he would mainly be digitalizing selected books from the Berlin research center. In fact, in the case of literature that relates to allusions and citations, the online edition is indeed linked to the server at the Wolfenbüttel Library.

---

[24] Schepers, p. 297.

[25] H. Hecht, E. Knobloch, S. Rieger: "Reihe VIII: Naturwissenschaftlich-medizinisch-technische Schriften. Ein neues Projekt im Rahmen der Akademie-Ausgabe", in: *VII. Internationaler Leibniz-Kongreß. Nihil sine ratione, Mensch, Natur und Technik im Wirken von G. W. Leibniz*, supplementary volume, ed. by H. Poser in connection with Chr. Asmuth, U. Goldenbaum, W. Li, Hannover 2002, p. 73–81.

The Berlin office was and is in close contact with editions that follow a similar conceptual framework, i.e., which also aspire to publication in book form and as an online edition:

1. The *edizione nationale* of the works of Francesco Maurolico, which is published under the direction of Pier Daniele Napolitani in Pisa. I belong to the steering committee (*comitato direttivo*).
2. The correspondence of the Danzig astronomer Johannes Hevelius, which is an edition planned as a German-French-Polish collaboration and which will be published under the auspices of the Académie Internationale d'Histoire des sciences. As the president of this Académie, I was the chair of the advisory board.

The first volume of Series VIII was published in 2009. The collaborators were Hartmut Hecht, Eberhard Knobloch, Alena Kuznetsova (†) and Sebastian Stork, assisted by Vladimir Kirsanov (†) and Anne-Lise Rey. A book presentation did not take place. The volume was received very positively. In Andreas Kleinert's review, he states:

> The reviewer is able to state, having viewed numerous sample texts, that the editors have worked with an accuracy and commitment to detail that can scarcely be surpassed and offer a convincing transcription for passages that are difficult to read. It is, above all, this part of the work that deserves the greatest commendation, and even more so by virtue of the fact that there are five languages represented in the volume.[26]

The second volume appeared in 2016, with particular thanks going to Hecht's successor Harald Siebert and the new co-worker Paolo Rubini. Thus, despite all difficulties, the current progress on Series VIII and the standard it has achieved justifies the statement that it is on the right track. Indeed, in 2007, the scientific commission of the eight academies of sciences in Germany evaluated all four Leibniz-Edition research posts, particularly the small research post in Berlin, very positively.

A.1 Epilogue

Vergil wrote in the Aeneid (Verse I, 33):

"Tantae molis erat Romanam condere gentem."

"Of such great effort it was to establish the Roman people."

[26]Review by A. Kleinert: "G. W. Leibniz: *Naturwissenschaft/iche, medizinische und technische Schriften. Erster Band 1668–1676. (Sämtliche Schriften und Briefe, Achte Reihe; l. Band)*. Ed. by the Leibniz-Edition Berlin of the Berlin-Brandenburg Academy of the Sciences. Revised by Hartmut Hecht, Eberhard Knobloch, Alena Kuznetsova (†) and Sebastian Stork. In collaboration with Vladimir Kirsanov (†) and Anne-Lise Rey. Berlin 2009", in: *Berichte zur Wissenschaftsgeschichte* 33 (2010), p. 329–330.

I would like to adapt the citation a little:

Tantae molis erat series septimam et octavam condere Leibnitianae editionis.

Of such great effort it was to establish the seventh and eighth series of the Leibniz-Edition.

Editorial remark

This article is the updated, enlarged English version of the paper: Eberhard Knobloch, Anmerkungen zu den Reihen VII und VIII der Leibniz-Edition, in: W. Li (ed.), *Komma und Kathedrale, Tradition, Bedeutung und Herausforderung des Leibniz- Edition*, Berlin: Akademie Verlag 2012, p. 95–113. (By kind permission of the publishing house de Gruyter).

References

A = Gottfried Wilhelm Leibniz. *Sämtliche Schriften und Briefe,* herausgegeben von der Berlin-Brandenburgischen Akademie der Wissenschaften und der Akademie der Wissenschaften zu Göttingen since 1923 (now Berlin: Walter de Gruyter; A VII, 1 means series VII, volume 1).

Albach, Horst. 1988. Die Akademie der Wissenschaften zu Berlin – Ein Experte fürs Allgemeine. In *Akademie der Wissenschaften zu Berlin* [The Academy of Sciences and Technology in Berlin] Jahrbuch/Yearbook 1987, 135–145. Berlin: Walter de Gruyter.

Folkerts, Menso. 2008. Die Leibniz-Edition zwischen Wissenschaft und Politik. In *Kosmos und Zahl, Beiträge zur Mathematik- und Astronomiegeschichte, zu Alexander von Humboldt und Leibniz*, ed. Hartmut Hecht et al., 23–45. Stuttgart: Franz Steiner Verlag (Boethius 58).

Gottschalk, Jürgen. 2000. Proposals for engineering improvements in mining in the Harz mountains. In *Gottfried Wilhelm Leibniz, The work of the great universal scholar as philosopher, mathematician, physicist, engineer*, ed. Karl Popp and Erwin Stein, 109–124. Hannover: Schlütersche.

Hecht, Hartmut, Eberhard Knobloch, and Simone Rieger. 2002. Reihe VIII: Naturwissenschaftlich-medizinisch-technische Schriften, Ein neues Projekt im Rahmen der Akademie-Ausgabe. In *VII. Internationaler Leibniz-Kongreß, Nihil sine ratione, Mensch, Natur und Technik im Wirken von G. W. Leibniz*, supplementary volume, ed. H. Poser in connection with Christoph Asmuth, Ursula Goldenbaum, Wenchao Li, 73–81. Hannover: Gottfried Wilhelm Leibniz Gesellschaft.

Holzemer, Karlheinz, and Istvan Szabó. 1978. Über das Mariotte-Leibnizsche Pendelproblem. *Humanismus und Technik* 22: 23–36.

Kleinert, Andreas. 2010. Review of G. W. Leibniz, Naturwissenschaftliche, medizinische und technische Schriften, Erster Band 1668-1676 (Sämtliche Schriften und Briefe, Achte Reihe, 1. Band), ed. by the Leibniz-Edition Berlin of the Berlin-Brandenburg Academy of Sciences and Humanities, revised by Hartmut Hecht, Eberhard Knobloch, Alena Kuznetsova and Sebastain Stork. In collaboration with Vladimir Kirsanov and Anne-Lise Rey. Berlin 2009. *Berichte zur Wissenschaftsgeschichte* 33: 329–330.

Knobloch, Eberhard. 1973. *Die mathematischen Studien von G. W. Leibniz zur Kombinatorik. Auf Grund fast ausschließlich handschriftlicher Aufzeichnungen dargelegt und kommentiert.* Wiesbaden: Franz Steiner Verlag (Studia Leibnitiana Supplementa 11).

———. 1976. *Die mathematischen Studien von G. W. Leibniz zur Kombinatorik, im Anschluss an den gleichnamigen Abhandlungsband zum ersten Mal nach den Originalhandschriften herausgegeben.* Wiesbaden: Franz Steiner Verlag (Studia Leibnitiana Supplemeta 16).

———. 1978. Übersicht über die unveröffentlichten mathematischen Arbeiten von Leibniz (1672–1676) mit einem Anhang über die ersten Ansätze zur algebraischen Indexbezeichnung während der Pariser Zeit. In Leibniz à Paris (1672–1676), *Symposion de la G. W. Leibniz-Gesellschaft (Hannover) et du Centre National de la Recherche Scientifique (Paris à Chantilly (France) du*

14 au 18 novembre 1976, Tome I: *Les sciences*, 3–43. Wiesbaden: Franz Steiner Verlag (Studia Leibnitiana Supplementa 17).

———. 1980. *Der Beginn der Determinantentheorie, Leibnizens nachgelassene Studien zum Determinantenkalkül*, Textband. Hildesheim: Gerstenberg Verlag (arbor scientiarum: Beiträge zur Wissenschaftsgeschichte, series B: Texte, vol. II).

———. 1983. Review of G. W. Leibniz, Specimen dynamicum, ed. and translated by Hans Günter Dosch, Glenn Most, Enno Rudolph, Latin-German. Hamburg 1982. *Annals of Science* 40: 501–504.

———. 1989. Leibniz und die Herausgabe seines wissenschaftlichen Nachlasses. In *Akademie der Wissenschaften zu Berlin* [The Academy of Sciences and Technology in Berlin] Jahrbuch/Yearbook 1988, 475–483. New York: Walter de Gruyter.

———. 2011. Die Kunst, Leibniz herauszugeben. *Spektrum der Wissenschaft* 9: 48–57.

Lack, Eva. 1988. Festakt aus Anlaß der Gründung der Akademie der Wissenschaften zu Berlin am 10. Oktober 1987. In *Akademie der Wissenschaften zu Berlin* [The Academy of Sciences and Technology in Berlin] Jahrbuch/Yerabook 1987, 129–130. Berlin: Walter de Gruyter.

Leibniz, Gottfried Wilhelm. 1906. *Nachgelassene Schriften physikalischen, mechanischen und technischen Inhalts*, ed. Ernst Gerland. Leipzig: B. G. Teubner (Reprint Hildesheim-Zürich-New York: Georg Olms Verlag, 1995).

———. 1976. *Ein Dialog zur Einführung in die Arithmetik und Algebra, nach der Originalhandschrift herausgegeben, übersetzt und kommentiert von Eberhard Knobloch*. Stuttgart-Bad Canstatt: Frommann-Holzboog.

———. 1982. *Specimen dynamicum*, ed. and Trans. Hans Günter Dosch, Glenn W. Most, Enno Rudolph. Hamburg: Felix Meiner Verlag.

———. 2000. *Hauptschriften zur Versicherungs- und Finanzmathematik*, ed. Eberhard Knobloch and J.-Matthias Graf von der Schulenburg with commentaries by Eberhard Knobloch, Ivo Schneider, Edgar Neuburger, Walter Karten, Klaus Luig. Berlin: Akademie Verlag.

Nowak, Kurt, and Hans Poser, eds. 1999. *Wissenschaft und Weltgestaltung, Internationales Symposium zum 350. Geburtstag von Gottfried Wilhelm Leibniz vom 9. bis 11. April 1996 in Leipzig*. Hildesheim: Georg Olms Verlag.

Schepers, Heinrich. 1999. Zur Geschichte und Situation der Akademie-Ausgabe von Gottfried Wilhelm Leibniz. In *Wissenschaft und Weltgestaltung, Internationales Symposium zum 350. Geburstatg von Gottfried Wilhelm Leibniz vom 9. Bis 11. April 1996 in Leipzig*, ed. Kurt Nowak and Hans Poser, 291–298. Hildesheim: Georg Olms Verlag.

The Edition of the Bernoulli Correspondence: A Historical Overview and Insights into the Most Recent Developments

Sulamith Gehr

Abstract

The aim of the online *Basler Edition der Bernoulli-Briefwechsel* (BEBB) is to make all as-of-yet unpublished letter exchanges of the seventeenth- and eighteenth-century mathematicians Daniel Bernoulli, Jacob II Bernoulli, Johann I Bernoulli, Johann II Bernoulli, Nicolaus I Bernoulli, Nicolaus II Bernoulli and Jacob Hermann available to the scientific community. Some of their correspondents were outstanding scientists. The edition of the Bernoulli correspondence has a long history that goes back to the eighteenth century, when the edition of the *Commercium epistolicum* between Johann I Bernoulli and Leibniz was prepared under the supervision of the Bernoullis. Despite the early interest in a scholarly edition of large parts of the Bernoulli correspondence and the repeated attempts by historians of mathematics to realize such a project, the undertaking experienced numerous setbacks. However, the digital age seems to provide the necessary conditions for a complete edition of the correspondence. In the context of the BEBB, around 1500 letters have been edited and published online in the last 10 years. In the next 10 years, another 2505 edited letters will be made available online. The present paper focuses on the main phases of the history of the Bernoulli letter edition and aims to give an overview of the on-going preparation of a modern online edition. This will include all related data, such as metadata, critical, emended transcriptions, digital facsimiles, commentaries, and indices, accessible through new technologies and tools that are currently being developed for the project.

S. Gehr (✉)
Bernoulli-Euler-Zentrum, University of Basel, Basel, Switzerland
e-mail: sulamith.gehr@unibas.ch

M. T. Borgato et al. (eds.), *Mathematical Correspondences and Critical Editions*, Trends in the History of Science, https://doi.org/10.1007/978-3-319-73577-1_3

1 Introduction

In this paper, I intend to give an overview of the different stages of the Bernoulli letter edition, with particular focus on the current project of a comprehensive online edition.[1] The Bernoulli letter Edition aims to produce an authoritative critical and fully annotated edition of the correspondence of the mathematicians and physicists Daniel I Bernoulli (1700–1782), Jacob I Bernoulli (1654–1705), Jacob II Bernoulli (1759–1789), Johann I Bernoulli (1667–1748), Johann II Bernoulli (1710–1790), Nicolaus I Bernoulli (1687–1759), Nicolaus II Bernoulli (1695–1726), and Jacob Hermann (1678–1733), a scholar closely associated with the Bernoullis. The Bernoullis and Hermann exchanged letters with over 400 correspondents, several of whom were world-renowned scientists of the seventeenth and eighteenth centuries.[2]

The history of the edition can be mainly divided into the following stages[3]:

1. 1745–1934: editions made prior to the foundation of the new Bernoulli-Edition;
2. 1935–1993: edition of four printed volumes of correspondence within the modern Bernoulli-Edition[4];
3. 2005–2015: edition of around 1500 letters within the *Basler Edition der Bernoulli-Briefwechsel* (BEBB) and their publication on the Bernoulli-wiki[5];
4. 2016–2018: transfer of the letters published on the Bernoulli-wiki and their metadata to the new platform "Bernoulli Euler OnLine" (BEOL);
5. 2018–2028: further edition and publication on BEOL (2505 letters).

2 The Editions Published Prior to the Foundation of the Modern Bernoulli Edition

In Early Modern Europe, letters counted among the most important vehicles of scientific communication and exerted a function similar to that of publications in scientific journals. In addition to reading them personally, their addressees often shared them with other scholars or even published them, entirely or partially,

[1] For their valuable advice and information, I wish to express my sincere thanks to Martin Mattmüller, Erwin Neuenschwander and Tobias Schweizer.

[2] A complete list of the correspondents can be found at http://www.ub.unibas.ch/bernoulli/index.php/Suchbaum. This and all further websites referred to in the following were last consulted on 12/31/2017.

[3] Individual phases of the history of the Bernoulli letter Edition between 1936 and 2005 are described in the introduction of Spiess (1955, pp. 9–85), and in Truesdell (1958, pp. 54–62), and Fellmann and Mikhajlov (1998, pp. 18–28). For the history of publication of the correspondence between the Bernoullis and Euler, see Neuenschwander (2002, pp. 103–105) and Nagel (2005, pp. 173–175).

[4] The edition of the correspondence between Johann I Bernoulli and Pierre Varignon is still in progress. The third volume is in preparation, and will be published in 2019.

[5] www.ub.unibas.ch/bernoulli

in journals.[6] The contemporary understanding of the scientific epistolary as an important aspect of scientific writing is visible, for instance, in a passage from Johann I Bernoulli's French autobiography.[7] Bernoulli says that his "assiduousness in writing" has not only led him to the production of a considerable amount of original scientific texts, but also to the establishment of a wide scientific correspondence:

> Cette assiduité d'écrire m'a procuré la connaissance de plusieurs savants du premier ordre, qui m'ont bien voulu honorer de leur correspondance. Ceux avec qui j'ai commercé le plus familièrement jusqu'à la fin de leurs jours, c'étaient Mr. le Mq. de l'Hospital, Mr. Leibnitz, Mr. Varignon, Mr. de Montmort, Mr. le chevalier Renau, Mr. de Tschirnhaus, Mr. Hermann, Mss. les frères Scheuchzer, Mr. Michelotti et plusieurs autres dont les noms ne me reviennent pas. Ce sont principalement Mss. de l'Hospital, Varignon, de Montmort, Michelotti qui voulurent bien me consulter comme leur oracle quand ils avaient des difficultés sur la sublime géométrie; aussi le premier de ces Messieurs donna rarement quelque chose au public, qu'il ne fut passé auparavant par mes mains, témoin grand nombre de ses lettres écrites à moi. Quant aux autres qui sont encore en vie et connus dans le monde savant, qui m'ont bien voulu honorer de leurs lettres, je n'en nommerais que quelques-uns, savoir le fameux Mr. Wolf, Mr. de Moivre, Mr. Burnet fils de Mr. l'Evêque de Salisbury, Mr. Craige, Mr. Cheynès, Mr. de Fontenelle, Mr. de Mairan, Mr. de Maupertuis, Mr. Clairaut, Mr. Poleni, Mr. de Crousaz, Mr. Cramer, Mr. Euler, Mr. Bilfinguer, etc. Quelques-uns de ces Messieurs sont encore jusqu'à-présent en correspondance avec moi. Si Mr. Newton eut vécu plus longtemps, je ne doute pas qu'il n'eût voulu lier avec moi une correspondance formelle.

This passage not only reveals how highly Johann I Bernoulli rated his correspondence with some of the most known "savants" of his time; it also highlights the importance that Bernoulli attached to the genre of scientific correspondence with respect to its function in scientific communication. While writing this autobiographical text, Bernoulli might already have had the edition of his own correspondence in mind.[8] Indeed, as early as 1719, some initial rumours about an edition of his writings had begun circulating.[9] At the beginning of 1731, a project to edit his letter exchange with Leibniz took shape. By this time, the mathematician Louis Bourguet (1678–1742) informed Johann I Bernoulli and Jacob Hermann of his plan to realize an edition of Leibniz's letter exchanges and received consent from both correspondents to publish their exchanges with the latter.[10] In the same year, Bernoulli and Hermann

[6]For the role of scientific correspondence in the eighteenth century, see Peiffer (1998).

[7]The text is printed in Wolf (1859, pp. 71–94).

[8]On the edition of the correspondence between Johann I Bernoulli and Leibniz, see Nagel (1989, pp. 167–174).

[9]Christian Wolff asks Bernoulli for information about this rumour in his letter of October 11, 1717 (Basel UB, Ms. L Ia 671, Nr.14*). In his answer of April 13, 1718 (Basel UB, Ms. L Ia 671, Nr.10), Bernoulli admits having already spoken with a publisher about the edition of his correspondence with Leibniz, but denies the rumours of an actual project dedicated to the edition.

[10]On Bourguet's plan for an edition of Bernoulli's and Hermann's letter exchanges with Leibniz, see Nagel (1994, pp. 525–528), and Bovet (1905, pp. 255–259). The correspondence of Bernoulli and Hermann with Bourguet, in which the topic of the edition of the Leibniz correspondence is

sent their manuscripts from Basel to Neuchâtel.[11] Because of financial reasons and health issues, Bourguet was not able to achieve his plan, and he returned the manuscripts to their owners in 1733.[12] With Hermann's death in the same year, every attempt to edit his exchange expired too; although his letters seem to have been sent back to Basel, they have unfortunately become lost. In the meantime, Johann I Bernoulli started a new attempt at an edition of his own correspondence with Leibniz, choosing Christian Kortholt (1633–1694) as the editor.[13] Kortholt planned to insert letters from the Bernoulli-Leibniz correspondence in the second volume of his collection, *Leibnitii Epistolae ad diversos*,[14] but for unknown reasons, this plan was never realized.

A third attempt, this time successful, to publish the Bernoulli-Leibniz correspondence was envisaged in 1740.[15] Information about this project is contained in the correspondence between Johann I Bernoulli, his son Johann II Bernoulli, the mathematicians Samuel König (1712–1757) and Gabriel Cramer (1704–1752), and the publisher Marc-Michel Bousquet (1696–1762). Johann I Bernoulli entrusted the editorship to Cramer, who was also a historian of mathematics and had already been responsible for the edition of Bernoulli's *Opera Omnia*.[16] The edition in two volumes appeared before the public in 1745, under the title *Virorum celeberr. Got. Gul. Leibnitii et Johan. Bernoullii Commercium philosophicum et mathematicum*.[17] Cramer was certainly an ideal choice for the task: he spent five months studying mathematics under Johann I Bernoulli, and therefore had profound knowledge of his mathematical thought. With his scientifically sound edition, characterised by the modern representation of mathematical formulae, he set a new standard for scholarly editions of scientific and mathematical texts. This earned him recognition as one of the first historians of mathematics specialising in editions of scientific writings.[18] Even so, it is understood that the *Commercium* does not comply with the requirements of contemporary scientific editions. For instance, Cramer left out many text passages, which he considered to be outside of the public interest,[19]

addressed, is housed at the Bibliothèque publique et universitaire de Neuchâtel (Suisse), Ms 1267 and Ms 1272. The letters will be published in 2020 as part of the BEBB.

[11]Johann I Bernoulli sent his manuscripts to Bourguet with his letter of June 11, 1731. See Nagel (1994, p. 528). In December 1731, Hermann sent a small parcel from Basel to Neuchâtel that probably contained his Leibniz manuscripts. See Bovet (1905, p. 258), and Nagel (1994, p. 527).

[12]Nagel (1994, p. 528).

[13]On Kortholt's project, see Nagel (1994, p. 529). The only sources of information on this project are a letter from Johann I Bernoulli to Bourguet of November 7, 1733, and a letter from Kortholt to Johann I Bernoulli of October 14, 1738.

[14]Kortholt (1734–1742).

[15]On the history of the *Commercium philosophicum et mathematicum*, see Nagel (1994, pp. 529–531).

[16]Bernoulli and Cramer (1742).

[17]Bernoulli et al. (1745).

[18]See Cantor (1898, p. 486).

[19]Bernoulli et al. (1745, pp. II–III).

producing what is, in effect, a censored edition. However, we also know from Cramer's correspondence with Johann II Bernoulli that, in doing so, he followed instructions from Basel. In a letter dated October 29, 1740, Johann II Bernoulli wrote to Cramer:

> Au reste, Monsieur, je crois superflu de vous avertir, de ne faire copier de ces lettres, que ce qui regarde les sciences et d'en omettre touttes les personalités, les particularités domestiques, enfin tout ce que vous jugerez que mon Pere ne souhaite pas qui soit publié. Vous trouverés aussi que bien des expressions ont besoin d'être corrigées parce qu'elles ne sont pas des meilleures ou adoucies parce qu'elles sont ecrites dans des lettres familières qu'on ne pensoit pas qui seroient jamais rendues publiques; c'est un article que nous vous recommandons très particulièrement.[20]

Despite the plan to edit other letter exchanges between Johann I Bernoulli and some of his primary correspondents, discussed in the abovementioned correspondence among Johann II, Bousquet and Cramer, no further edition of the vast epistolary exchanges of the Bernoullis and Hermann was attempted during the correspondents' lifetime. When Johann I Bernoulli died in 1748, the lion's share of his scientific manuscripts came under the possession of his son, Johann II, who, in turn, handed them over to his son, Johann III Bernoulli (1744–1807), director of the observatory and of the mathematical class of the Akademie der Wissenschaften in Berlin.

After several unsuccessful attempts to edit some of the major correspondences of his ancestors, beginning with that between Johann I Bernoulli and the Marquis de L'Hôpital,[21] Johann III Bernoulli put the letters up for sale in 1796[22] and sold the most important part of the letter collection to the Royal Swedish Academy of Sciences in Stockholm in 1797. The remaining part was acquired in 1799, along with further manuscript collections from the Bernoulli estate and a collection of manuscripts belonging to Johann Heinrich Lambert (1728–1777), by the Herzogliche Bibliothek of Gotha. The letters were kept safe in these libraries, yet they remained unpublished, and were thus, in fact, hardly accessible to the scientific community. After Johann III Bernoulli's death in 1807, the existence and location of these letters fell into oblivion for more than 50 years.[23]

Besides the letter collections sold to Stockholm and Gotha, there were other collections with substantial parts of the Bernoulli correspondence, the most important of which were conserved at St. Petersburg and Hannover. The scientific edition of these collections began in the first half of the nineteenth century.

In 1826, Paul Heinrich Fuss (1798–1855), a great-grandson of Leonhard Euler (1707–1783), became permanent secretary of the St. Petersburg Academy of

[20]Genève, BGE, Ms. Suppl. 384, fo. 14r-15v.

[21]For an overview of Johann III Bernoulli's plans, see Spiess (1955, pp. 27–31).

[22]The sales offer of approximately 1700 letters was made anonymously in Bernoulli Jo III (1796).

[23]A detailed account of the rediscovery of the Bernoulli manuscripts can be found in Spiess (1955, pp. 31–38).

Sciences. Driven by his interest in the scientific heritage of his ancestor, he viewed the manuscript collections kept in the Academy's archive. There, he found the letter exchanges of several figures of scientific interest that had been active in the Academy.[24] Among others, he came across 185 letters written by Johann I, Nicolaus I, Nicolaus II and Daniel Bernoulli to Leonhard Euler (1707–1783), Christian Goldbach (1690–1764) and his father Nicolaus Fuss (1755–1826). Fuss fully recognised the scientific importance of the letters. In 1841, he made a report to the Academy on the results of his investigation, with the proposal of preparing an edition of the material, a project he then started with the Academy's consent. Hoping to receive information about the location of the letters that were missing from the Academy's collection, he published his proposal publicly.[25] As a positive effect, the scientific community became aware of his project; nevertheless, he did not manage to discover the location of the missing letters. Fuss's edition appeared in 1743 in two volumes, with the title *Correspondance mathématique et physique de quelques célèbres géomètres du XVIIIème siècle.*[26] His work was very well received among the scientific community and revived interest in research on the history of mathematics. Regrettably, though, the edition was incomplete, as the volumes were missing the parts of the letter exchanges that Fuss had not been able to trace, which were kept in unknown locations in European archives. Additionally, 46 more letters from the exchange between Johann I and Daniel Bernoulli with Euler were being kept in St. Petersburg in a separate section of the archive.[27] The incompleteness of the edition, the lack of documentation of the editorial methods adopted and of commentaries, as well as the contemporary need for modern editorial standards led to a new edition of these letters in the frame of the Euler-Edition.[28]

Another important edition project that involves substantial parts of the Bernoulli correspondence was realized by the mathematician and historian of mathematics Carl Immanuel Gerhardt (1816–1899). Gerhardt edited almost all known letters exchanged between Leibniz and Jacob I, Johann I, Nicolaus Bernoulli and Jacob Hermann, and published them between 1855 and 1859 in four volumes of *Leibnizens mathematische Schriften*.[29] The edition was carried out on the basis of the manuscript collections now kept at the Gottfried Wilhelm Leibniz-Bibliothek in Hannover, which included the biggest portion of the letters sent by the Bernoullis and the drafts of the letters sent to them by Leibniz. In his introductions, Gerhardt

[24]At present, the manuscripts are located at the St. Petersburg branch of the Archive of the Russian Academy of Science.

[25]Fuss (1842).

[26]An account of the history of this edition is given in Fuss (1842, 1843 I, pp. IXX–XXXV) and in Fellmann and Mikhajlov (1998, pp. 21–23).

[27]Fellmann and Mikhajlov (1998, p. 23).

[28]These letters are edited in Fellmann and Mikhajlov (1998, 2016).

[29]Gerhardt (1855–1856) and (1859). For an overview and description of Gerhardt's editorial work, see Hess (1986). The edition of Leibniz's mathematical correspondence with the Bernoulli is addressed on pp. 35–38.

mostly does not explicitly outline the editorial principles underlying his edition. An exception is the introduction to the edition of the correspondence between Johann I Bernoulli and Leibniz,[30] in which he points out the main shortcomings of the *Commercium*, such as the lack of large passages due to the censoring efforts of the editors. In the same place, Gerhardt admits the incompleteness of his source material and the need to resort to the *Commercium* for the letters whose manuscript sources were inaccessible to him.[31] Despite the great merits of Gerhardt's editing work, an edition of the Bernoulli-Leibniz correspondence that is in accordance with today's editorial standards is therefore still a desideratum. Such an edition is being implemented in the context of the *Akademie-Ausgabe* of Leibniz's works and correspondence.[32]

With their work, Fuss and Gerhardt testify that, in the nineteenth century, the European community of historians of science still had an interest in the Bernoulli correspondence. In Switzerland, it was mainly the astronomer Rudolf Wolf (1816–1893) who focussed on the Bernoulli letters, while doing research for his Bernoulli biographies.[33] Wolf used the source material known to him as a basis for his studies, in the case of Johann I Bernoulli, mainly the part of the still unpublished correspondence with the brothers Johann Jakob (1672–1733) and Johannes (1684–1738) Scheuchzer, which was kept at the university library of Zürich. Wolf also contributed to the rediscovery of the letters that remained forgotten for several decades in the libraries of Stockholm and Gotha. In 1858, he learned about the existence of the Gotha manuscripts, and in 1877, about the location of the Stockholm manuscripts through his correspondence with Hugo Gyldén (1841–1896), director of the Stockholm observatory.[34] These discoveries led to a new attempt to edit the Bernoulli correspondence. The mathematician Gustaf Eneström (1852–1923), who became librarian of the Royal Library in Stockholm in the same year, delved into the Bernoulli papers kept in its collection. In 1880, he edited three unpublished letters from Leonhard Euler to Johann I Bernoulli.[35] Another promoter of the edition of the Bernoulli manuscripts in Stockholm was the mathematician Magnus Gösta Mittag-Leffler (1846–1927). With respect to the large amount of text material that had to be elaborated upon and the estimated high costs, Mittag-Leffler tried to find a partner institution for the project in Switzerland. In 1886, he contacted the physicist Eduard Hagenbach-Bischoff (1833–1910) with the offer that the whole Bernoulli estate kept at Stockholm would be donated to the University Library at Basel, if Basel, in return, would finance the printing of the edited letters. As editor, he recommended

---

[30] Gerhardt (1855, p. 132).

[31] The original letters sent by Leibniz to the Bernoullis were, at the time, kept at the University Library of Basel.

[32] See below (p. 12).

[33] Wolf (1858–1862).

[34] The discovery is reported in Wolf (1876).

[35] Eneström (1880).

Gustaf Eneström, who was apparently willing to do the work on a voluntary basis.[36] Despite the excellent conditions offered, the answer from Basel was negative, as funding of the printing expenses was not possible. Therefore, the original plan of an integral edition of the letters was set back again. The letters edited in this phase by Gustaf Eneström consisted only of the remaining part of the still unpublished letters from the correspondence between Johann I Bernoulli and Leonhard Euler kept at Stockholm.[37] These were just 34 letters of the approximately 1500 in the Stockholm collection. Nevertheless, the engagement of Gustaf Eneström did not stop here. In the 1920s, he accepted the redaction of the volume of the Bernoulli-Euler correspondence within the frame of the Euler-Edition led by Ferdinand Rudio. His edition was planned as volume III, 12 of the *Opera omnia Leonhardi Euleri*,[38] but could not be realised due to Eneström's death in 1923. The edition within the *Opera omnia* was implemented only in 1998 and 2016 with the two volumes edited by Emil Fellmann and Gleb Mikhajlov.[39]

In the meantime, the Bernoulli letters kept at Gotha also came into public awareness. Their new discovery is due to the professor of mathematics Paul Stäckel (1862–1919), who, in 1899, was searching for the manuscripts of Johann Heinrich Lambert (1728–1777) that had been sold to the Herzogliche Bibliothek Gotha by Johann III Bernoulli, together with the Bernoulli collection.[40] In 1916, the historian of mathematics Karl Bopp (1877–1934) edited Lambert's *Monatsbuch*. In the introduction to the book, he printed an inventory of the letter manuscripts kept at Gotha. Three of Karl Bopp's students edited some of the letters in this collection.[41] Karl Wollenschläger published the 17 letters exchanged between Johann I Bernoulli and Abraham de Moivre,[42] Otto Julius Rebel (1910–ca. 1934) edited 11 letters of the Bernoulli-L'Hôpital correspondence,[43] and Ernst Jakob Fedel edited 12 letters between Johann I Bernoulli and Pierre Varignon (1654–1722).[44]

In 1926, another small edition was put out by the historian of economics Wilhelm Stieda (1852–1933), who published 16 letters sent by Johann II Bernoulli to Henri Alexandre de Catt (1725–1795), Frederick the Great (1712–1786), and Prince Henry of Prussia (1726–1802).[45]

---

[36]See Spiess (1955 pp. 40–41).

[37]Eneström (1903–1905) and (1906).

[38]Fellmann and Mikhajlov (2016, p. 16).

[39]Fellmann and Mikhajlov (1998; 2016).

[40]Spiess (1955, pp. 44–45).

[41]For an overview, see Spiess (1955, pp. 9–85).

[42]Wollenschläger (1933). The need for a re-edition of this correspondence, due to the inadequate standard of the Wollenschlägers edition, is pointed out in Spiess (1955, pp. 47–48). Moivre's correspondence will be edited anew within the BEBB.

[43]Rebel (1934).

[44]Fedel (1934).

[45]Stieda (1926).

3 The Bernoulli-Edition at Basel

Although the interest of historians of science in the Bernoulli correspondence was lively and constant through the centuries, and initiatives to start a comprehensive edition were repeatedly taken, the letter manuscripts remained, to the greatest extent, undisclosed. One main hindrance to the project's advancement was the lack of sufficient financial funding. Perhaps the most decisive factor to lead to a change was the rise of a new understanding of history of science as a discipline that preserves and discloses the national cultural heritage with a focus on outstanding figures of national interest. The establishment of the Euler Edition[46] on the occasion of the second centenary of Leonhard Euler's birth in 1907 is a good example of how this programmatic context enhanced the fundraising of a long-term edition project. It was indeed in the wake of the example of the Euler Edition that the mathematician Otto Spiess (1878–1966), then professor of mathematics at the University of Basel, initiated the Bernoulli Edition. With his project, Spiess aimed to make the whole corpus of letters and works of the mathematicians and physicists of the Bernoulli family and of Jacob Hermann accessible to the scientific community.[47] Spiess laid the ground for the edition by organizing an institutional framework indispensable for the execution of such a long-term project.[48] In addition, he coordinated the repatriation of most known Bernoulli manuscripts and the acquisition of the Gotha manuscripts by the University Library of Basel through funding from private donors.[49]

Spiess reports that the proper editorial work started in 1935.[50] From the beginning, the focus was on the edition of letters, since, in contrast to the scientific works, these had remained unpublished for the most part. Once the manuscripts were reunited in Basel, Spiess began with some preparatory work. First of all, he tracked the remaining letters scattered all over Europe, compiled an inventory of all known letters sent or received by the Bernoullis and Jacob Hermann and prepared an editorial plan for a comprehensive Bernoulli edition.[51] On this basis, he began, with a small team, to transcribe almost all letter manuscripts that were accessible to him, producing far more than 10,000 typewritten pages. In the meantime, he recorded

[46] An overview of the Euler-Edition is given in Kleinert (2015, 2017), and Kleinert and Mattmüller (2007).

[47] For more detailed information on Spiess, see Neuenschwander (2013) and Dauben and Scriba (2002, p. 526).

[48] In 1935, Spiess managed to bring the project under the patronage of the *Naturforschende Gesellschaft Basel* and to create a foundation supervised by the Bernoulli-Kommission. See Spiess (1955, p. 49).

[49] The University Library of Basel obtained the manuscripts conserved at the Royal Academy of Stockholm on a loan basis in 1935 for the purpose of their edition, and finally acquired them in 1965. See Spiess (1955, p. 29, 50), and Fellmann and Mikhajlov (1998, p. 25).

[50] Spiess (1955, p. 10).

[51] Spiess (1955, p. 75).

the names mentioned in the letters in a card index and in inventories of single letter exchanges.[52]

The first volume of the edition, containing 162 letters from the correspondence between Jacob I Bernoulli and Johann I Bernoulli, Johann I Bernoulli and the Marquis de L'Hôpital, as well as some smaller early correspondences of Johann I Bernoulli, was edited by Spiess and published by the *Naturforschende Gesellschaft* in 1955.[53] The letters were reproduced in the original language and commented upon in German. Besides the above-mentioned introductory section on the history of the Bernoulli manuscripts and their edition, the volume included an exposition of the editorial principles and methods adopted, a chronological inventory of the edited letters, a catalogue of the printed works by L'Hôpital, a bibliography of the literature mentioned in the letters and of that consulted for the commentaries, and indices listing scientific problems, subjects, and names.[54] The volume follows clear editorial rules and can be considered the first modern edition of the Bernoulli letters. Spiess also planned, and entrusted to the historian of science Pierre Costabel (1912–1989), the edition of the Bernoulli-Varignon correspondence in 1938.

After Spiess's death in 1966, a new team of scholars continued the work on the Bernoulli Edition under the guidance of the professor of mathematics Joachim Otto Fleckenstein (1914–1980),[55] who was already a member of the editorial team of both the Bernoulli Edition and the Euler Edition. Fleckenstein shifted the attention away from the letters and put the main focus on the edition of the works of the Bernoullis. Fleckenstein had already worked on the Bernoulli-Varignon letter edition under Spiess's direction, but he died before the first volume could appear. His successor, the professor of mathematics David Speiser (1926–2016), who directed the edition from 1980 to 2001, reassigned the project to Pierre Costabel.[56] Costabel prepared the first two volumes in collaboration with the historian of mathematics Jeanne Peiffer. The volumes containing 158 letters from the Bernoulli-Varignon correspondence and 9 letters by other authors appeared in 1988 and 1992.[57] Moreover, in 1993, André Weil (1906–1998) edited the complete correspondence of Jacob I Bernoulli (with the exception of the exchange with Leibniz, consisting of 49 letters) in one volume, with contributions by Clifford Truesdell (1919–2000) and Fritz Nagel.[58] The four extant volumes of the Bernoulli-Edition (comprising

[52]Spiess (1955, p. 58). The typewritten transcriptions prepared by Spiess still form the basis of the Bernoulli letter edition. They are conserved, together with the card index, at the Bernoulli-Euler-Zentrum in the University Library of Basel.

[53]Spiess (1955, pp. 95–485).

[54]Spiess (1955, pp. 87–92).

[55]See Costabel and Peiffer (1988, p. IX).

[56]On the history of the correspondence between Bernoulli and Varignon, see Costabel and Peiffer (1988, pp. 3–4).

[57]Costabel and Peiffer (1988–1992).

[58]Weil (1993).

369 letters altogether) all follow the structural model, the methods and the editorial standards set by Spiess in 1955, with a few exceptions, such as the working language.

4 The Euler- and Leibniz-Edition and Other Modern Editions Prepared Independently from the Bernoulli-Edition

In his outline of the 1955 edition, Otto Spiess established that the correspondences of the Bernoullis with Leonhard Euler and Gottfried Wilhelm Leibniz should be edited, respectively, within the Euler and Leibniz Editions. In the following section, I shall therefore give a short overview of the state of the Bernoulli correspondence in these editions. After a first attempt to edit the Bernoulli-Euler correspondence was interrupted by the death of Eneström in 1923,[59] a further step was taken at the Archive of the St. Petersburg Academy of Science by I. I. Ljubimenko. In 1937, she published an inventory of the correspondence of the members of the Academy with some newly discovered letters of Daniel Bernoulli in the appendix.[60] In the late 1950s, Gleb K. Mikhajlov also worked at the same Archive, viewing the Euler and Bernoulli manuscripts and preparing the edition of the epistolary between the Bernoullis and Leonhard Euler in common agreement with Otto Spiess in Basel. The establishment of a Swiss-Russian cooperation between Gleb K. Mikhajlov and the historian of mathematics Emil A. Fellmann (1927–2012) finally led to the accomplishment of the editorial work.[61] In 1998, the Euler Edition published a volume, edited by Fellmann and Mikhajlov, containing the correspondence of Johann I Bernoulli and Nicolaus I Bernoulli with Leonhard Euler (respectively, 38 and 11 letters) and some officials of the Imperial Academy of Sciences of St. Petersburg (12 letters).[62] The remaining part of the Bernoulli-Euler correspondence, consisting of the letters that Daniel I, Johann II and Johann III Bernoulli exchanged with Leonhard Euler (112 letters), Johann Albrecht Euler (33 letters), Nicolaus Fuss, and the officials of the Academy at St. Petersburg (93 letters), was prepared by the same editors and appeared in 2016. The editorial standards of these volumes can be compared with those of the volumes of the Bernoulli letter edition: they follow almost the same rules, which are described in the introduction of the two volumes.

The edition of the Leibniz correspondence is being prepared at the Leibniz-Archiv Hannover as section III (Mathematischer, naturwissenschaftlicher und technischer Briefwechsel) of the series *Leibniz, Gottfried Wilhelm: Sämtliche Schriften und Briefe*.[63] Also known under the designation *Akademie-Ausgabe*, this historical-

[59]See above (p. 8).

[60]Ljubimenko (1937). See Fellmann and Mikhajlov (2016, p. 16).

[61]Fellmann's work is credited in Neuenschwander (2018).

[62]Fellmann and Mikhajlov (1998).

[63]On the history of the Leibniz edition see inter alia Wahl (2013), Folkerts (2008), Lorenz (2007), Poser (2000), and Schepers (1999).

critical, fully annotated edition was initiated in 1901 and stands under the patronage of the Berlin-Brandenburgische Akademie der Wissenschaften and the Akademie der Wissenschaften zu Göttingen. The correspondence is presented there in chronological order, so that the letters exchanged with a single correspondent are not grouped into one volume, but rather spread out over several volumes. The correspondence with the Bernoullis starts in 1672, and is therefore contained in volume 4 onwards.[64] The volumes that have appeared so far contain the letters exchanged by the Bernoullis and Hermann with Leibniz from 1672 to 1705, which is about one-half of the 406 surviving letters of this exchange. So far, the edition has been released in book form, but lately, online publishing has also been adopted. Some edition volumes are also available as a download of free pdf-files interlinked with the *Persons and Correspondence Database* of the Leibniz edition, which has been accessible on the web since 2017.[65] Over the years, the following editors have been responsible for the project: Heinz-Jürgen Hess, James G. O'Hara, Herbert Breger, Charlotte Wahl, Ralf Krömer, Heike Sefrin-Weis, Uwe Mayer and Michael Kempe.

Besides the correspondences with Euler and Leibniz, some smaller correspondences, such as those between Jacob Hermann and some Italian scientists and between Nicolaus II Bernoulli and Jacopo Riccati (1676–1754), have been edited independently outside the Bernoulli, Euler and Leibniz Editions. These editions were implemented in the context of research on the role of the Bernoullis in the divulgation of the infinitesimal calculus in Italy, for the most part by the historians of mathematics Clara Silvia Roero and Silvia Mazzone.[66]

5 The Basler Edition der Bernoulli-Briefwechsel (BEBB)

Between 1993 and 2007, the Bernoulli Edition in Basel continued its work with the preparation of an online inventory of the Bernoulli correspondence, providing the metadata of all known letters based on the inventory and name index prepared by Otto Spiess. Fritz Nagel extended the Spiess inventory and index through the addition of codicological descriptions of the letters, content summaries of a large number of letters, bibliographical information on the letters already published, lists of all persons mentioned, and information on the existence of different versions of letters and their location. The data are accessible online in the *Basler Inventar der Bernoulli-Briefwechsel* (BIBB) via the manuscript catalogue of the

[64]Leibniz-Archiv, Leibniz-Forschungsstelle Hannover (eds) (1976–). Volumes 4-8 were published between 1995 and 2015. Volume 9 is still in progress, but a preliminary version is already available on the webpage of the Leibniz edition (http://www.gwlb.de/Leibniz/Leibnizarchiv/Veroeffentlichungen/III9.pdf).

[65]Berlin-Brandenburgische Akademie der Wissenschaften/Akademie der Wissenschaften zu Göttingen (eds) (2017–) Personen- und Korrespondenz-Datenbank der Leibniz-Edition (https://leibniz.uni-goettingen.de/pages/index). At present, the database is not yet completely accessible to the public; the remaining parts will be released in stages.

[66]The published letters are edited in Mazzone and Roero (1992, 1997), and Grugnetti (1986).

University Library of Basel and can be searched according to criteria similar to those used for the online library catalogue.[67] This inventory provided the basis for the subsequently planned online-edition of the Bernoulli letters.

In 2007, Fritz Nagel and I started the project *Basler Edition der Bernoulli-Briefwechsel (BEBB)*.[68] The first aim of the project was to bring together the already existing material on an online platform and to build from this ground an open-access database providing an integral edition of all unpublished Bernoulli letters.

For this purpose, the University Library of Basel[69] created a Bernoulli-wiki, hosted by the library together with the editorial team. There, the edited letters could be published and further elaborated upon. The texts on the Bernoulli website were encoded in the MediaWiki syntax, which, due to its simplicity, was well suited to the collaborative approach of the edition.

In comparison to the traditional printed edition, the publication on the Bernoulli-wiki offered many advantages:

- The metadata contained in the BIBB are far more detailed than those that can be provided in the header of a traditional printed edition. While only a selection of the full metadata is rendered in the headline of the considered letter, the user can reach the full dataset recorded in the inventory through a link.
- In the printed volumes, a number is assigned to each letter to facilitate its citation. This is, however, only applicable within the volume itself, as the numeration is not continuous between the volumes. By contrast, the online edition assigns a unique number to each letter that can thus serve as its absolute identifier.

Some of the advantages of the BEBB are specific to the electronic form of publication:

- Greater flexibility in the planning of the edition. The work can be started without the need to determine the fixed order of the letters or to distribute the text corpus on volumes.
- Absence of space restrictions on the publication platform, and therefore no limits to an integral edition of the letters.
- The possibility of adding more versions of the text, such as diplomatic transcriptions or translations.
- The possibility of processing the edition in stages and with different levels of elaboration.

[67]It can be consulted at http://aleph.unibas.ch/F/func=option-update-lng&file_name=find-b&p_con_lng=GER&local_base=bernoulli_edition

[68]The edited letters can be consulted at http://www.ub.unibas.ch/bernoulli/index.php/Briefe_im_Volltext.

[69]The Bernoulli-wiki was developed by Andreas Bigger, at the time, the coordinator of the digitalisation team of the University Library.

- The possibility of supplementing digital images of the original manuscripts.[70]
- An improved representation of the network character of the correspondence, due to the network structure of an online edition and the possibility of interlinking the different resources available on the web.
- Flexibility of the edition: the possibility of adding new materials, emending mistakes and updating the commentaries.
- Search tools that permit structured or full-text searches.
- Open and easy access.

Each letter was published as a separate webpage that comprises the emended transcriptions, a footnote apparatus with commentaries and an abridged version of the letter metadata extracted from the BIBB through an interface. Thumbnails of the images of the manuscript pages positioned at the top of the page and smaller icons positioned in the text to indicate the beginning of the relative manuscript page were linked to high-resolution images of the corresponding manuscript. The Bernoulli-wiki is searchable by full text search. Additionally, a more structured search is available in the interlinked inventory, where a search mask modelled on that offered in the HAN/Aleph catalogue of the University Library is available. This includes search categories such as author, addressee, date, correspondence, mentioned person or location of the manuscript.

During the first phase of the project, lasting from 2007 to 2011, Fritz Nagel and I edited 712 letters belonging to the correspondence of Johann I Bernoulli. The main aim of the BEBB was to make the hitherto unpublished texts accessible to the scientific community in a reliable and citable way as quickly as possible. Therefore, the focus of the editorial work was on the compilation of the text. The editorial guidelines that the editors of the printed volumes complied with were partly renewed based on a comparison with the current guidelines of other letter editions, such as the Leibniz and Euler Editions.[71] In this phase, the letters of Johann I Bernoulli with 81 correspondents were edited.[72] The major epistolary exchanges were those with John Arnold (ca. 1688–17xx) (11 letters), Georg Bernhard Bilfinger (1693–1750) (50 letters), William Burnet (1688–1729) (30 letters), Alexis Claude Clairaut (1713–1765) (8 letters), Jean Pierre de Crousaz (1663–1750) (44 letters), Bernard le Bovier de Fontenelle (1657–1757) (19 letters), Jacob Hermann (79 letters), Pierre Louis Moreau de Maupertuis (1698–1759) (95 letters), Johann Burckhardt Mencke (1674–1732) (30 letters), Pierre Rémond de Montmort (1678–1719) (40 letters), Giovanni Poleni (1683–1761) (31 letters), Burchard de Volder (1643–1709) (13 letters), and Christian Wolff (1679–1754) (97 letters).

[70]The presentation of digital images of the manuscript is particularly useful in editions of mathematical texts, since the editorial representation of mathematical formulae or figures sometimes implies a loss of information. For the problems related to the representation of mathematical figures in editions, see Beeley (2007, p. 106).

[71]The guidelines containing a detailed documentation of the editorial rules can be consulted at http://www.ub.unibas.ch/bernoulli/images/e/e0/Richtlinien_2011.pdf

[72]For more detailed information on the correspondents, see www.ub.unibas.ch/bernoulli

Between 2011 and 2015, Fritz Nagel and I, with the temporary support of Elena Mazzei, Damaris Gehr and other collaborators, continued the editorial work on the BEBB with the publication of the Bernoulli-Scheuchzer correspondence. The corpus of this correspondence comprises 829 letters that Johann I, Nicolaus I, Nicolaus II and Daniel Bernoulli, as well as Jacob Hermann, exchanged with the brothers Johann Jakob and Johannes Scheuchzer. In contrast to the first phase of the BEBB, this project was planned as a fully annotated edition. In order to address the exceptionally wide range of subjects treated in the letters at a certain level of competence, experts from several specialist fields were invited to contribute more or less extensive thematic commentaries.[73] The Bernoulli-Scheuchzer edition can therefore be called an interdisciplinary annotated edition and represents a first step in the direction of a collaborative online edition.

The third and final stage of the edition of the Johann I Bernoulli-Varignon correspondence (95 letters) was initiated in 2016 and is being edited by Jeanne Peiffer and the editorial team in Basel.[74] It will be published in printed form as volume 3 of the Bernoulli-Varignon correspondence (volume 4 of the series *Der Briefwechsel von Johann Bernoulli*, within the overall edition *Die gesammelten Werke der Mathematiker und Physiker der Familie Bernoulli*) and in the form of an open-access online database within the BEBB. In the same period, Fritz Nagel added scientific commentaries to the correspondence of Johann I Bernoulli and Christian Wolff, which has already been published online as part of the BEBB.

6 Continuation of the BEBB Within Bernoulli-Euler OnLine (BEOL)

In the meantime, the Bernoulli letter edition experienced a reorganisation phase following the foundation of the Bernoulli-Euler-Zentrum in 2012.[75] The reunion of the Bernoulli-Forschungsstelle Basel and the Euler-Archiv in the new centre gave new impulses for a collaborative approach to the previously separate Euler and Bernoulli editions and entailed the discussion of a new common basis for both editions. The first big step in this direction was taken with the planning of

[73]The Bernoulli Scheuchzer edition benefited from the cooperation of the following experts: Daniel Bernoulli, professor of geology, ETH Zürich; Ulrich Gäbler, professor of church and dogma history, University of Basel; Ernst Jenni, professor of Old Testament and Semitic linguistics, University of Basel; Paul Michel, professor for older German literature, University of Zürich; Rebekka Schifferle, cultural historian, University of Basel; Martin Rickenbacher, historian of geography, Swisstopo Bern; Andreas Verdun, historian of astronomy, University of Bern.

[74]Jeanne Peiffer participated in the publication of the previous two volumes of the correspondence, which appeared in 1988 and 1992 as volumes 2 and 3 of *Der Briefwechsel von Johann Bernoulli*.

[75]The Bernoulli-Euler-Zentrum was established by the University of Basel in 2012 and is hosted on the premises of the Basel University Library. The pivotal aims of the Centre are the maintenance of the heritage of the "Basel school" of mathematicians in the seventeenth and eighteenth centuries, the edition of their works and correspondence, and the promotion of international research cooperation. More information can be found online at: https://bez.unibas.ch

the Bernoulli-Euler OnLine project (BEOL), initiated by Hanspeter Kraft of the Bernoulli-Euler-Zentrum and Lukas Rosenthaler of the Digital Humanities Lab Basel (DHLab).[76] BEOL is a collaborative and interdisciplinary project combining editorial scholarship and digital humanities. With this project, the editorial team of the BEZ aims to continue the editorial work in an online format sharing the same platform, ontologies and technologies. In this context, new edition projects are already being implemented. The DHLab, in turn, aims at the realisation of a generic virtual research environment (VRE), in which specific methods and functionalities needed for an interoperable, long-term online edition of source texts from the Early Modern history of mathematics and science are implemented.[77] The collaboration of the BEZ and the DHLab provides the framework for discussing the real needs of the editorial teams and represents a testing ground for the tools that are being developed. Four concrete editorial projects have been used as test runs: the web-based presentation of volume IVA/4 of Leonhard Euler's *Opera Omnia*, which was published as a book in 2015,[78] the "genuinely digital" edition of Jacob I Bernoulli's notebook *Meditationes* elaborated by Martin Mattmüller, the presentation of the last part of the Bernoulli-Varignon correspondence, which will be published as a book and simultaneously integrated into the platform,[79] and the import of the letters already published within the Bernoulli-wiki. The results of the first phase of BEOL are now accessible in a beta release.[80]

With the import into BEOL, the BEBB will benefit from several significant improvements:

1. All data now scattered over the BIBB and the Bernoulli-wiki will be brought together in a single place and processed with the same tools and methods.
2. The data will be searchable with search tools elaborated within the platform and designed according to the needs of the BEBB.
3. The VRE permits generating and processing structured data through semantic mark-ups, in analogy to the handling of text in TEI/XML, but with improved flexibility. Biographical, bibliographical and other metadata will be organised in ontologies that are shared with the other projects on the platform.

---

[76]The work on BEOL is funded by the Swiss National Foundation and currently stands under the direction of Lukas Rosenthaler and Helmut Harbrecht, the actual director of the BEZ.

[77]The VRE used here is based on the Knora/SALSAH platform developed by Lukas Rosenthaler at the DHLab. Knora stands for Knowledge Organization, Representation, and Annotation. It is an open, modular, extensible and flexible platform on which the data are stored and processed. In order to represent data, Knora uses industry standards such as the Resource Description Framework (RDF). A description of the platform with a graphical representation of its architecture is given in SAGW (2015).

[78]Lemmermeyer and Mattmüller (2015).

[79]The Bernoulli-Varignon edition is a hybrid edition, having been elaborated in parallel for publication in print and online. For more information on the project, see above (pp. 15–16).

[80]A link to the beta release can be found on the BEZ website (https://www.bez.unibas.ch/beol.php).

4. Different stages of elaboration, such as diplomatic and normalized versions, can be generated from one file and published in parallel, offering a detailed documentation of the editorial changes and simplifying the scholarly apparatus of footnotes in the meantime.
5. As the Knora/SALSAH platform is being developed within the Swiss Data and Service Center for the Humanities (DaSCH) initiated by the Swiss Academy of Humanities and Social Sciences with the purpose of developing solutions to the problem of long-term availability of research data in the humanities, the data security will be improved.
6. The platform enables different access levels. This simplifies the further elaboration of the material that had already been published on it. The public can access the definitive results, while the editorial teams can continue working on it in the background.
7. Other than on the Bernoulli-wiki, formulae are represented as text rather than images and can be encoded, compiled and checked easily in the VRE in LaTeX or MathML.
8. Due to a versioning system and the possibility of defining text passages that have been referred to, a precise referencing of the letters is now possible.
9. In this VRE, the edition process, publication, consultation, and scholarly re-use of the edited material all take place in a common virtual space. On this basis, a user-friendly online collaboration on complex edition projects with high scholarly standards will be possible.

The next phase of the BEBB within BEOL, planned for a period of 10 years, consists mainly in the edition of 2505 letters. These will be distributed among the single letter exchanges as follows: Johann I Bernoulli (331 letters), Daniel I Bernoulli (387 letters), Jacob II Bernoulli (42 letters), Johann II Bernoulli (1073 letters), Nicolaus I Bernoulli (467 letters), Nicolaus II (39 letters), Jacob Hermann (166 letters). The working phase also includes the fact that the letters that have already been edited in the Bernoulli-wiki will be further processed on the new platform. This work, which will take place in several stages, will mainly consist in structuring the text data and commenting upon the letters. In this stage, the edition of three new letter exchanges will be elaborated: 115 letters exchanged between Johann I Bernoulli and Jean Jacques Dortous de Mairan (1678–1771), prepared by Fabienne Schwizer, 109 letters exchanged between Johann I Bernoulli and Pietro Antonio Michelotti (1663–1740), prepared by me, and 78 letters exchanged by the Bernoullis and Hermann with Louis Bourguet (1678–1742), prepared by Fritz Nagel and me.

For this phase, the BEBB received the support of the Swiss National Foundation, which has recently committed itself to the funding of long-term edition projects. Starting in 2019, the patronage of the project will be taken over by the Swiss Academy of Humanities and Social Sciences.[81]

[81] Immenhauser (2017, p. 27).

Bibliography

Beeley, Ph. 2007. Edition und Rezeption. Zur Rolle der Akademie-Ausgabe in der Gegenwärtigen Diskussion zur Philosophie des G. W. Leibniz in den angelsächsischen Ländern. In *Editionen – Wandel und Wirkung*, ed. A. Sell, editio/Beihefte 25, 93–108. Tübingen.

Berlin-Brandenburgische Akademie der Wissenschaften, Akademie der Wissenschaften zu Göttingen eds. 2017–. Personen- und Korrespondenz-Datenbank der Leibniz-Edition. https://leibniz.uni-goettingen.de/pages/index

Bernoulli, Ja. Meditationes, Annotationes, Animadversiones Theologicae & Philosophicae. University Library of Basel, Ms L Ia 3.

Bernoulli, Jo. 1859. [Autobiography 1741]. In *Biographien zur Kulturgeschichte der Schweiz*, ed. R. Wolf, vol. 2, Zürich, 71–94.

Bernoulli, Jo, and G. Cramer, ed. 1742. Opera omnia, tam antea sparsim edita, quam hactenus inedita, Lausannae/Genevae (M.-M. Bousquet), 4 vols.

Bernoulli, Jo, G.W. Leibniz, and G. Cramer, ed. 1745. Virorum celeberr. Got. Gul. Leibnitii et Johan. Bernoullii Commercium philosophicum et mathematicum, Lausannae/Genevae (M.-M. Bousquet), 2 vols.

Bernoulli, Jo. III 1796. Nachricht wegen eines Briefwechsels Joh. Bernoulli's. *Archiv der reinen und angewandten Mathematik* 5(2): 3–4 (cover).

Bovet, P. 1905. Louis Bourguet, Son projet d'édition des oeuvres de Leibniz. In *Congrès International de Philosophie, IIe session, Genève 1904, Rapports et comptes rendus*, Genève, 252–263.

Cantor, M. 1898. Vorlesungen über Geschichte der Mathematik, vol. 3, Leipzig.

Costabel, P., and J. Peiffer, eds. 1988–1992. Der Briefwechsel von Johann I Bernoulli, vols 2–3 (Der Briefwechsel mit Pierre Varignon, vol 1 1692–1702 and vol 2 1702–1714). Berlin.

Dauben, J.W., and Ch.J. Scriba, eds. 2002. Writing the History of Mathematics: Its Historical Development. Berlin.

Eneström, G., ed. 1880. Trois lettres inédites de Jean Ier Bernoulli à Léonard Euler tirées de la correspondance de Jean Ier Bernoulli gardées dans la Bibliothèque de l'Académie Royale des Sciences de Stockholm. In *Bihang till Kongl. Svenska Vetenskaps-Akademiens Handlingar*, 5(21), Stockholm, 5–20.

———., ed. 1903–1905. Der Briefwechsel zwischen Leonhard Euler und Johann I Bernoulli. In: *Bibliotheca Mathematica* 1903 4(3): 344–388; 1904 5(3): 248–291; 1905 6(3): 16–87.

———. 1906. Der Briefwechsel zwischen Leonhard Euler und Daniel Bernoulli. *Bibliotheca Mathematica* 7 (3): 126–156.

Fedel, E.J., ed. 1934. Der Briefwechsel Johann (I) Bernoulli-Pierre Varignon aus den Jahren 1692 bis 1702 in erläuternder Darstellung. Dissertation, University Heidelberg.

Fellmann, E.A., and G.K. Mikhajlov, eds. 1998. Leonhardi Euleri Opera Omnia, vol IVA/2: Der Briefwechsel von Leonhard Euler mit Johann I Bernoulli und Niklaus I Bernoulli, Basel.

———, eds. 2016. Leonhardi Euleri Opera Omnia, vol IVA/3: Leonhardi Euleri commercium epistolicum cum Daniele, Johanne II, Johanne III Bernoulli ..., Basel.

Fleckenstein, J.O., and D. Speiser, eds. 1969–1999. Die Werke von Jakob Bernoulli. Im Auftrag der Naturforschenden Gesellschaft in Basel und der Otto Spiess-Gesellschaft, Basel.

Folkerts, M. 2008. Die Leibniz-Edition zwischen Wissenschaft und Politik. In *Kosmos und Zahl*, eds. H. Hecht et al. 23–45. Stuttgart.

Fuss, P.H. 1842. Extrait du procès verbal de l'Académie Impériale des Sciences de St.-Pétersbourg du 24 Septembre (6 Octobre). *Journal für die reine und angewandte Mathematik* 23: 196–198.

———., ed. 1843. Correspondance mathématique et physique de quelques célèbres géomètres du XVIIIème siècle ..., vol 2. St.-Pétersbourg.

Gerhardt, C.I., ed. 1855–1856. Leibnizens mathematische Schriften. 1. Abtheilung, vol 2, 3 and 2. Abtheilung, vol 3. Halle (Briefwechsel zwischen Leibniz, Jacob Bernoulli, Johann Bernoulli und Nicolaus Bernoulli) and 1. Abth. vol 4, Halle 1859 (Briefwechsel zwischen Leibniz, Wallis, Varignon, Guido Grandi, Zendrini, Hermann, und Freiherrn von Tschirnhaus).

Gehr, S., and F. Nagel, eds. 2008a–. Basler Edition der Bernoulli-Briefwechsel, Basel. http://www.ub.unibas.ch/bernoulli

———, eds. 2008b–2011. Editorische Richtlinien zu Transkription, Textkonstitution und Kommentierung der Bernoulli-Briefwechsel, Basel. http://www.ub.unibas.ch/bernoulli/images/e/e0/Richtlinien_2011.pdf

Grugnetti, L. 1986. L'Equazione di Riccati: un carteggio inedito tra Jacopo Riccati e Nicola II Bernoulli. *Bollettino di Storia delle Scienze Matematiche* 6 (2): 45–82.

Hess, H.-J. 1986. Karl Immanuel Gerhardt. Ein großer Leibniz-Editor. In: Beiträge zur Wirkungs- und Rezeptionsgeschichte von Gottfried Wilhelm Leibniz (Studia Leibnitiana Supplementa, vol XXVI), 27–64. Stuttgart.

Immenhauser, B. 2017. Transfer von Editionen vom Schweizerischen Nationalfonds zur SAGW. *Bulletin SAGW* 4: 26–27.

Kleinert, A. 2015. Leonhardi Euleri Opera Omnia: Editing the works and correspondence of Leonhard Euler. *Prace Komisji Historii Nauki PAU* XIV: 13–35.

———. 2017. Hundert Jahre russisch-schweizerische Zusammenarbeit bei der Euler-Gesamtausgabe. In *Deutsch-russische Zusammenarbeit wissenschaftlicher und kultureller Institutionen vom 18. zum 20. Jahrhundert*, eds. I. Kästner, and M. Schippan (Europäische Wissenschaftsbeziehungen vol 14), 213–224. Aachen.

Kleinert, A., and M. Mattmüller. 2007. Leonhardi Euleri Opera Omnia: a centenary project. *Newsletter of the European Mathematical Society* 65: 25–31.

Korthold, Christian, ed. 1734–1742. Viri illustris Godefridi Guil. Leibnitii epistolae ad diversos ... Lipsiae.

Leibniz-Archiv, Leibniz-Forschungsstelle Hannover, eds. 1976–. Sämtliche Schriften und Briefe (Akademie–Ausgabe), Reihe III: Mathematischer, naturwissenschaftlicher und technischer Briefwechsel, Berlin.

Lemmermeyer, F., and M. Mattmüller, eds. 2015. Leonhardi Euleri Opera Omnia, vol IVA/4: Correspondence of Leonhard Euler with Christian Goldbach. Basel.

Ljubimenko, I.I., ed. 1937. *Ученая корреспонденция Академии наук XVII века: Научное описание 1766–1782* [Uchenaya korrespondenciya Akademii nauk XVIII veka: Nauchnoe opisanie 1766–1782]. Moskva/Leningrad.

Lorenz, S. 2007. Auferstehung eines Leibes dessen Glieder wunderbarhlich herum zerstreuet sind. Leibniz-Renaissancen und ihre editorischen Reflexe. In *Editionen – Wandel und Wirkung*, ed. Sell A. (editio/Beihefte 25), 65–92. Tübingen.

Mazzone, S., and C.S. Roero, eds. 1992. *Guido Grandi – Jacob Hermann*. Carteggio. Firenze.

———. 1997. *Jacob Hermann and the diffusion of the Leibnizian calculus in Italy (1707–1713)*. Firenze.

Nagel, F. 1989. Der Briefwechsel zwischen Johann Bernoulli und Leibniz. Zur Geschichte der Basler Handschriften. *Studia Leibnitiana, Sonderheft* 17: 167–174.

———. 1994. Schweizer Beiträge zu Leibniz-Editionen des 18. Jahrhunderts. Die Leibniz-Handschriften von Johann Bernoulli und Jacob Hermann in den Briefwechseln von Bourguet, Cramer, König und Kortholt. In *Leibniz und Europa. VI. Internationaler Leibniz-Kongress*, 525–533. Vorträge, Hannover.

———. 2005. Die Bernoulli-Edition. In *Geschickte Leute, die was praestiren können ... Gelehrte aus Basel an der St. Petersburger Akademie der Wissenschaften des 18. Jahrhunderts*, eds. F. Nagel, and A. Verdun, 173–175. Aachen.

———. 2007–. Bernoulli-Briefinventar (BIBB). Basel. https://aleph.unibas.ch/F/?func=option-update-lng&file_name=find-b&p_con_lng=GER&local_base=bernoulli_edition

Neuenschwander, E. 2002. Switzerland. In *Writing the History of Mathematics: Its Historical Development*, eds. J. W. Dauben, and Ch.J. Scriba, 103–105. Basel.

———. 2013. Spiess, Otto. Historisches Lexikon der Schweiz. http://www.hls-dhs-dss.ch/textes/d/D43136.php

———. 2018. In memoriam Emil Alfred Fellmann (1927–2010). *Archives Internationales d'Histoire des Sciences* 66 (177): 501–516.

Peiffer, J. 1998. Faire des mathématiques par lettres. *Revue d'histoire des mathématiques* 4 (1): 143–157.

Poser, H. 2000. Langzeitvorhaben in der Akademie. Die Geschichte der Leibniz-Edition zwischen Kaiserreich und geteiltem Deutschland. In *Die Preußische Akademie der Wissenschaften zu Berlin 1914–1945*, ed. W. Fischer, 375–389. Berlin.

Rebel, O.J., ed. 1934. Der Briefwechsel zwischen Johann (I) Bernoulli und dem Marquis de L'Hospital in erläuternder Darstellung. Dissertation, University Heidelberg.

Schepers, H. 1999. Zur Geschichte und Situation der Akademie-Ausgabe von Gottfried Wilhelm Leibniz. Wissenschaft und Weltgestaltung, 291–298. Hildesheim.

Schweizerische Akademie der Geistes- und Sozialwissenschaften (SAGW), ed. 2015. *Final report for the pilot project "Data and Service Center for the Humanities" (DaSCH)*, 25–26. Bern. Online at http://www.sagw.ch/dms/sagw/liste_der_publikationen/wissenschaft/Final_report_pilot-project_DaSCH/document/Final_report_pilot%20project_DaSCH.pdf

Spiess, O., ed. 1955. *Der Briefwechsel von Johann Bernoulli*, vol 1. Basel.

Stieda, W. 1926. Johann [III] Bernoulli in seinen Beziehungen zum preussischen Herrscherhause und zur Akademie der Wissenschaften. Abh. d. preuss. Akademie der Wissenschaften, 1925, Phil.-Hist. Kl. Nr. 6, Berlin.

Truesdell, C. 1958. The New Bernoulli Edition. *Isis* 49: 54–62.

Wahl, Charlotte. 2013. Leibniz' mathematische Notation und der Druck. In *Medienwandel/Medienwechsel in der Editionswissenschaft* (editio/Beihefte 35), ed. A. Bohnenkamp-Renken, 41–59.

Weil, A. ed. 1993. *Der Briefwechsel von Jacob Bernoulli*. Basel.

Wollenschläger, K., ed. 1933. *Der mathematische Briefwechsel zwischen Johann I Bernoulli und Abraham de Moivre*. Basel.

Wolf, R. 1858–1862. Biographien zur Kulturgeschichte der Schweiz, 4 vols, Zürich.

———. 1876. Notizen. Die Correspondenz von Johannes Bernoulli. *Vierteljahrsschrift der Naturforschenden Gesellschaft in Zürich* 21 (4): 384–386.

D'Alembert's Mathematical Correspondence: Beyond the Formal Description of Networks

Irène Passeron

Abstract

D'Alembert (1717–1783) corresponded with some of the greatest mathematicians of his time, Leonhard Euler, Gabriel Cramer, and Joseph Louis Lagrange. This correspondence sheds light on the scientific controversies and epistemological issues of the day. It also clarifies the organization of the academic world in the middle of the eighteenth century, despite its lacks and losses. It allows us to determine the precise various statuses of a "letter," from the most public to the very private. We will question the relevance of a network epistolary representation, by inserting mathematical problems into the context of other forms of scientific communication: published treatises, academic reports, periodicals, and the *Encyclopédie* (1751–1772), the main medium for D'Alembert's work and for the Enlightenment. We will then focus on the relationships between Euler and D'Alembert, inserted into the overlapping of Paris-Berlin antagonisms and alliances.

Amongst the many different approaches that correspondences open us up to, two lines will be put forward here: to question the meaning that sets down a graphical representation of epistolary networks and, in parallel, to describe controversies by interweaving the various levels of interpretation suggested by the analyses of other written networks.

To speak of a network concerning mathematical correspondence may seem both trivial and inadequate: the triviality of a graphical representation of this

I. Passeron (✉)
CNRS, Sorbonne Université, Institut de Mathématiques de Jussieu-Paris Rive Gauche, Paris Cedex 05, France
e-mail: irene.passeron@imj-prg.fr

M. T. Borgato et al. (eds.), *Mathematical Correspondences and Critical Editions*, Trends in the History of Science, https://doi.org/10.1007/978-3-319-73577-1_4

correspondence as a mapping of geographical networks,[1] and the inadequacy between a focus on this mode of expression and the diversity of the modes of both exchanges and mathematical works.

By examining this tension, and by using the example of D'Alembert's correspondence, which is better known today,[2] we wish to clarify some of the parameters of the notion of network and draw attention to what can be done so that it no longer remains a static and inert material, but may, on the contrary, become a fertile ground for further research and readings.

For the last 30 years, perceptive studies carried out on scientific correspondence networks[3] have enabled us to clarify, in different times and places, the three functions of scientific correspondence: information (direct or indirect), legitimization of the respective positions of speakers, and mediation with third parties.[4]

Despite keeping it as a model, the notion of a "network" goes further than the simple graph of interconnections or even the structural frameworks of institutional links (for example, academic) and the local spheres of exchange (family or high society). Conversely, epistolary networks are often used to shed a contrasting light on other known records, both academic and literary.

We must therefore question the very nature of an epistolary relationship and the specific character of the semantic weight we attribute to it. Indeed, the notion of an "epistolary network" only makes sense if we are capable of identifying the parameters of the information, the impact, diffusion and variation of which we want to measure. Therefore, we must possess an accurate, up-to-date and reliable inventory and compare this network with other vectors of scientific exchange.[5]

---

[1]The success of mapping correspondence lies in the fact that it provides a synthetic vision (on a map—modern or period, the correspondents being represented by localised dots and the letters by arrows) and, less often, a dynamic (in time), by placing these periodic exchanges within the context of other epistolary relations. It remains a challenge to map uncertainty and shortcomings.

[2]*D'Alembert. Œuvres complètes*, volume V/1, *Inventaire analytique de la correspondance*, Irène Passeron, in collaboration with Anne-Marie Chouillet and Jean-Daniel Candaux, 2009. Volume V/2, *Correspondance 1741–1752*, I. Passeron, in collaboration with Jean-Daniel Candaux, Alain Cernuschi, Frédéric Chambat, Michelle Chapront-Touzé, Christian Gilain, Alexandre Guilbaud, Guillaume Jouve, Françoise Launay, Marie-Laure Massot, François Prin, Christophe Schmit. Paris: CNRS Éditions, 2015. All of the information and references on the correspondence are taken from the collective works of the Groupe D'Alembert to which we refer. The *Inventaire* has benefited from the preliminary works undertaken by Martha Rezler, Gilles Maheu and John Pappas (especially J. Pappas 1989).

[3]See Beaurepaire et al. (2006) and Beaurepaire (2014).

[4]See Passeron et al. (2008).

[5]See A. Guilbaud's article in the present book and our attempt to provide a new way of browsing a mix of print and manuscript documents: http://dalembert.academie-sciences.fr/encyclopedie/Dossier_Affaire_Tolomas/index.php

1 D'Alembert's Mathematical Correspondence as a Part of His Activity

In order to discover which mathematical ideas and crucial scientific issues emerge from D'Alembert's currently extant correspondence, we first of all need a broader mapping of his activities and more precise biographical data than that revealed in the letters, all written after the Minister's letter notifying him of his admission to the Académie Royale des sciences when he was 23.[6]

Jean Le Rond, "newly born" when he was "found" on the steps of the church St-Jean-Le-Rond on October 16th, 1717, was baptised after the name of the patron saint of that very church. This was a name of which he was unaware for the part of his youth spent with his wet nurse,[7] his boarding school master and at the Collège Mazarin, thinking at the time that his name was "d'Aremberg."[8] He obtained a Master of Arts degree (maître ès arts) in 1735 under the name of Jean Le Rond, and it is only before entering the Académie that he took, from then on, the name of D'Alembert.

It is therefore Jean Le Rond D'Alembert, scholar with a complex identity,[9] member of the most prestigious European Academies,[10] who died in Paris on October 29th, 1783, at the "Château du Louvre," in the apartment to which he was entitled as Permanent Secretary of the Académie française.

As we must have knowledge of his scientific work (his treatises, academic memoirs, participation in the Académie awards and his expert reports) and analyse his publishing strategies, a third cartography is necessary, that of the social connections his position in the enlightened world enabled him to establish at various stages of his life.

We will discuss the context in which his discussions took place with the Basel native Leonhard Euler[11] (during Euler's time in St. Petersburg and Berlin) and the Genevan Gabriel Cramer.[12]

---

[6]D'Alembert (2015), letter 41.01 from Maurepas to D'Alembert, May 13, 1741, p. 3–4.

[7]See F. Launay (2010).

[8]See F. Launay (2012).

[9]For this complexity, see I. Passeron (2009).

[10]Académie royale des sciences of Paris (1741), Academy of Berlin (1746), Royal Society of London (1748), Académie française (1754), Academy of Stockholm (1755), Institute of Bologna (1755), Imperial Academy of Saint Petersburg (1764), and many others throughout Europe.

[11]Leonhard Euler (1707–1783), *princeps mathematicorum,* was born in Basel. After his Saint Petersburg period, from 1727 to 1741, he took up a position at the Berlin Academy (1741), where he spent 35 years, before returning to Russia. During his very fruitful Berliner period, he was elected "associé étranger" of the Académie royale des sciences of Paris (1755) and won a great number of Academy Prizes. At least 40 letters were exchanged between both scholars.

[12]Gabriel Cramer (1704–1752) was a Genevan mathematician, not to be confused with Voltaire's publisher of the same name, Gabriel Cramer (1723–1793). We know that 25 letters were exchanged between Cramer and D'Alembert, of a very friendly nature, and we also have evidence that some are missing (D'Alembert 2015, p. xli).

Finally, a mapping of the entanglement of the Republic of Letters and the Republic of Sciences, which was still in its early stages, will enable us to identify the part played by D'Alembert as he became renowned for solving mathematical problems, or the forms of interactions existing between these resolutions and his epistemological positions, expressed not only, for example, in the *Encyclopédie*, for which he wrote the well-known *Discours préliminaire*, but also in his roughly 1700 articles on physics and mathematics.[13]

2 The Characteristics of D'Alembert's Correspondence

Now that these biographical points have been made, we can thereby evaluate the significance of the extant correspondence of D'Alembert: 2300 letters sent or received, none of which were from relatives (which is partly explained by the fact that he had been abandoned and unrecognised, as well as being childless). However, the definition of an epistolary corpus must be specified, because the very notion of a "letter" has different facets, from the private letter, passed from hand to hand, to the ostensible letter,[14] or even the fake letter.[15] Additionally, further evaluation has led to further doubts: we counted up to 450 correspondents, but their identities are sometimes uncertain, and may even mask multiple authors. Finally, in what is probably the most difficult aspect to assess, one that is too often obscured by the inventories, we need to determine the way in which these 2300 letters and 450 interlocutors are representative of D'Alembert's correspondence, which implies that we need to discover why and how letters have been kept, gathered, destroyed or lost.[16]

Unlike others, D'Alembert did not keep an epistolary register, and we might even say that he was not at all interested in doing so. What is obvious, however, is that we only have a fraction of his correspondence, and that only exchanges with correspondents who were, for various reasons, "privileged," such as Voltaire[17] and

[13]For specific studies on D'Alembert's articles, see (online) the file "D'Alembert et l'*Encyclopédie*" in *RDE 21* (Chouillet 1996), *Les branches du savoir dans l'*Encyclopédie, *RDE 40–41* (Leca-Tsiomis and Passeron 2006) and Gilain (2010), Guilbaud (2012), Schmit (2014).

[14]An "ostensible", or conspicuous letter is a letter written for the purpose of being shown or written knowing it would be shown (see D'Alembert 2015, Introduction, Privé *versus* Public, pp. xiv–xvii). Some of these, the most "ostensible" (from A43.10 to A83.01), are listed in D'Alembert (2009, Appendice, pp. 518–547). We have listed 2300 private and ostensible letters, excluding the fake and apocryphal ones.

[15]For fake and apocryphal letters, see D'Alembert (2009, p. xv and pp. 549–551) and D'Alembert (2015, Introduction, Lettres apocryphes, p. xvi).

[16]See D'Alembert (2015), Introduction, pp. xxi–xxiii.

[17]The correspondence with Voltaire begins in 1746 and comprises over 527 letters (known so far) until the death of the latter. See D'Alembert (2009), Introduction, pp. xxviii–xxix.

Frederick II,[18] or important figures within mathematics, such as Euler, Cramer and Lagrange,[19] have been rather well preserved. The density of these written records must not, however, overshadow the many oral discussions he had with Diderot,[20] Condillac,[21] Clairaut[22] and De Gua[23] that were not followed up in writing.

D'Alembert often wrote for the purpose of cultivating intellectual and amicable relations disrupted by distance (as was the case with Cramer after he left Paris in 1748), but also to follow academic activities, and even support them in cases of priority disputes (as with Euler, whom he never met) or to clarify his reasoning (with Lagrange).

Therefore, the only means to evaluate the way in which the letters are purported to differ from academic relations and information transfers through periodicals and prints is to compare their substance with what we learn from accounts of institutional life and the various circulations of printed matter.

D'Alembert's extant correspondence is very unevenly distributed:

Apart from his nomination letter as "adjoint" at the Académie royale des sciences in 1741, none of the letters written before 1746, the year he turned 29, have been preserved. This gap is largely due to his status as a "bastard," *nec pater nec res*,[24] and even more to the fact that he was childless. He had, however, more than one adoptive family: his foster family, the Rousseaus,[25] a family of glaziers with whom

---

[18]The correspondence with Frederick II (1712–1786, King of Prussia from 1740) begins in 1746 and comprises over 285 letters (known so far), until D'Alembert's death. See D'Alembert (2009), Introduction, p. xxvi–xxvii.

[19]Joseph Louis (Giuseppe Luigi) Lagrange (1736–1813), was a member of the Berlin Academy (1756) and of the Académie royale des sciences of Paris (1772). Their correspondence begins in 1759 and comprises over 172 letters, until D'Alembert's death. See D'Alembert (2009, Introduction, pp. xxviii–xxxi).

[20]In regard to his friend and colleague Denis Diderot (1713–1784), only one letter is known (D'Alembert 2009, letter 65.39). Like many Parisians, D'Alembert and Diderot ("Da and Di" as Voltaire said in his letter to Damilaville on April 8th, 1765) probably sent short letters to each other, all of which were lost.

[21]Also a friend from his early years, Etienne Bonnot de Condillac (1714–1780) had a great influence on D'Alembert's philosophy. No letters have been found, despite evidence of the existence of at least one (D'Alembert 2009, pp. xxxiv–xxxv).

[22]Alexis Claude Clairaut (1713–1765) was not a friend, but rather a stimulating rival, which is also a good circumstance under which to send and receive letters, although only a few letters are known of (D'Alembert 2009, letters 59.08, 59.09, 64.24, 64.25, 64.30 and ostensible letters A61.06, A62.05, A62.01, A62.06).

[23]Very little is known about the relationship between D'Alembert and the editor of the first project of the *Encyclopédie*, Jean Paul De Gua de Malves (1710–1786), except that they were close colleagues in the 1740s. No letters between the two academicians remain, an unsurprising fact, given that De Gua's correspondence and D'Alembert's correspondence of the 1740s have both disappeared.

[24]"Neither father nor fortune": to understand how these words gave rise to the "Affaire" with Jesuit Father Tolomas, see the interface mentioned note 6.

[25]D'Alembert's well known wet-nurse, "Mme Rousseau", was not very famous before the publication of F. Launay's study (Launay 2010): she was born Etiennette Gabrielle Ponthieu

D'Alembert lived in Le Marais until he reached the age of 48, and the family who served as his official guardians, the Destouches,[26] from whom he received a pension and with whom he was very close. The social status of the Rousseaus and the fact that the members of the Destouches family were simply benevolent guardians explains why D'Alembert did not write to them or, if he did, which was probably the case with the Destouches family at least, why these exchanges were not kept on either side. Furthermore, in 1746, he had already written two major works,[27] and there must have been exchanges, with Maupertuis,[28] for instance, one of his supporters at the Académie des sciences who had, at the time, left for Berlin. Therefore, a number of D'Alembert's letters were not kept, and were most likely discarded while he was still alive.

Besides, another imbalance is due to the fact that we have more letters sent by D'Alembert than those received by him: the scholar kept fewer of his letters than his correspondents did.[29] What is more, D'Alembert's fame as Secretary of the French Académie and as the "leading light" of French philosophers[30] ensured the preservation of certain of his letters from the eighteenth century to the present.

If we now examine the chronological distribution of the letters and correlate it with their actual content, we notice a connection between their volume and the scholar's official career. Indeed, it increases at the time of the *Encyclopédie*, except during the mysterious 1759 "gap,"[31] and reaches its peak during the activities of the salon that he hosted together with Mlle de Lespinasse[32]; it then remains substantial, thanks to his envied status as Permanent Secretary.

More specifically, during the early years, this correspondence was largely fuelled by his exchanges with Euler. These stopped due to profound disagreements, and even disputes, as was the case in 1751, and it was only in the 1760s that the

(1682–1775), and when D'Alembert was born in 1717, she had taken the name "Mme Gerard" after her first marriage; she then took the name of her second husband, the glazier Rousseau, in 1726.

26Louis Camus Destouches (1667–1726), one of Fenelon's friends, was the first to be in charge of the young "Jean d'Aremberg". When he died, his brother Michel, and finally Michel's widow (in 1731), continued to provide for him (1200 livres tournois every year), guiding his first steps in the world (Launay 2012).

27*Traité de dynamique* (D'Alembert 1743), which contains the famous "D'Alembert's principle" and *Traité de l'équilibre et du mouvement des fluides* (D'Alembert 1744).

28Pierre Louis Moreau de Maupertuis (1698–1759), "the man who flattened the Earth", led an expedition to Lapland to determine the shape of the Earth (1736–1737), an issue at stake in the Newtonian-Cartesian controversy. He was then invited by Frederick the Great to be the first President of the Prussian Academy of Science (Terrall 2002).

29See D'Alembert (2015, Introduction, pp. xxi–xxiii).

30As Chevalier de Roubin said: "puisqu'avec raison on vous regarde comme le flambeau de l'europe, c'est vers vous qu'il faut aller pour etre éclairé" (D'Alembert 2009, letter 73.84).

31See Fig. 1.

32Julie Jeanne Eleonore de Lespinasse (1732–1776), an illegitimate child like D'Alembert, became famous after she broke her friendship with Mme Du Deffand (1764) and set up her own "salon", "rue Saint-Dominique", where she lived with D'Alembert until her death.

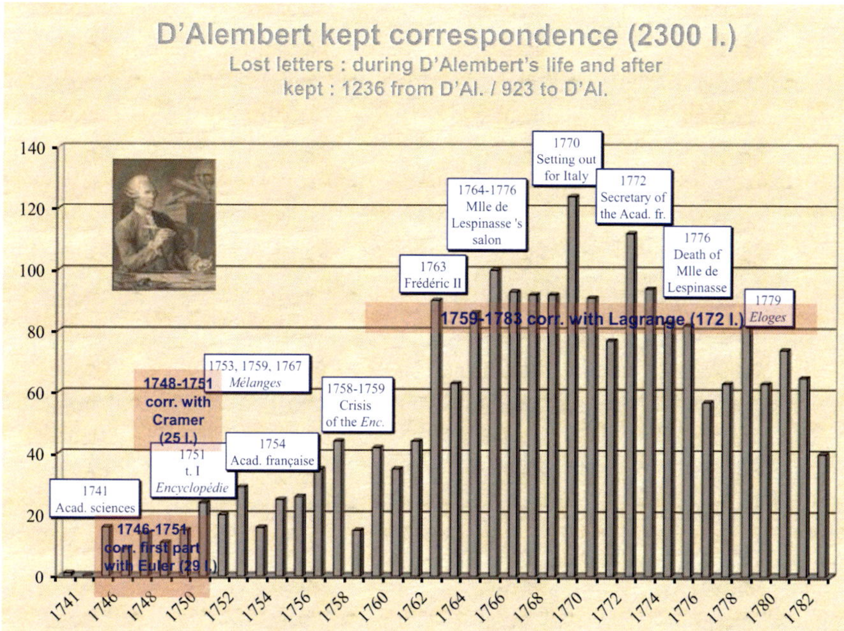

Fig. 1 Kept correspondence from and to D'Alembert

correspondence was resumed. The 30 letters exchanged between 1746 and 1751 (out of 40 in total) are thus the most important, as they shed a completely new light on the works of both scholars during that period: the fundamental theorem of algebra, the vibrating strings,[33] the theoretical crisis over Newton's theory of gravitation, hydrodynamics, all elements that contributed to the genesis of the views of both scholars, views they developed in their printed memoirs in a way that smooths over the debates.

We are also lucky to possess, as a counterpoint, the discussions that both had with the Genevan scholar Gabriel Cramer. Cramer's stay in Paris in 1747–1748 gave rise to a very close friendship between the two mathematicians, most likely *via* the salons of Mme Geoffrin[34] and Mme Du Deffand.[35] But it was also a friendship between philosophers: the theory of infinity, the theory of music, a large number of issues brought them together and led to fruitful exchanges, the echoes of which can

[33]In particular, a recently discovered letter from Euler, D'Alembert (2015, pp. 39–47, letter 46.12).

[34]Mme Geoffrin, born Marie Thérèse Rodet (1699–1777), was linked with the Genevan patricians, the Saladins. Her "salon" was probably where D'Alembert met Cramer during his stay in Paris.

[35]Mme Du Deffand, born Marie de Vichy-Chamron (*c.* 1697–1780), was, at the beginning, a very dear friend of D'Alembert's, and supported him in 1754 in his election to the Académie française. Upset after her break with Mlle de Lespinasse, she was, at the same time, angry with D'Alembert and never forgave him for the choice he had made.

be found even in the *Discours préliminaire* and in the articles of the *Encyclopédie*.[36] Cramer's death in January 1752, of course, put an end to these exchanges.

It is only with Lagrange that D'Alembert was once more to have a genuine scientific exchange, over a long period of time and with growing intensity, from 1759 until his death.

The figure represents what we call "private letters," even if there is no universal definition of such letters. D'Alembert wrote at a period during which the channels of expression were increasingly diversified through periodicals, but even though the gradation of existing intermediaries makes it difficult, we can still distinguish, in most cases, between the letters that were "ostensible" and those that were "private."[37]

It is not always easy to identify the sender or the recipient of a letter: it could be an institution[38] or it could also be more than one person. For instance, when Mlle de Lespinasse dictated a letter to D'Alembert, which was destined for Condorcet,[39] the "secretary" was no longer "permanent" but "private," and he took the liberty of including personal jokes.[40] In addition, some people wrote letters that they knew, or else pretended not to know, would be used publicly by, for example, being read in the salons, being copied or even printed.

Finally, as far as some of the letters are concerned, we only know that they were written to a woman or to an unidentified collaborator in the *Encyclopédie*, or we have no certainty as to the identity of either the sender or the recipient.

We had to leave out all writings (notes, certificates or receipts) mistakenly considered as letters by the previous publishers, in order to exclude, but nevertheless report, literary frauds or sheer fabrications, often of slanderous intent. On the other hand, the "Extraits des différentes Lettres de M. d'Alembert à M. de la

[36]For instance, the article "Courbe" of the *Encyclopédie ou Dictionnaire raisonné des sciences, des arts et des métiers*, written by D'Alembert, quotes and praises Cramer's book (Cramer 1750): "ouvrage très-complet, très-clair & très-instructif, & dans lequel on trouve d'ailleurs plusieurs méthodes nouvelles" (*Encyclopédie*, vol. IV, 1754, p. 388a).

[37]See note 15.

[38]*A fortiori* if the sender or the recipient is the permanent secretary of an Academy, it is often impossible to draw a clear distinction between what falls under an individual word and what falls under an academic word.

[39]Marie Jean Antoine Nicolas de Caritat, marquis de Condorcet (1743–1794), was D'Alembert's spiritual son. Unfortunately, their correspondence was lost. See Nicolas Rieucau's contribution in the present book.

[40]Mlle de Lespinasse dictating, D'Alembert writing: "vous avez encore tort de faire de la géométrie comme un fou, de souper comme un ogre et de ne pas plus dormir qu'un lièvre. Vous croyez bien que ce n'est pas mon secrétaire qui dit cela"; "mon secrétaire ne sait jamais ni ce qu'il dit, ni ce qu'il fait—(pure bêtise de dire cela: cette pensée est du secrétaire)"; "il est très incommode de dicter à un homme aussi *admirable* que mon secrétaire, qui fait d'aussi beaux mémoires à l'Académie, ou qui est aussi maussade *à la maison*"; "mon revêche secrétaire veut bien écrire à mon bon Condorcet"; "mon secrétaire et moi, nous vous écrivons en commun"; "le secrétaire vous embrasse et trouve qu'*en voilà assez*. Ce mot est son cachet et vous y reconnaîtrez sa grâce enchanteresse." (D'Alembert 2009, letters 69.46; 69.50; 69.55; 71.77; 72.33; 74.43).

Grange" (*Mémoires de Turin*, 1765), which we also had to eliminate from this correspondence, although they are worthy of publication as part of D'Alembert's works, were but a stratagem he came up with, in connivance with Lagrange, to publish mathematical texts elsewhere than in the Paris or Berlin academic memoirs.

The great variety and quantity of documents that we possess demonstrate that the boundaries between the public and the private spheres are blurred, and that they are representative of the various positions occupied by D'Alembert: as co-director of the *Encyclopédie*; member of the Académie royale des sciences; Permanent Secretary of the Académie française; but also recipient, as well as sender, of epistles, and therefore both a "public" and a private figure.

3 The Characteristics of D'Alembert's Publications and Manuscripts

D'Alembert's correspondence is closely linked to the printed word[41] and to academic activity.[42] Even the few purely amicable or social letters that have reached us, sometimes miraculously, bear witness to this connection.[43] Indeed, the tumultuous news that they relate is closely connected to literary and political debates.[44]

[41] This fact illustrates the growing influence of the printing world, even over the growing culture of the private sphere (on this, see Chartier and Martin 1991). For instance, D'Alembert gets acquainted with Euler by sending him his works: "Monsieur de Maupertuis m'a remis tant Votre lettre que Vos ouvrages" (D'Alembert 2015, letter 46.12, p. 39). But, on the other hand, a book can also be linked to an acutely topical issue, when he writes to Cramer: "M. Diderot, mon intime ami, que vous connoissés de reputation, s'est avisé de donner au Public une lettre sur les aveugles, ou il y a d'excellentes choses *sed non erat his locus*", telling his friend living in quiet Geneva that Diderot has been imprisoned in La Bastille following the publication of the *Lettre sur les aveugles à l'usage de ceux qui voyent*, a controversial issue due to his atheistic materialism (D'Alembert 2015, letter 49.09, pp. 227–228).

[42] Almost nothing is known about the debates, if not disputes, which enlivened the academic sessions in Paris or Berlin, but allusions are numerous, for example, when D'Alembert blames Euler for his partiality when examining his paper proposed for the Berlin prize: "Je scai aussi par une voie tres sure qu'il s'est passé bien des *vilainies* sur ce sujet" (D'Alembert 2015, letter 51.15, pp. 348–349).

[43] In a short letter (a "billet" hand-delivered between addresses in Paris) written to his friend, la marquise de Crequÿ, (1714–1803) D'Alembert expresses this proximity in a witty remark: "Je m'amuse à vous ecrire, à condition que c'est pour vous seule, j'ay pourtant assez d'ouvrage: quatre epreuves à corriger, un avertissement a achever, l'Errata du second volume à composer, les Jesuites à batonner, les jansenistes à fustiger... " (D'Alembert 2015, letter 51.24, pp. 364–365).

[44] The letters give public information ["Comme j'allais fermer ma lettre il m'est arrivé trois prêtres [...] Ils m'ont beaucoup parlé du livre de Buffon et de celuy de Montesquieu, que la Sorbonne veut condamner" (D'Alembert 2009, letter 53.07 to Mme de Crequÿ)] and unofficial information ("A propos de President, le Montesquieu m'a ecrit une assez jolie lettre. Il ne veut point de *democratie et despotisme* mais il est bien tenté de prendre l'article *Goût*. Vous ne vous en seriés jamais douté ny moy non plus.") (D'Alembert 2009, lettre 53.26 to Duché, about Montesquieu's contribution to the *Encyclopédie*).

Treatises	Académie des sciences	Académie française	*Encyclopédie*
	1741: « adjoint »		
1743: *Traité de dynamique*			
1746-1752: works on winds, moon, precession, fluids	1746: « associé »		1751: vol. I *Discours préliminaire*
1752 : *Elémens de musique*		1753 : *Mélanges*	1752 : vol. II 1753 : vol. III
1754-56 : *Système du monde*		1754: member	1754 : vol. IV 1755 : vol. V
	1756: « pensionnaire surnuméraire »		1756 : vol. VI 1757 : vol. VII
		1759 : *Mélanges* + 2 vol.	1759: condamnation
1761: *Opuscules*, t. I-II		1762 : Cath. II's proposition from 1763 : Fréd. II	
	1765: « pensionnaire »	1765-7 : *Jésuites*, *Mél.*+1 vol.	1765: vol. VIII-XVII
	1769: « directeur »		
		1772: « secrétaire perpétuel »	
	1773: Condorcet « secrétaire perpétuel »		1776-7: Supplément
1780: *Opuscules*, t. VII-VIII		1779 : Eulogies	

Fig. 2 Bio-bibliographic landmarks

If we consider his printed works, we can divide them into four major categories, which are punctuated with biographic landmarks and which differ in their dissemination and audience (Fig. 2).

The first is that of the great scientific treatises. Theses treatises, written between 1743 and 1756, are all centred around a theme: dynamics (1743), fluids (1744), the general cause of winds (1746), the lunar theory (1747–1749), the precession of equinoxes (1749), music (1752) or the world system (1754–1756); and they involve key mathematical issues: partial differential equations of motion and equilibrium, the fundamental theorem of algebra, complex numbers, series. Later on, the *Opuscules* were published (nine volumes in total, although the last remains unpublished) which comprise a collection of various memoirs and in which we find further discussions on these subjects.

In the second category, we find the works related to his activities at the Académie des sciences: the texts he published in his annual *Mémoires*, his expert reports and his role as a member of the prize commission.[45]

The third category is linked to his literary activity, which gained momentum throughout his life and was at its peak after his nomination as Permanent Secretary

[45]These internal texts will be published in volume III/11 "D'Alembert académicien des sciences" of *D'Alembert. Œuvres complètes*, along with the speeches read out on special sessions of the Paris Academy of Sciences.

of the Académie française in 1772. Many of these works address issues relative to language and its formalisation in the Dictionnaire de l'Académie française. He also wrote numerous eulogies,[46] which formed a kind of "catechism" of good scholarly practice.

The fourth category, frequently interacting with the previous ones, is centred on his writings for the *Encyclopédie*,[47] essentially from 1746 to 1759 (the moment when D'Alembert withdrew from the enterprise).

4 D'Alembert, Euler and Cramer Between 1748 and 1752

Having established this groundwork, we can describe more precisely how mathematical information was restored and circulated between D'Alembert, Euler and Cramer from 1748 until 1752, just before and just after the publication of the first volume of the *Encyclopédie*.

If we place these exchanges on a map of Europe, we notice that mathematical correspondence (pure or mixed mathematics), which was the theatre of numerous controversies between the years 1746 and 1752, revolves around two main interlocutors, and therefore two destinations. Apart from his subsequent correspondence with Lagrange, it is during that period that D'Alembert's mathematical exchanges intensify: with Euler in Berlin and with Cramer in Geneva. The intensity of these exchanges enables us to estimate the number of letters that have been lost at approximately 30, but the remaining 55 letters are still very instructive.

Apart from reading the correspondence with the secretary of the Berlin Académie, Formey, having background knowledge of the correspondence of its President, Maupertuis, would be desirable, as he played an important part in D'Alembert's early career at the beginning of the 1740s.[48] It is through him that D'Alembert came into contact with the Berlin Académie of which he then became a member, thanks to the prize he received for his *Réflexions sur la cause générale des vents* in 1746 (the only one he ever received, because the one he applied for next, the prize on resistance of fluids, caused his dispute with Euler). Unfortunately, most of Maupertuis's correspondence has been destroyed, but the few elements that

[46]These eulogies contributed to his success, as he read a lot of them in the French Academy public sessions, such as the eulogy for Massillon read on August 25th, 1772, or the eulogy for Bossuet read in 1775: "cet endroit a été saisi avec transport par le public, qui a applaudi à enfoncer la salle", as reported by Julie de Lespinasse to Condorcet (Passeron 2009).

[47]He wrote numerous articles in the *Encyclopédie ou Dictionnaire raisonné des sciences des arts et des métiers* about physics and mathematics, and the famous "Discours préliminaire", as well as controversial non- scientific articles: "Collège", on Jesuit education, and "Genève", on Genevan pastors' deism. For more details about these contributions, see the ENCCRE project: http://enccre.academie-sciences.fr/

[48]A letter found recently, from D'Alembert to Father Jacquier, shows that, from 1745 on, Maupertuis asked the French scholar to come to Berlin. See also D'Alembert (2015, Introduction, pp. lii–liii and lxxxi–lxxxii).

remain enable us to see that, in 1752, the negotiations between D'Alembert and Euler on whether or not the memoirs or rectifications should be published in Berlin passed him, and to some extent Formey, by.

D'Alembert and Euler had indeed ceased to communicate by the end of 1751, the former being sickened by what he thought were bad practices on the part of the Berlin mathematician, the latter frightened by the encyclopaedist and showing little interest in exchanging with him.[49] It is obvious that the strong friendship, if we can call it thus, between D'Alembert and Frederick II did not contribute to making D'Alembert agreeable in the eyes of the right-minded and true believer Euler. In the meantime, in Geneva, Cramer did his best to deal with these two great mathematicians.

If we take into account all of the information provided by the bibliographical analysis, the canvas of the mathematical controversies between D'Alembert and Euler during these years appears to be a junction and an overlapping of various antagonisms and alliances.

At first, their exchanges were connected to academic prizes, whether from Berlin or from Paris. It is thanks to D'Alembert's vote that, in 1748, Euler[50] won, amongst many others, the prize related to Jupiter's satellites in Paris. At the time, the points of friction between both scholars were not yet explicit. This mathematical exchange takes its full meaning in the dual context of the French and Berlin academies, *via* Maupertuis,[51] and in D'Alembert's opposition to the French Court, which drew him closer to Berlin as early as 1746.[52] This rapprochement is illustrated by the very flattering dedication of the *Cause des vents* to Frederick the Great, which D'Alembert had carefully worked out in his letters to the Marquis d'Adhemar.[53]

In parallel, between 1747 and 1749, the well-known[54] controversy over the movement of the lunar apsis was in full swing. There was no open rupture

[49] About the ideological differences and scientific focal points between D'Alembert (or Condorcet) and Euler, see Gilain (2013) and D'Alembert (2015, Introduction, § IV, VI and VIII).

[50] At this time, Euler was a well-known mathematician, a member of the Berlin Academy since 1741 and "commissaire" (member of the committee that awards the prizes). He competed prizes of the Académie des sciences, where D'Alembert was also "commissaire", and on the other side, D'Alembert competed Berlin prizes.

[51] Maupertuis was a member of the Académie des sciences from 1723 to 1746, when he left Paris for Berlin, to become President of the Berlin Academy. Frederick II subsequently requested of him that he recruit the best European scholars.

[52] D'Alembert thought he was treated better by Frederick II than by his own king. He was very proud to win the first prize given by the new Berlin Academy, in 1746, and quite angry to be underestimated in France.

[53] Antoine Honneste François, marquis d'Adhemar (c. 1710–1785), was a cavalry captain, friend of the encyclopedists, who left the army in 1752 to become grand-maître at the court of Frederick II's sister, Wilhelmine, Margravine of Bayreuth.

[54] Thanks to the works of Michelle Chapront-Touzé, which made possible the publication of D'Alembert's Lunar theory: see D'Alembert (2002) and also D'Alembert (2006) for the important question of precession of equinoxes.

(except vis-a-vis Buffon[55]), but rather a fruitful debate between D'Alembert, Euler and Clairaut, although disputes over priority were being hatched. Many questions of pure mathematics underlying the works on the general cause of winds (1746) and on celestial mechanics were raised and discussed: the way to solve the so-called "fundamental theorem of algebra,"[56] which put Euler in opposition to D'Alembert, the relevance of examples and counter-examples, up to the principle of the method itself—algebraic for Euler, analytic for D'Alembert. There again, the epistolary voice of the friend, Cramer, came as a counterpoint, and that of the rival, Clairaut, expressed itself through the *Mémoires* de l'Académie.

Nevertheless, the main reason for the breakdown between Euler and D'Alembert was probably the award for the determination of the fluid resistance laws: as we can see in the correspondence, D'Alembert was convinced that Euler was responsible, as "commissaire" in Berlin, for many "villainies," which prevented him from receiving the prize.[57]

Finally, the publication at the end of 1750 of Cramer's work, *Introduction à l'analyse des lignes courbes algébriques*, was used by D'Alembert to echo and orchestrate his opposition to Euler. Indeed, D'Alembert thought that he had discovered the mistake made by De Gua[58] before Euler, when De Gua [?] affirmed, in his work published in 1740, that no algebraic curve could have a cusp of the second kind. D'Alembert also claimed to be the first to demonstrate analytically the existence of such singular points.

Moreover, his change of position regarding De Gua, although he had written a positive review of his work in 1740, can also be understood in the context of the early stages of the *Encyclopédie*.[59] Indeed, De Gua managed the enterprise from June 1746—at the time, it was just a project centered around translating the two volumes of the *Cyclopaedia*—but in October 1747, the booksellers dismissed him, judging that his work was insufficient, and so they signed a contract with D'Alembert and Diderot, who thereafter became the editors of the *Encyclopédie*.[60]

These beginnings were already tumultuous: Diderot was at the Bastille and Formey was claiming payment for the papers he had given to the enterprise, which De Gua was taking time to pay. D'Alembert intervened in all of these developments, and consolidated his status as an academician who could not be overlooked, despite his unorthodox opinions, particularly in the eyes of Jesuits.

[55]At this time, the relations between the mathematician D'Alembert and the naturalist Buffon (1707–1788) were friendly, but not close.

[56]See D'Alembert (2007) and Christian Gilain's analysis.

[57]The affair began with the report of the prize in May 1750, and ended with D'Alembert's letter 51.15: see D'Alembert (2015, pp. cxvi–cxx).

[58]See n. 24.

[59]As D'Alembert tells Cramer: "c'est un homme qui se plaint de tout le monde, parce que tout le monde a à se plaindre de luy" (D'Alembert 2015, letter 51.02, p. 313).

[60]For the "prehistory" of the *Encyclopédie*, see Wilson (1985, pp. 73–82).

It is to this type of reconstruction of the ideological landscape that the publication of the *Œuvres complètes* is dedicated. A reconstruction made possible by a detailed analysis of a correspondence, which is at times incomplete, and that makes sense only in light of the whole body of writings that inspired it.

Acknowledgement I would like to thank Sophie Bond for the translation from the French.

References

Beaurepaire, Pierre-Yves, Jens Häseler, and Antony McKenna, eds. 2006. *Réseaux de correspondance à l'âge classique (XVIe–XVIIe siècle)*. Saint-Etienne: Publications de l'Université de Saint-Etienne.

Beaurepaire, Pierre-Yves, ed. 2014. *La Communication en Europe de l'âge classique au siècle des Lumières*. Paris: Belin.

Chartier, Roger, and Henri-Jean Martin. 1989–1991. Histoire de l'édition française, 2nd edition, 2. Paris: Fayard et Cercle de la librairie.

Chouillet, Anne-Marie, ed. 1996. *D'Alembert et l'Encyclopédie. Recherches sur Diderot et sur l'Encyclopédie* 21: 70–178.

Cramer, Gabriel. 1750. *Introduction à l'analyse des lignes courbes algébriques*. Genève: Frères Cramer et Philibert.

D'Alembert, Jean Le Rond. 1743. *Traité de dynamique*. Paris: David l'aîné. 1758, 2nd edition. Paris: David.

———. 1744. *Traité de l'équilibre et du mouvement des fluides*. Paris: David l'aîné. 1770 2nd edition. Paris: Briasson.

———. 1746. *Réflexions sur la cause générale des vents*. 1747. Berlin: Haude & Spener (published in november 1746).

———. 2002. *Œuvres complètes. Série I. Vol. 6. Premiers textes de mécanique céleste (1747–1749)*, ed. M. Chapront-Touzé. Paris: CNRS Editions.

———. 2006. *Œuvres complètes. Série I. Vol. 7. Recherches sur la précession des équinoxes et sur la nutation de l'axe de la terre dans le système newtonien*, eds. M. Chapront-Touzé and J. Souchay. Paris: CNRS Editions.

———. 2007. *Œuvres complètes. Série I. Vol. I/4a. Textes de mathématiques pures 1745–1752*, ed. C. Gilain. Paris: CNRS Editions.

———. 2009. *Œuvres complètes. Série V. Vol. 1. Inventaire analytique de la correspondance 1741–1783*, eds. I. Passeron, A.-M. Chouillet, and J.-D. Candaux. Paris: CNRS Editions.

———. 2015. *Œuvres complètes. Série V. Vol. 2. Correspondance générale 1741–1752*, eds. I. Passeron, J.-D. Candaux, A. Cernuschi, F. Chambat, M. Chapront-Touzé, C. Gilain, A. Guilbaud, G. Jouve, F. Launay, M.-L. Massot, F. Prin, and C. Schmit, Paris: CNRS Editions.

Gilain, Christian. 2010. La place de l'analyse dans la classification des mathématiques: de l'*Encyclopédie* à la *Méthodique*. *Recherches sur Diderot et sur l'Encyclopédie* 45: 109–128.

———. 2013. Sur l'expression "mathématiques des Lumières". In *Raison et passions des Lumières*, eds. Simone Mazauric and Pierre-François Moreau, 213–226. Paris: L'Harmattan.

Guilbaud, Alexandre. 2012. À propos des relations entre savoirs théoriques et pratiques dans l'*Encyclopédie*: le cas du problème de la résistance des fluides et de ses applications. *Recherches sur Diderot et sur l'Encyclopédie* 47: 207–242.

Launay, Françoise. 2010. D'Alembert et la femme du vitrier Rousseau, Etiennette Gabrielle Ponthieux (c. 1683–1775). *Recherches sur Diderot et sur l'Encyclopédie* 45: 100–106.

———. 2012. Les identités de D'Alembert. *Recherches sur Diderot et sur l'Encyclopédie* 47: 243–289.

Leca-Tsiomis, Marie, and Irène Passeron, eds. 2006. *Les branches du savoir dans l'Encyclopédie. Recherches sur Diderot et sur l'Encyclopédie* 40–41: 7–13.

Passeron, Irène, Siegfried Bodenmann, and René Sigrist, eds. 2008. *La République des sciences. Dix-Huitième siècle 40*. Paris: La Découverte.
Passeron, Irène. 2009. D'Alembert: construction d'une identité scientifique au XVIIIe siècle. *Jahrbuch für Europäische Wissenschaftskultur* 4 (2008): 11–34.
Schmit, Christophe. 2014. Les articles de mécanique de l'*Encyclopédie,* ou D'Alembert lecteur de Varignon. *Recherches sur Diderot et sur l'Encyclopédie* 47: 225–258.
Terrall, Mary. 2002. *The Man Who Flattened the Earth. Maupertuis and the Sciences of the Enlightenment*. Chicago: The University of Chicago Press.
Wilson, Arthur Mc Candless. 1985. *Diderot*. New York: Oxford University Press (French trad. *Diderot. Sa vie est son œuvre*. Paris: Laffont Ramsay).

The Digital Edition of D'Alembert's Correspondence

Alexandre Guilbaud

Abstract

In conjunction with the preparation of a critical edition in printed form of the complete works of D'Alembert, a group has been working for several years now to produce a digital edition of the correspondence of the famous French scholar and encyclopaedist based on the following principles: various means of consultation and search tools in the corpus, development of the many surviving manuscripts, renewal of the circulation within the corpus, the rich critical apparatus being formed and many other works by the author, and provision of useful and relevant research tools for the more specialized reader. This article describes the concrete results of the work and states the difficulties encountered, as well as the long-term prospects of development envisaged.

A digital edition entitled "D'Alembert en toutes lettres" is being developed within the scope of a project aimed at establishing the first critical edition of the *Œuvres complètes* of Jean le Rond D'Alembert (1717–1783),[1] including, above all, the paper edition of all active and passive correspondence of the renowned French scientist, philosopher and encyclopaedist.

Initiated about 20 years ago, the research works of I. Passeron, in collaboration with A.-M. Chouillet and J.-D. Candaux, have led to the publication of an analytical inventory (D'Alembert 2009) summarizing the content and presenting in detail

[1] See http://dalembert.academie-sciences.fr

A. Guilbaud (✉)
Sorbonne Université, Institut de Mathématiques de Jussieu-Paris Rive Gauche, CNRS, Université Paris-Diderot, Paris 5, France
e-mail: alexandre.guilbaud@imj-prg.fr

M. T. Borgato et al. (eds.), *Mathematical Correspondences and Critical Editions*, Trends in the History of Science, https://doi.org/10.1007/978-3-319-73577-1_5

the localization and the source of almost 2300 letters D'Alembert exchanged with about 450 correspondents between 1741 and 1783.[2] Some months ago, the volume collecting the letters exchanged between 1741 and 1752 (D'Alembert 2015), presented and annotated by a large team of historians of sciences and ideas, was published. Ten further volumes will be published in the future, with the rest of the letters organized according to chronological periods.

Hence, the edition "D'Alembert en toutes lettres" is closely connected with a printed publication process, which itself is related to current research works. In accordance with Passeron, the online version of the correspondence, freely accessible at the web link http://dalembert.academie-sciences.fr/Correspondance, aims, firstly, to highlight these research dynamics, offering a continuously enhanced edition updated with the new findings (such as, for instance, the recently found letters), and secondly, to supplement the content of the printed edition thanks to access to effective navigation and search tools, as well as to the material traces of the letters, which constitute a crucial aspect of the corpus.

I will first explain how the digital media and the materials at our disposal enabled us to meet these different challenges. I will then highlight what we learned from this first step of editorial development and outline some prospects for future research works, as we currently consider them within the more general context of digital humanities.

1 The Functionalities of the Edition

The first advantage of the digital media pertains to the new search capacities—which, in the printed format, are often reduced to a table of contents and an index—offered to the reader for the purpose of searching a corpus. These search capacities themselves depend on the level of description of the edited documents.

The search functions incorporated into the interface "D'Alembert en toutes lettres" are largely based on the digital version of the analytical inventory of the correspondence, in the form of a data table, which has been continuously updated by I. Passeron and F. Prin for a period of roughly 15 years, and whose printed volume, published in 2009, with the exception of its introduction and appendices, may be considered as more of a snapshot. This table indicates the descriptive items given for each letter, including those given at the time of the 2009 publication:

- The reference number of the letter in the inventory of D'Alembert's correspondence, made up of an initial number (ranging between 41 and 83) referring to the year and of a second number corresponding to the chronological order of the letter during the given year;
- The date of the letter, whether provided by the source or inferred from research works;

[2]See the contribution by I. Passeron (2018).

- The name of the correspondent, whether explicitly named or specifically researched;
- The places from where and to where the letter was sent, when known;
- The material description of the source of the letter, which may belong to one of the three following categories: a manuscript (always given by default when it was found), a printed document or a sales catalogue mentioning this letter;
- The place where the manuscript source of this letter is kept, when known;
- The list of the known edited versions of this letter;
- Another manuscript source, if any (e.g., another existing copy, a draft, etc.);
- The incipit of the letter;
- The summary of the letter's content.

These different items, called metadata, define the search capacities of our digital edition, as well as the different information types that we can provide the reader with for each letter. At the present time, thanks to the search tool of the interface "D'Alembert en toutes lettres", simple or cross-queries can be made on the three following items—dates, correspondents, place where the letter was sent from/to—locating a letter according to its inventory number or making a full-text search in the incipit and summary of the 2300 letters referenced in this table.

From an editorial point of view, these new search means are only one facet of a more crucial advantage offered by the digital format, i.e., its dynamic nature. Contrary to a printed version, frozen at a given moment, the new media and its related display and search functions enable us to sort out, reorder and update the information in real time. Our digital edition works directly from a database, which is constantly updated with the most recent research results. That is why the interface "D'Alembert en toutes lettres" allows us at any time to read and search an updated version of the analytical inventory of the correspondence, including, for instance, the description of the last letters discovered (around 30 since 2009) or the latest information relating to the dating of given documents (the preparation works for the first volume of correspondence led to the re-dating of a dozen out of the 117 edited letters). The digital edition thus offers an elegant solution to the classical tension between two conflicting temporalities: the strongly discontinuous temporality of a printed edition and the ever-evolving temporality of research.

"D'Alembert en toutes lettres" is not only a searchable inventory of the correspondence, rich as it may be, but also has an interface allowing access to the very text of a given number of letters, while informing the reader on the status—especially the material status—of the documents read and, when possible, providing access to the original documents.

Started in 2013, the transcriptions are being made available step by step and in coordination with the publication of the annotated volumes of the correspondence. The digital edition of the texts of the letters directly benefits from the thorough collation work done for the preparation of the printed edition, in order to offer the most reliable information in both formats. Two hundred of the 278 transcribed letters currently available on the interface belong to D'Alembert's 1741–1755 correspondences, i.e., to the two volumes of annotated correspondence. The first (corresponding to the years 1741–1752) has just been published and the second (corresponding to the years 1753–1757) is currently being prepared. Even though this editorial chain does not permit us to propose strictly similar versions in both formats (the letters are notably published without critical notes on the interface, in order to leave the scoop to the paper editions, and without indicating the manuscript corrections, as we will explain below), it has, however, been possible to profit from the LaTeX data input language chosen for the printed edition to offer a digital transcription of the letters, including a reliable and perfectly readable reproduction of the many mathematical formulae punctuating D'Alembert's exchanges with Cramer and Euler during this period.[3]

Furthermore, the dynamic role of the display provides the reader with the possibility of combining the information in many ways: since the inventory data and the texts of the letters are gathered on the same page, the former enables an understanding of the latter. This is particularly true in the case of the information regarding the sources of each letter.

A letter from D'Alembert may indeed take very diverse forms. In some cases, it can take the form of the original autographed handwritten version sent by the author, or of an autographed draft or a copy of the letter—which is called a "minute" when it is made before sending the letter. Sometimes, only a printed version of the letter was found, published during the lifetime of the author—this is the case with the so-called "public letters" (*lettres ostensibles*)[4]—or after his death, in a later edition or re-edition of his works. More indirect proof of the existence of the letter can also exist, either as a simple mention or partial citation thereof (in another letter, in the reply from a correspondent, in a book by the author or by one of his contemporaries), or as a trace, in a sales catalogue, before it disappeared into a private collection.

Our aim for the digital edition of the correspondence was to inform the reader about the source of the letter, whose transcription was being consulted, and any other existing versions thereof. Each page featuring the text of a letter also contains the list of its material versions. From this list, the reader is given the possibility of identifying the source from which the text offered has been collated and consulting its other versions when they are available online.

Last but not least, the "D'Alembert en toutes lettres" interface provides access to the original manuscripts of the transcribed letters. In the case of a handwritten version of a letter, when it was possible to obtain both a high quality digital scan thereof and the rights to reproduce it online, the reader has the opportunity to compare the transcription of each folio with the original document and enlarge the size if necessary (this is currently possible for 135 out of the 278 transcribed letters). This kind of consultation not only facilitates greater access to the original

[3]The edition of the formulae is ensured thanks to the Opensource Mathjax module (https://www.mathjax.org/), able to process LaTeX code fragments encapsulated within XHTML documents. In terms of reading, the results obtained with Mathjax are up to the performance of the renowned mathematical edition language.

[4]See the definition of *lettre ostensible* by I. Passeron in D'Alembert (2009, pp. 519–520).

sources, often preserved in libraries and not accessible to a wide audience, but also offsets the imperfections of the chosen linearized transcription system by enabling the reader to view the text in its original state. Moreover, our aim to improve the material dimension of the documents highlights yet another way to browse the correspondence: the consultation of the letters according to conservation fund. This browsing mode was compiled by presenting the content and history of each fund, when the libraries themselves possess this information.

Finally, this first digital version of the interface offers several modes of access to and consultation of the correspondence, which are complementary to the paper edition. This version is user-friendly, since, having moved beyond the home page, the next page (see Fig. 1) provides for selection from among the 3000 letters of the correspondence inventory according to various selection criteria: on the left-hand side, a direct selection can be made per year, per correspondent, per reception or sending place or per conversation fund, and on the right-hand side, more precise (and even cross-) selection is available, thanks to the search engine of the interface. The selected letters are then displayed at the centre of the page and the incipit, as well as a summary of each letter, can be consulted, combined with the source used for collating the data when the transcription is available and the history of the fund consulted as to when the manuscript was made available online.

If the letter is transcribed, a second page can be accessed (see Fig. 2) via a web link showing, on the right-hand side, the previous selection and, on the left-hand side, the list of the material versions and the summary of the letter searched, whereas the centre of the page reveals the text of the letter preceded by the following information: the first banner indicates the number of the letter in the

Fig. 1 Screenshot of the interface "D'Alembert en toutes lettres". An example of a letters' selection

Fig. 2 Example of a transcribed letter

inventory of D'Alembert's correspondence, the name of the correspondent (inferred by research), the place from which the letter was sent and where it was received, when known, as well as the related date (inferred or not); the second banner gives access to the material description of the transcribed document, its history, how to quote it, and, when necessary, the critical information enabling the reader to date it.

If the manuscript of the letter is consultable, the reader can access it via a simple link ("see the manuscript") or the folio number retained in the right margin of the transcribed text. A third page (see Fig. 3) will then appear on the screen, with the original manuscript on the left and its transcription next to it on the right, all of these being completed by commands that can be used to scroll through the document, folio after folio, and by a reminder on the essential information pertaining to the document.

2 Account of the Experience and Development Perspective

Having presented the objectives of the edition "D'Alembert en toutes lettres" and the functions implemented to meet them, we feel that it would be of interest to provide the reader with some information on the difficulties we faced and the lessons we learned from them, in the hope that such an account of our experience could be useful to other edition groups working in the field of Digital Humanities.

Some of the constraints we faced were due to the specific context of this project: this critical and annotated edition of D'Alembert's *Œuvres complètes* necessarily

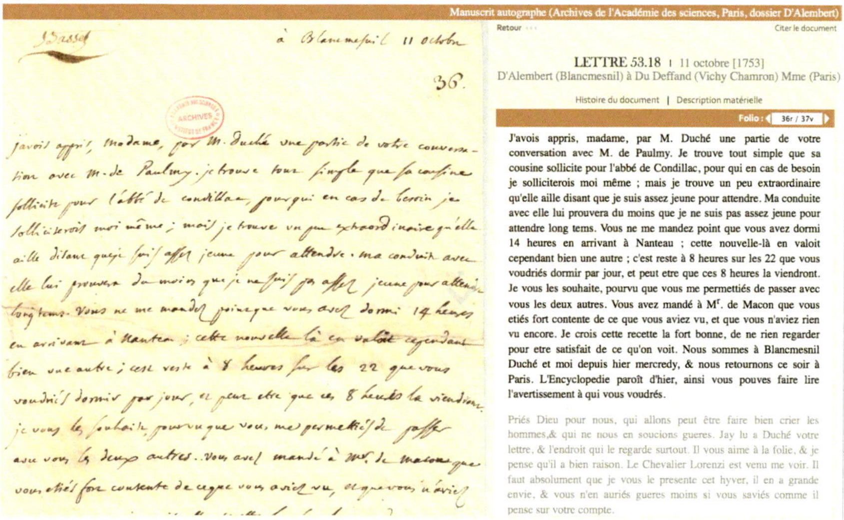

Fig. 3 Example of a transcribed letter with the original manuscript

implied an editorial process depending strongly on choices made beforehand, from a completely different perspective.

The correspondence inventory database was originally designed for use in a printed format and would have been structured in a very different way if we had developed it to be used in a digital format. As an instance, the data, which, until some months ago, had been all mixed up in the database, had to be separated into those coming directly from the letter sources and those inferred by the researchers. This was necessary in order to differentiate, both in terms of display and for the future inventory search, between the information from the primary documents on the one hand and the outcomes of the critical work (which now show up in square brackets on the interface) relating to the dating of the letters[5] and the identification of the correspondents on the other. Other changes, like the introduction of the notion of a "material version" of the letter in the edition's data structure, required more complicated technical work, i.e., the development of a specific database complementary to the inventory database. Given the scope of the work and our limited resources, we can only work on this as the publication of the transcriptions is put online, thus highlighting the need for a tool with many features for searching

[5]Sometimes partial, the square brackets can indicate the day and/or the month and/or the year.

among the material versions of the letters, which is essential in our editorial policy,[6] and could open search opportunities regarding conservation funds, edition, etc.

As in any digital edition project, we are, of course, facing the problems of data survivability and interoperability, in relation to their storage and format. The relevant format will particularly ensure their compatibility with other databases and other edition projects, as well as their visibility on the Internet at a time when the Semantic Web is developing.

The data of the Correspondence and the interface "D'Alembert en toutes lettres" are currently hosted at IN2P3 under the domain name of the French Academy of Sciences, which offers several guarantees. Yet, the format issue raises more questions, since the first digital works made within the scope of the *Œuvres complètes*, partially inherited by "D'Alembert en toutes lettres", historically favoured the use of a different encoding format from the XML-TEI, which is nowadays considered the best guarantee of data survivability and interoperability. At that time, the objective was to edit the *Eloges*, composed by D'Alembert when he was Permanent Secretary of the French Academy. We often have several preliminary handwritten versions of them (*avant-textes* in French), as well as drafts requiring a genetic digital edition capable of showing the writing process of these *Eloges* from the various materials available. These issues led to the development of the software named ORIGAMI (meaning a tool for IT scholars for the genesis of preliminary versions from manuscripts to prints—in French: Outil de Recherche Informatique sur la Genèse des Avant-textes du Manuscrit à l'Imprimé). Since they are inserted into a dedicated encoding format, this tool provides a dynamic visualization of the various handwritten correction campaigns (additions, erasures, replacements) D'Alembert brought to his texts (Barrellon and Guilbaud 2014).

Maintained in "D'Alembert en toutes lettres", this editorial chain, as well as the initial encoding format, has both the advantage of allowing the edition of handwritten corrections of the letters on the interface (particularly because several of them have preliminary versions and the transcriptions made for the paper edition already include most of this information), and the disadvantage of a non-standard digital transcription and enrichment language. We are now working on features that will allow us to export our data and metadata to XML-TEI, in order to ensure the interoperability of our edition with the main digital platforms of the national and international communities in the field of Humanities.[7]

Controlling this important question of the encoding format, as well as the various languages enabling the development of a web interface, requires specific

[6]On the other hand, these difficulties remind us that the development of a digital edition implies the previous definition of an editorial policy, embodied here by a specific structure of information, which will be difficult to amend as soon as the data have already begun to be organized according to this structure, i.e., edited.

[7]Contrastingly, the data of the ENCCRE project (Edition Numérique Collaborative et CRitique de l'*Encyclopédie*—the Collaborative and Critical Digital Edition of the *Encyclopédie*), whose first version will be online in 2017, were primarily encoded in XML-TEI. See Guilbaud et al. (2014) and the website http://enccre.academie-sciences.fr

engineering skills. Hence, the engineer becomes a valuable partner of the researcher in Humanities who wants to develop a digital edition. Furthermore, the increasing opportunities offered by digital media in terms of the quantity of publishable information and possible editorial functions make the question of the organization of the information proposed to the reader more complex. As a consequence, we also require other skills responding specifically to these new ergonomics and design issues.

These various needs raise several types of difficulty: there are financial difficulties, such as the fact that gathering all of the necessary skills requires significant funding, which only a few projects obtain; difficulties related to the technical sustainability of the editions developed, since a digital edition requires constant maintenance and updating in a political context, favouring short-term employment of engineers and/or technicians; but also methodological difficulties, as this new context assumes the development of a constructive dialogue between two types of stakeholder, the researcher and the engineer, whose issues and work practices are very different.

Hence, the researcher willing to contribute to Digital Humanities must adapt to a new work context, which is particularly time-consuming: the more complex work required to construct the text, the necessary search for funding, the necessary—but fascinating—interactions with new stakeholders, as well as the new intellectual property issues, to mention but a few of the challenges at stake.

Like several other projects, the digital edition of D'Alembert's correspondence relies heavily on volunteer work, raising again, from a less technical point of view, the essential question of the sustainability of this type of work.

Finally, the emergence of what is generally called Digital Humanities raises questions that have a strong impact on the research and edition methods. The enhanced analytical capacity offered by IT tools combined with the availability of a huge number of digitalized books on the web favours the development of new approaches, among which is the undeniable temptation to use massively statistical analysis tools: in extreme cases, it has, for instance, taken the form of "distant reading." As opposed to "close reading," this method encourages digital analysis of large corpuses without having to read them (Moretti 2013). Although the advantages that new technologies provide for corpus analysis cannot be denied, I feel it is useful to remember that any statistical analysis requires, on the one hand, the definition of a rigorous methodology beforehand, which takes into account the features of the corpus analysed, and on the other hand, the availability of sufficiently enriched and verified data that ensures the relevance of the results obtained. Unfortunately, we have no choice but to note that these obvious scientific rules are sometimes broken.[8]

From an editorial point of view, the provision of corpuses and search tools on the web naturally requires the same caution. The digital edition of D'Alembert's correspondence presently offers no means of conducting full-text searches in the

[8]This is the case in the article Allen et al. (2010) on the computerized detection of the sources of the *Encyclopédie*. For a critical analysis of this article, see Leca-Tsiomis (2013).

very content of the letters, as most of the transcriptions are still missing—the summaries, however, are freely searchable, as they have been made available by I. Passeron. Acting on the same principle, our present goal is to complete the reconstruction of the data related to the material versions of the letters before activating the corpus search based on this criterion (per conservation fund, per printed edition, etc.).

These limits implicitly define the future development perspectives for the edition, which will be in line with the previous works. Following the same logic, we will continue to add the manuscripts as soon as we have obtained the rights to publish them on the interface. The correspondence data export feature to XML-TEI is another priority of the project: the current work on the metadata of the inventory will soon be extended to the genetic information already integrated into the transcriptions in order to propose, in the near future, a dynamic visualization mode of the corrections made on the manuscripts of the letters, similar to the one we started to implement for D'Alembert's *Eloges*.

Finally, we are planning the edition of chronologic and thematic dossiers that will enable us to propose, as with the interface we designed for the Tolomas Affair within the scope of the ENCCRE project,[9] an annotated consultation tool of a set of letters selected on the basis of its links with certain aspects of D'Alembert's work, such as his position as editor of the *Encyclopédie*, or a specific scientific dispute (with Euler on logarithms of imaginary numbers or on negative pressure, with Euler and Lagrange on the issue of vibrating strings, etc.). Here again, the objective will be to find the best editorial expression among the new possibilities offered by the digital media, the promotion of original historical documents, as well as any new information brought to light by the ongoing research into the correspondence.

References

Allen, Timothy, Charles Cooney, Stéphane Douard, Russel Horton, Robert Morrissey, Mark Olsen, Glenn Roe, and Robert Voyer. 2010. Plundering philosophers: identifying sources of the *Encyclopédie*. *Journal of the Association for History and Computing* 13 (1). http://hdl.handle.net/2027/spo.3310410.0013.107.

Barrellon, Vincent, and Alexandre Guilbaud. 2014. ORIGAMI: première démonstration et perspectives de développement. In *Claude Simon. Les Vies de l'archive*, 211–224. Dijon: Editions universitaires de Dijon.

D'Alembert, Jean Le Rond. 2009. *Inventaire analytique de la correspondance (1741–1783). Œuvres complètes*, série V, vol. 1, ed. Irène Passeron, Jean-Daniel Candaux, and Anne-Marie Chouillet. Paris: CNRS Éditions.

———. 2015. *Correspondance générale 1741–1752. Œuvres complètes*, série V, vol. 2, ed. Irène Passeron, J.-D. Candaux, A. Cernuschi, F. Chambat, M. Chapront-Touzé, C. Gilain, A. Guilbaud, G. Jouve, F. Launay, M.-L. Massot, F. Prin, and C. Schmit. Paris: CNRS Editions.

Guilbaud, Alexandre, Irène Passeron, Olivier Ferret, and Vincent Barrellon. 2014. Éditer l'*Encyclopédie* au 21e siècle: un projet d'édition numérique, critique et collaborative. *Dix-Huitième Siècle* 46: 153–166. https://doi.org/10.3917/dhs.046.0153.

[9]http://dalembert.academie-sciences.fr/encyclopedie/Dossier_Affaire_Tolomas/

Leca-Tsiomis, Marie. 2013. The use and abuse of the digital humanities in the history of ideas: how to study the *Encyclopédie*. *History of European Ideas* 39 (4): 467–476. https://doi.org/10.1080/01916599.2013.774115.

Moretti, Franco. 2013. *Distant reading*. New York: Verso.

Passeron, Irène. 2018. D'Alembert's mathematical correspondence: beyond the formal description of networks. In *Mathematical Correspondences and Critical Editions*, ed. Maria Teresa Borgato, Erwin Neuenschwander, and Irène Passeron. Cham: Springer.

La correspondance scientifique de Condorcet : un aperçu

Nicolas Rieucau

Abstract
Cette brève étude consacrée à la correspondance scientifique de Condorcet, qui demeure dans sa majeure partie inédite, s'articule autour de trois aspects. Il s'agit dans un premier temps de circonscrire cette correspondance en tentant de la définir et d'en examiner la nature. L'identité des principaux correspondants scientifiques de Condorcet est ensuite évoquée. Enfin, les problèmes d'identification, de prospection et de datation de la correspondance scientifique de Condorcet sont considérés, en comparaison de ceux qui interviennent pour le reste de son corpus épistolaire.

Les *Œuvres de Condorcet* (1847–1849) demeurent encore aujourd'hui la principale référence utilisée pour ses écrits. Il s'agit pourtant d'une publication incomplète, en premier lieu parce qu'elle omet la plupart des travaux scientifiques de Condorcet.[1] Cette insuffisance est aussi reflétée par les lacunes affectant la correspondance : moins de 200 lettres, dans leur très grande majorité sans aucune tonalité scientifique, sont colligées dans cette publication. Vers 2010, l'Équipe *Inventaire Condorcet* a décidé de concentrer ses efforts sur l'établissement d'un répertoire le plus exhaustif possible de la correspondance de l'encyclopédiste. À ce jour, ont été recensées plus de 2100 lettres dont les originaux, lorsqu'ils subsistent, sont répartis dans plus de

[1] À ce sujet, voir P. Crépel (2009), N. Rieucau (2005, 2009).

N. Rieucau (✉)
Université Paris VIII (LED) et Équipe *Inventaire Condorcet*, Paris, France
e-mail: nicolas.rieucau@univ-paris8.fr

M. T. Borgato et al. (eds.), *Mathematical Correspondences and Critical Editions*, Trends in the History of Science, https://doi.org/10.1007/978-3-319-73577-1_6

130 lieux de conservation dans le monde.[2] Or, non seulement plus de la moitié de ces lettres demeurent inédites mais, de surcroît, elles sont principalement de nature scientifique.

Il est difficile de mesurer l'ampleur exacte de la correspondance que Condorcet a entretenue. À la différence de certains de ses contemporains, il ne tenait pas de registre dans lequel auraient été consignés ses échanges épistolaires.[3] Mais ceux-ci furent assurément bien plus importants que la correspondance dont nous avons aujourd'hui une trace.[4] Les estimations que nous donnons ici doivent donc être impérativement considérées avec réserve, sans compter que notre inventaire n'est pas arrivé à son terme et qu'une découverte ultérieure de documents en assez grand nombre demeure toujours possible, en particulier sur le marché des autographes.

Cela étant, la présence importante de lettres scientifiques dans le corpus déjà repéré reflète assez bien l'activité intellectuelle de Condorcet. Rappelons qu'il fut en effet avant tout un savant. Il soutint une thèse d'analyse en 1760 et continua activement ses travaux de mathématiques pures jusqu'au début des années 1780 ; alors que ceux de mathématiques appliquées, entamés dès la fin des années 1760, se poursuivirent jusqu'aux dernières années de sa vie. D'autre part, l'essentiel de sa carrière intellectuelle s'est déroulée à l'Académie royale des sciences de Paris, la plus grande institution scientifique du monde à son époque. Condorcet y entra en mars 1769 comme adjoint mécanicien, avant d'être nommé, en mars 1773, secrétaire adjoint puis, en août 1776, secrétaire perpétuel, poste qu'il occupera jusqu'à la Révolution.

Nous entendons ici ne fournir qu'une modeste vue d'ensemble de la correspondance scientifique de Condorcet. Après avoir tenté de définir cette correspondance, nous envisagerons sa nature et qui sont les correspondants de Condorcet. Nous examinerons ensuite les problèmes relatifs à la constitution de l'inventaire de la correspondance scientifique, au regard de ceux rencontrés pour établir celui du reste du corpus épistolaire de Condorcet.

1 Qu'est ce que la "correspondance scientifique" de Condorcet ?

Cette interrogation conduit en réalité à deux questions : comment définir une correspondance et quand peut-on considérer que son objet relève des sciences ? Dans les deux cas, aucune réponse tranchée ne peut être avancée.

Pour ce qui concerne la première question, il est possible de mentionner plusieurs exemples de documents dont le statut demeure problématique, à partir du

[2]Consulter à ce propos la base de données disponible à l'adresse http://alpha.inventaire-condorcet.com/Inventaire/Correspondance. Une carte des lieux de conservation se trouve par ailleurs à l'adresse http://alpha.inventaire-condorcet.com/Map/index.php

[3]Sur ce point, voir Équipe *Inventaire Condorcet* (2014, p. 17).

[4]*Ibid.*, p. 20.

moment où l'on envisage une correspondance comme un ensemble de messages écrits. De nombreux cas épineux s'appliquent ainsi à Condorcet[5] comme à ses contemporains[6] : lettres envoyées ou reçues individuellement ou collectivement au titre d'une fonction institutionnelle, lettres rédigées de la part d'un tiers ou sous sa dictée, billets, messages sur un manuscrit destinés à un copiste ou un imprimeur, notes envoyées sans messages particuliers, dédicaces adressées à un tiers sur un ouvrage, faire-part, reconnaissances de la bonne réception d'un versement ou d'un écrit, lettres ostensibles,[7] libelles ou mémoires publiés sous forme de lettres dans un périodique, épîtres, adresses, avis, lettres apocryphes . . .

Il est déjà acquis, en l'état actuel de notre travail sur l'inventaire de la correspondance de Condorcet, que les textes édités sous le titre de "lettres", les épîtres ou tous les écrits relevant du genre épistolaire au sens large seront signalés mais constitueront plusieurs catégories à part. Si on comptabilise près d'une cinquante d'écrits en la matière, ils ont avant tout une tonalité politique et peu nombreux sont ceux d'ordre scientifique.[8]

En revanche, en raison des activités de Condorcet à l'Académie des sciences, entreprendre de définir sa correspondance savante impose d'être confronté en particulier à deux des cas énoncés ci-dessus, à savoir les reçus relatifs à un versement ou à un écrit et les lettres relevant d'une fonction institutionnelle. Nous exclurons les reçus : si les documents de ce type ont par exemple été publiés dans les correspondances de Laplace[9] et de Lavoisier[10] avec Condorcet, il nous paraît excessif de considérer qu'ils détiennent un statut épistolaire. Nous comprendrons au contraire les lettres liées aux fonctions de Condorcet à l'Académie des sciences. Presque la moitié de la totalité des lettres à ce jour repérées peuvent être considérées comme relevant de cette catégorie, ce qui représente environ un millier de pièces. Les quatre cinquièmes d'entre elles sont constitués de lettres reçues par Condorcet compte tenu de ses activités académiques. Elles sont pour la plupart aujourd'hui conservées à Paris, aux Archives de l'Académie des sciences (désormais "AAds").

À première vue, considérer que ces dernières lettres ont effectivement été destinées à Condorcet ne va cependant pas toujours de soi. C'est le cas des lettres envoyées à l'Académie sans que l'on y trouve une trace explicite attestant qu'elles

[5]Voir aussi *ibid.*, pp. 15–16 et A. Magnan (2014, pp. 142–143).

[6]Consulter en particulier les multiples exemples relatifs à D'Alembert dans I. Passeron (2009, pp. xiii–xvii).

[7]Rappelons que cette expression n'est pas anachronique. Elle est d'ailleurs employée par Condorcet lui-même. Voir *Condorcet à Voltaire* (28 nov. 1776), dans Voltaire, *Correspondence and related documents* (Besterman D 20431). L'usage d'une telle expression doit être préféré à celui de "lettres ouvertes" ou de "lettres publiques" : des lettres initialement privées (et non "ouvertes" ou "publiques") peuvent par la suite avoir circulé de mains en mains, ou avoir été publiées par leur expéditeur ou leur destinataire.

[8]Parmi ces derniers, le plus notable est *Le marquis de Condorcet à M. D'Alembert sur le système du monde et sur le calcul intégral* (1768).

[9]*Correspondance de Pierre-Simon Laplace (1749–1827).*

[10]*Œuvres de Lavoisier – Correspondance.*

sont adressées à Condorcet. Ce cas intervient lorsqu'il s'agit de certains billets remis en mains propres ou, beaucoup plus souvent, lorsque les enveloppes – ou le papier plié puis cacheté qui en tenait lieu – comportant le nom du destinataire ont été jetées ou détruites, tandis que le contenu des lettres ne révèle pas d'indications sur ce nom. Le nombre de ces lettres est imposant puisqu'il s'élève à près de 400, soit près de la moitié du corpus de la correspondance académique de Condorcet. Ces lettres débutent la plupart du temps par "Monsieur" et on est *a priori* bien en peine de savoir si l'expéditeur connaît réellement Condorcet, d'autant que ce dernier n'y est pas systématiquement désigné par le titre de marquis.[11]

Même si des cas incertains demeurent, plusieurs éléments permettent souvent de déduire que ces lettres s'adressent à tout le moins au représentant de l'institution et donc à Condorcet, ou que ce dernier a dû "traiter" ces lettres, précisément compte tenu de ses fonctions de secrétaire perpétuel. La consultation du *Plumitif des séances* de l'Académie des sciences est à cet égard précieuse puisque Condorcet y mentionne fréquemment qu'il a fait lecture de telle ou telle lettre lors d'une réunion de l'institution. Son écriture peut aussi figurer sur certaines missives, souvent pour désigner les rapporteurs du mémoire que l'expéditeur avait joint à son courrier.

Le statut de la plupart des lettres lues lors de séances académiques mérite en particulier qu'on s'y attarde. Il s'agit de lettres qui en l'occurrence sont finalement destinées à l'ensemble des académiciens. Certains correspondants, s'ils adressent leur courrier à Condorcet, font d'ailleurs allusion à une telle destination.[12] Mais on doit inversement considérer que les lettres explicitement adressées à "Messieurs les académiciens" le sont également à celui qui les représente, et donc à Condorcet. Preuve en est le fait que, dans les cas où une adresse complète figure sur la lettre, une mention allographe indiquant que Condorcet en est le destinataire est très souvent ajoutée aux côtés de cette adresse.[13]

D'une manière générale, la correspondance de Condorcet liée à ses activités académiques ne doit toutefois pas être envisagée comme une correspondance ès qualités au sens strict. Certaines lettres revêtent il est vrai un caractère institutionnel mais aussi intime, le départ entre les deux étant impossible à effectuer. Cette dualité se manifeste dès les premières lettres que Condorcet signe en tant que secrétaire de l'Académie. Dans une lettre à Franklin par exemple, en date du 2 décembre 1773, à la suite de plusieurs questions principalement relatives à la minéralogie et la météorologie, un hommage personnel est rendu par Condorcet à l'Américain

[11]Ce titre disparaîtra d'ailleurs progressivement sous la Révolution. Sur ce point, voir N. Rieucau (2013, pp. 702–703).

[12]Citons par exemple l'extrait d'une lettre de Gasté : "permettés [. . .] que je m'adresse a Vous, et par Vous Monsieur le Marquis a Messieurs de l'Academie", *Gasté à Condorcet* (28 mars 1784), AAds, pochette de la séance du 31 mars 1784.

[13]Par exemple, à l'adresse indiquée par Délmas "A Messieurs/Messieurs De lacademie/des Sciences/À Paris" succède, d'une autre main, l'inscription "Pour M. de Condorcet/Rue de Bourbon", *Délmas à Condorcet* (8 fév. 1792), AAds, pochette de la séance du 18 fév. 1792.

qualifié de "promethée moderne", en référence à ses travaux sur l'électricité.[14] Encore plus éloquente est une lettre que Condorcet envoie à Haller le 14 octobre 1777.[15] Condorcet signe là aussi sa lettre en tant que secrétaire de l'Académie des sciences et y évoque l'*Éloge de van Swieten* (1773) prononcé par Fouchy, mais l'essentiel de son courrier consiste à demander à Haller du crottin de chamois afin de soigner le cancer du sein dont souffre sa mère[16] ! On pourrait ainsi multiplier les exemples. Ils sont également nombreux s'agissant de la correspondance passive de Condorcet. On se contentera de renvoyer à ce sujet aux lettres que Lagrange lui adresse, où tournures proprement amicales et égards personnels sont mêlés à des considérations scientifiques et/ou relatives à la vie académique. Un passage de l'une de ces lettres, en date du 1er octobre 1774, synthétise bien cette ambivalence : "j'aime vos ouvrages, et comme ceux d'un des premiers savants du siècle, et comme ceux d'un de mes meilleurs amis".[17]

Concernant la seconde question que nous avons posée – à partir de quand doit-on considérer que l'objet d'une correspondance est scientifique ? – l'une des difficultés de son traitement réside justement dans le fait que les échanges épistolaires de Condorcet peuvent mêler considérations personnelles et savantes. Nous n'y reviendrons donc pas. Mais deux autres difficultés, qui n'exigent pas un fort développement, peuvent aussi être avancées.

Premièrement, une lettre peut contenir des passages proprement scientifiques et/ou relevant de l'activité académique de Condorcet, mais il est également possible qu'elle aborde des thèmes relevant d'autres disciplines. Le cas de la correspondance entre Condorcet et Turgot[18] est à ce titre particulièrement exemplaire : au sein d'une même lettre portant sur la chimie ou l'astronomie, ou bien renvoyant à des écrits et événements académiques de toute sorte (rapports et mémoires discutés en séance, nominations, débats et querelles internes ...), il est fréquent de trouver une ou plusieurs considérations relatives à la jurisprudence, la politique ou encore l'économie politique.

Deuxièmement, il existe des lettres qui traitent de l'application des connaissances ou d'une démarche scientifiques à des objets relevant des sciences morales et politiques. Ces lettres se trouvent donc à mi-chemin entre plusieurs champs de la connaissance humaine, et pas seulement des sciences au sens strict. On peut estimer

[14]On trouvera une analyse de la correspondance entre Condorcet et Franklin dans M. Albertone (2014, pp. 84–87).

[15]Bern, Burgerbibliothek, N. Albrecht von Haller, 105.10.

[16]La mère de Condorcet décèdera le 19 décembre de l'année suivante. Il est à noter que le "remède" consistant à appliquer du crottin de chamois sur le sein pour en soigner le cancer est demeuré peu connu. Il n'est en particulier pas signalé dans l'ouvrage de D. Droixhe (2015) consacré aux thérapies du cancer du sein au XVIIIe siècle. D. Droixhe nous a par ailleurs confirmé qu'il n'avait pas connaissance d'un tel traitement.

[17]*Œuvres de Lagrange*, vol. xiv, p. 28.

[18]Rappelons que la majeure partie de cette correspondance a été publiée par Ch. Henry en 1883. Certaines lettres se trouvent aussi, notamment, dans B. Bru et P. Crépel (1994, § 1.3).

qu'une grande partie de la correspondance avec Windischgrätz, sur l'application de l'analyse aux transferts de propriété,[19] détient ce statut. De même, il est possible de renvoyer à la correspondance que Condorcet entretient sur les canaux avec Turgot, Trudaine de Montigny et les commissaires des États de Bretagne.[20] Nous pensons également aux échanges de Condorcet avec Lacroix[21] relatifs aux questions de population, avec Duvillard à propos des problèmes d'assurance et de finances,[22] ainsi qu'aux lettres que Condorcet adresse à Verri au sujet de l'application du calcul mathématique à l'économie politique.[23]

2 Quelle est la nature de la correspondance savante de Condorcet et qui sont les correspondants ?

La nature de la correspondance savante de Condorcet couvre deux principaux aspects. Comme nous l'avons laissé entendre, elle concerne des échanges institutionnels et/ou proprement scientifiques.

Pour ce qui regarde le cadre institutionnel, près d'un cinquième de la totalité de la correspondance savante est relatif à des nominations de savants à divers rangs académiques, envoyés par la Maison du roi.[24] S'y ajoutent des courriers – nettement moins nombreux – adressés par Condorcet au même titre, principalement à des scientifiques étrangers, ainsi que les lettres de remerciements de ces derniers.[25] On dispose par ailleurs de plusieurs dizaines de lettres liées à l'impression ou à l'envoi de textes académiques, notamment les volumes de l'*Histoire de l'Académie*

[19]Cette correspondance, qui demeure inédite, a été analysée par M. Ondo-Grečenkovà dans plusieurs contributions. Voir par exemple M. Ondo-Grečenkovà (2007, 2009).

[20]Pour un commentaire, comprenant près d'une trentaine de lettres inédites, consulter É. Szulman (2014, 2019).

[21]Ces échanges ont été publiés et commentés par R. Taton (1959).

[22]Paris, Bibliothèque nationale de France, NAF 20576, f. 261–278. Cet ensemble comporte treize lettres. L'une d'entre elles a été publiée et commentée par G. Thuillier (1997, pp. 449–451).

[23]Une édition récente de ces lettres se trouvent dans B. Bru et P. Crépel (1994, § 1.4) où elles sont aussi commentées. À ce sujet, voir également N. Rieucau (1997, chap. 2, sect. 2.2).

[24]À ce titre, et en l'état actuel de nos recherches, nous avons en particulier répertorié environ 150 lettres d'Amelot et du baron de Breteuil, successivement ministres de la Maison du roi (12 mai 1776–18 nov. 1783 et 18 nov. 1783–26 juil. 1788). Ces lettres sont conservées dans les pochettes des séances, aux AAds. Leurs minutes se trouvent aux Archives nationales, dans la sous-série O^1.

[25]Voir par exemple : *Condorcet à Van Swinden* (15 sept. [1777]), Leiden, Universiteitsbibliotheek, BPL 755 ; *Hunter à Condorcet* (19 fév. 1782), AAds, Dossier biographique Hunter ; *Condorcet à Bergman* (13 mars 1782), dans *Torbern Bergman's Foreign Correspondence*, vol. 1, p. 21 ; *Condorcet à Bonnet* (25 mai 1783) et *Bonnet à Condorcet* (31 mai 1783), Bibliothèque de Genève, Ms Bonnet respectivement 86, f. 64–65 et 76, f. 97–98 ; *Wargentin à Condorcet* (20 juin 1783), AAds, Dossier biographique Wargentin ; *Priestley à Condorcet* (16 [?] mars 1784), AAds, Dossier biographique Priestley ; *Condorcet à Black*, 20 mars 1789, dans *The correspondence of Joseph Black*, vol. II, pp. 1016–1017.

royale des sciences et ceux dits des *Savants étrangers*.[26] Une quinzaine de lettres, à mi-chemin entre la correspondance institutionnelle et celle réellement scientifique, sont des courriers échangés par Condorcet afin de se procurer des renseignements sur les savants dont il doit composer les éloges académiques.[27] Le reste de la correspondance institutionnelle concerne des sujets divers et plus ponctuels, parfois importants, tels que ceux relatifs aux tentatives de réformes entreprises par Condorcet au milieu des années 1770 pour rapprocher les académies provinciales de l'Académie royale des sciences de Paris,[28] parfois anecdotiques, comme les demandes relatives à la nomination de commissaires pour un mémoire ou celles concernant la certification de rapports académiques.

Quant à la nature scientifique de la correspondance de Condorcet, celle-ci couvre une part importante des connaissances et des techniques savantes de son temps. Un premier groupe, composé de plus de 300 lettres, à peu près également réparties entre celles reçues et envoyées, concerne par ordre d'importance les disciplines suivantes : calcul intégral, calcul des probabilités, hydraulique, chimie, métrologie. À l'exception de la chimie, ces disciplines sont aussi celles auxquelles Condorcet consacrera des ouvrages, des articles, des mémoires ou des rapports. Un second groupe, de plus de 150 lettres principalement reçues, concerne des disciplines moins fréquemment évoquées dans la correspondance de Condorcet et qui ne sont d'ailleurs pas (tout comme la chimie), ses domaines intellectuels de prédilection : météorologie, géodésie, minéralogie, géologie, optique, botanique ou encore agronomie. Concernant les arts, ceux dont il est question dans la correspondance de Condorcet relèvent essentiellement de sa correspondance reçue en raison de son poste de secrétaire de l'Académie. Cela n'est guère surprenant sachant que ce dernier n'était pas à proprement parler un inventeur de techniques. En l'occurrence, les machines hydrauliques et (à partir des années 1780) les ballons dirigeables sont abordés dans plus d'une cinquantaine de lettres reçues par Condorcet, ce qui représente un corpus plus imposant que celui relatif à l'ensemble des autres arts envisagés dans le reste de sa correspondance passive. Par ordre d'importance et sans être exhaustif, ces arts sont les suivants : paratonnerres, ponts, fours, boussoles, cabestans, baromètres, compas, montres.

---

[26] En se limitant à ces deux derniers cas, on peut citer, entres autres, les lettres suivantes : *Condorcet à Trudaine de Montigny* [1773–1774], London, British Library, RP 5959/2 ; *Amelot à Condorcet* (15 fév. 1777), AAds, pochette de la séance du 19 fév. 1777 ; *Condorcet aux académiciens de Toulouse* [début 1782], Archives de l'Académie des sciences, inscriptions et belles-lettres de Toulouse, vol. "correspondance" 80159, pièce 78 ; *Anisson Duperron à Condorcet* (1er déc. 1782), AAds, pochette générale de l'année 1782 ; *Castilhon à Condorcet* (6 mars 1785), AAds, pochette de la séance du 16 avr. 1785.

[27] Pour ne donner qu'un exemple d'une correspondance assez fournie, à propos de l'*Éloge de Linné* (1779), voir *Wargentin à Condorcet* (12 fév. 1778), Paris, Bibliothèque de l'Institut de France, Ms 876, f. 59–60 ; *Condorcet à Wargentin* (14 avr. 1778), Stockholm, Centrum för vetenskapshistoria, sans cote ; *Condorcet à Linné fils* (15 août 1778), London, Linnean Society, Linnean Correspondence, vol. 3, f. 89–90.

[28] À ce sujet, voir K. M. Baker (1975, pp. 66–75), ainsi que A. Chassagne et P. Crépel (2019).

Venons-en maintenant aux correspondants scientifiques de Condorcet, dont certains noms ont déjà été signalés lors de l'examen des questions précédentes. Leur nombre s'élève à plus de 250. Nous n'allons évidemment pas les énumérer ici[29] et il faut par ailleurs conserver à l'esprit que leur correspondance avec Condorcet est souvent lacunaire.[30] On y relève, et à cela rien de surprenant, le nom de savants plus ou moins prestigieux. Ainsi peut-on citer Adanson, Bicquilley, Borda, Bossut, Cotte, D'Alembert, Dionis du Séjour, Duvillard, Guyton de Morveau, A. L. de Jussieu, Lacroix, Laplace, La Tourette, Lavoisier, Monge, J.-É. et J.-M. de Montgolfier, Ratte, Seignette, ou encore Vicq d'Azyr. Pour ce qui concerne les étrangers, il est possible d'évoquer Struick à Amsterdam, Jean II Bernoulli à Bâle, Lagrange à Berlin, Haller à Berne, Canterzani et Malvezzi à Bologne, van Swinden à Franeker, Bonnet, Le Sage et H.-B. de Saussure à Genève, Pezzi à Gênes, Banks, Blagden et Priestley à Londres, Frisi à Milan, Franklin à Philadephie, L. et J. A. Euler, Fuss ainsi que Lexell à Saint-Pétersbourg, Wargentin à Stockholm, ou encore Bergman à Uppsala.

Mais ce qu'il y a *a priori* de plus frappant si l'on examine les correspondants scientifiques attestés de Condorcet, n'est pas tant l'absence de certains d'entre eux[31] que la présence de près d'une centaine de "petits" correspondants souvent issus des provinces françaises[32] et provenant d'horizons fort divers : érudits locaux, savants aux recherches demeurées confidentielles, ou personnages en marge des réseaux sociaux connus. Cette particularité s'explique par le fait que les correspondants en question souhaitent s'adresser à l'Académie des sciences afin de faire reconnaître leurs travaux ou leurs découvertes et, moins fréquemment, afin de la solliciter sur un problème qui se pose à eux.

3 Les problèmes d'identification, de prospection et de datation de la correspondance scientifique

En comparaison du reste de la correspondance de Condorcet, ces problèmes se posent d'une manière spécifique s'agissant de sa correspondance scientifique, et cela compte tenu des fonctions académiques de l'encyclopédiste.

[29]Ils sont tous répertoriés au sein de la table "Personnes" (http://alpha.inventaire-condorcet.com/Inventaire/Expediteurs_Destinataires), sous la responsabilité de F. Launay, de notre base de données.

[30]Le nombre de lettres, pour chaque correspondant, est précisé à l'adresse mentionnée à la note précédente. Le corpus le plus important est constitué par 22 lettres reçues de Lagrange. C'est un nombre relativement faible, non seulement parce qu'aucune lettre de Condorcet à Lagrange n'est parvenue jusqu'à nous, mais aussi parce que des lettres de Lagrange sont manquantes.

[31]Parmi les proches de Condorcet, on peut évoquer les noms d'Arbogast, Cabanis, Keralio et Girault de Keroudou.

[32]À ce sujet, voir N. Rieucau (2019b) où les correspondants en question sont toutefois rangés sous la catégorie plus large des "non-académiciens".

Les "petits" correspondants auxquels nous avons précédemment fait allusion sont en particulier difficiles à identifier car leur signature – bien que déterminante *in fine*, nous le verrons bientôt, par la façon dont elle est écrite – n'a fréquemment qu'une valeur informative limitée : prénom absent (car l'usage était à l'époque de ne pas le mentionner dans les lettres, pas plus d'ailleurs que dans les éventuels imprimés), cas d'homographie, patronyme parfois incomplet (dans le cas des nobles) et/ou qui n'est pas signalé dans les dictionnaires ou les encyclopédies biographiques généralistes ou spécialisés etc. L'identification de ces correspondants exige alors des recherches approfondies,[33] en s'appuyant non seulement sur le contenu de la missive mais aussi, lorsqu'elle est présente, sur la mention de leur adresse et de leur profession. Des hypothèses une fois formulées sur plusieurs personnages potentiels, le but ultime est de parvenir à retrouver leur signature autographe sur des documents manuscrits de nature administrative, habituellement utilisés en généalogie, tels que les actes notariés, les registres paroissiaux et les actes d'état civil. Pour les correspondants parisiens ou ayant séjourné à Paris, l'examen des cartes de sûreté, rendues obligatoires à partir de septembre 1792 jusqu'au moins la fin de l'année 1793, peut également se révéler précieux. Sur les registres de délivrance qui sont conservés aux Archives nationales, dans la sous-série F/7, figurent en effet la signature, les noms, prénoms, âge, profession et adresse des détenteurs des cartes de sûreté, ainsi que leur lieu de naissance, la date de leur arrivée à Paris, et leur éventuelle adresse précédente dans la capitale.

Les problèmes de prospection des pièces de la correspondance de Condorcet résultent de leur dispersion assez importante. Quatre raisons principales expliquent cette dispersion[34] : la négligence de Condorcet quant à la gestion de sa correspondance ; le sort chaotique de ses papiers lors de sa proscription, à partir de juillet 1793, puis après sa mort ; la circulation, depuis les premières décennies du XIXe siècle, de plusieurs centaines de pièces sur le marché des autographes ; enfin la multiplicité des destinataires des lettres de Condorcet.

Cette quatrième et dernière raison concerne en particulier la correspondance académique. En sa qualité de secrétaire de l'Académie des sciences, et comme nous l'avons suggéré quand nous avons évoqué les correspondants de Condorcet, ce dernier entretenait une relation épistolaire avec de nombreux scientifiques français et étrangers. Paradoxalement, il est plus aisé de mener la prospection hors de France plutôt qu'en France. Les lettres de Condorcet ont souvent, en effet, été adressées à des sommités scientifiques étrangères : leur correspondance est en général assez

[33]Nous devons la description qui suit de ces recherches à Françoise Launay. Qu'elle en soit ici remerciée. On trouvera d'ailleurs dans F. Launay (2019) une illustration de ce genre d'investigation poussée, en l'occurrence menée pour identifier les correspondants suivants (nous ajoutons entre crochets droits les prénoms et fragments de patronymes rétablis) : [Jean Baptiste] Paroisse, [Charles Etienne] Magnin, [Joseph Antoine] de Sauteiron [de St Clément], [Jean Claude] Rivey, [Pierre] Remy, [Gilles François de] Segondat, [Michel Jacques] Chapaux, [Gilles Nicolas Jean] Mancel, [Joseph René] de Gasté, [Jean Michel] Cavé, [Jean Antoine] Lobgeois, [Pierre Joseph] Geouffre d'Aurussac.

[34]Pour un exposé développé, voir Équipe *Inventaire Condorcet* (2014, pp. 16–19).

bien connue et donc relativement facile à chercher, d'autant que les sommités en question étaient fréquemment attachées à de grandes sociétés savantes soucieuses de la conservation de leurs archives. La situation est nettement plus contrastée pour la correspondance de Condorcet en France, et en premier lieu celle envoyée dans les provinces françaises. C'est manifestement en grande partie compte tenu de cette différence que nous avons, à ce jour, trouvé à peine plus d'une trentaine de lettres scientifiques adressées par Condorcet dans cette dernière zone géographique, tandis que le corpus identifié à l'étranger est *grosso modo* deux fois plus important.

Dans les deux cas, cela étant, les moyens de prospection sont sensiblement les mêmes : examen des catalogues imprimés ou électroniques d'archives ou de bibliothèques, enquête auprès des conservateurs, recherche dans les catalogues de vente, exploration *in situ* de certains fonds . . .

Si les problèmes d'identification et de prospection se posent de façon plus aigüe pour la correspondance scientifique de Condorcet que pour le reste de sa correspondance, l'inverse se présente en revanche quant aux questions de datation. Deux raisons peuvent être avancées. Premièrement, les lettres reçues représentent approximativement les deux tiers de la correspondance scientifique de Condorcet, contre seulement un tiers, environ, de sa correspondance non scientifique. Or, sachant que les lettres reçues par Condorcet sont nettement mieux datées que celles qu'il envoie, la part plus importante de sa correspondance passive en matière scientifique minimise les problèmes de datation. Deuxièmement, la correspondance scientifique de Condorcet, envoyée ou reçue, revêt fréquemment, nous l'avons vu, un aspect institutionnel. Elle est par conséquent mieux datée que celle relevant d'un cadre strictement privé. On observe ainsi que plus des trois quarts de la correspondance scientifique de Condorcet est correctement datée, alors qu'inversement plus des trois quarts de sa correspondance non scientifique est incomplètement ou non datée.

Néanmoins, et comme en matière de prospection, les méthodes de datation dont nous usons demeurent similaires quelle que soit la nature des lettres envisagées. En contrepoint de l'étude de leur contenu, y compris de l'adresse et des inscriptions allographes lorsqu'elles figurent, il convient parfois d'examiner d'autres lettres du corpus voire celles échangées par des tiers. À cela rien d'original. La particularité de notre recherche est, en revanche, de s'appuyer sur une analyse systématique et approfondie des caractéristiques matérielles de la correspondance, telles que le type de papier utilisé (souvent filigrané) et les cachets de cire.[35]

4 Conclusion

Laissée à l'écart de ses *Œuvres* publiée au milieu du XIXe siècle, la correspondance scientifique de Condorcet a par la suite bénéficié de plusieurs éditions ponctuelles, lors de la publication de la correspondance de certains de ses contemporains ou

[35]Pour une présentation détaillée, voir N. Rieucau (2013).

à l'occasion de contributions exclusivement consacrées à l'encyclopédiste.[36] Il demeure, rappelons-le, que la majeure partie de cette correspondance de Condorcet demeure encore aujourd'hui inédite, tandis que les éditions exhaustives de la correspondance des principaux hommes de sciences de son temps – les Bernoulli, D'Alembert, L. Euler, Franklin, Laplace ou encore Lavoisier – sont en revanche aujourd'hui disponibles ou en passe de l'être.

De façon générale, nul doute que la diffusion de la correspondance savante de Condorcet permettrait de mieux comprendre son œuvre et son action scientifiques. Ce faisant, il s'agirait de réévaluer sa figure de savant en tant que telle, souvent occultée au profit de l'image, quasi symbolique, du philosophe des Lumières défenseur des idéaux de 1789 et finalement victime de la Terreur, rédigeant au seuil de la mort son *Esquisse d'un tableau historique des progrès de l'esprit humain*.[37] À la tête de l'Académie royale des sciences de Paris, Condorcet a exercé un rôle capital dans le développement des sciences et des réseaux savants dans les dernières décennies de l'Ancien Régime puis sous la Révolution. Diffuser sa correspondance permettrait d'en mesurer, de façon décisive, l'ampleur.

Bibliographie[38]

Albertone, Manuela. 2014. "Condorcet et ses correspondants américains : de grands interlocuteurs et de petites découvertes", dans Nicolas Rieucau (dir.), pp. 83–107.

Baker, Keith Michael. 1975. *Condorcet. From natural philosophy to social mathematics*, Chicago, University of Chicago Press ; trad. fr. de Michel Nobile : *Condorcet, raison et politique*, Paris : Hermann, 1988.

Bergman, Torbern. 1965. *Torbern Bergman's Foreign Correspondence*, vol. I : *Letters from foreigners to Torbern Bergman*, eds. Göte Carlid & Johan Nordström. Stockholm : Almqvist & Wiksell.

Black, Joseph. 2012. *The correspondence of Joseph Black*, eds. Robert G. W. Anderson, & Jean Jones. Farnham : Ashgate, 2 vol.

Bru, Bernard, et Crépel Pierre. 1994. *Condorcet. Arithmétique politique. Textes rares ou inédits (1767–1789)*. Paris : INED.

Cardinali, Sandro, et Pepe Luigi. 1989. "Les rapports de Condorcet avec l'Institut de Bologne et Alfonso Bonfioli Malvezzi", dans Pierre Crépel et Christian Gilain (dir.), *Condorcet mathématicien, économiste, philosophe, homme politique*, 443–449. Paris : Minerve.

Chassagne, Annie, and Crépel Pierre. 2019 (à paraître). "Deux projets inédits de Condorcet pour stimuler l'activité des académies dites "de province"", dans Nicolas Rieucau (dir.) (2019a).

Condorcet Marie Jean Antoine Nicolas Caritat de. 1768. *Le marquis de Condorcet à M. D'Alembert sur le systême du monde et sur le calcul intégral*, dans Condorcet, *Essais d'analyse*, 3–93. Paris : impr. de Didot, t. I.

[36] La plupart de ces éditions sont répertoriées en bibliographie.

[37] Sur ce point, voir N. Rieucau (2009, pp. 7, 57–58).

[38] Figurent ci-dessous, outre celles citées dans le texte, la plupart des références comprenant ou constituant des éditions de la correspondance scientifique de Condorcet. Nous ajoutons une sélection de références commentant cette correspondance. Ne sont en revanche pas reprises les références manuscrites ou électroniques intégralement citées dans le corps de notre étude.

———. *Œuvres de Condorcet*, [É. O'Connor], F. Arago, F. Génin, F. Isambert, A. O'Connor, (éds.), Paris : Firmin Didot, 1847–1849, 12 t.

———. 1779. *Éloge de Linné*, dans *Histoire et Mémoires de l'Académie royale des sciences*, année 1778, [partie Histoire], Paris : Imprimerie royale, 1781, pp. 66–84.

———. 1795. *Esquisse d'un tableau historique des progrès de l'esprit humain*, dans Jean-Pierre Schandeler et Pierre Crépel (dir.), *Condorcet, Tableau historique des progrès de l'esprit humain. Projets, Esquisse, Fragments et Notes (1772–1794)*, 223–459. Paris : INED, 2004.

Corvaglia, Raffaele. 2007. *La corrispondenza di R.G. Boscovich con tre scienziati francesi et l'analisi del suo metodo di lavoro attraverso alcuni scritti matematici*, Tesi di laurea magistrale, Università degli studi di Torino.

Crépel, Pierre. 1996. "Une curieuse lettre de Borda à Condorcet et un non moins curieux article du Journal encyclopédique", dans Éric Brian et Christiane Demeulenaere-Douyère (dir.), *Histoire et mémoire de l'Académie des sciences – Guide de recherches*, 325–337. Londres : Technique & documentation.

———. 2009. "Ce qu'Arago a fait et n'a pas fait de l'œuvre de Condorcet", *Les Arago acteurs de leur temps*, Archives départementales des Pyrénées-Orientales, 355–366.

Crépel, Pierre, et Orain Arnaud. 2002. Une lettre (probablement inédite) de Condorcet à Frisi. *Recherches sur Diderot et l'Encyclopédie* 33 : 206–210.

Droixhe, Daniel. 2015. *Soigner le cancer auXVIII^e siècle. Triomphe et déclin de la thérapie par la ciguë dans le* Journal de médecine. Paris : Hermann.

Durry, Marie-Jeanne. 1959. *Autographes de Mariémont*, Deuxième partie, t. II : De Marchangy à Victor Hugo. Paris : Nizet.

Équipe *Inventaire Condorcet*. 2014. "L'inventaire analytique et matériel de la correspondance de Concorcet : enjeux et méthodes", dans Nicolas Rieucau (dir.), pp. 15–28.

Euler, Leonhard. "Correspondance Euler-Condorcet", *Opera Omnia* (1911–), Série I VA, vol. 9, Birkhäuser ed. (à paraître).

Fouchy, Jean-Paul Grandjean de. 1773. *Éloge de M. le baron Van Swieten*, dans *Histoire* [*et Mémoires*] *de l'Académie royale des sciences*, année 1772, 1ère partie, [partie Histoire], 114–128. Paris : Imprimerie royale, 1775.

Franklin, Benjamin. 1959–. *The Papers of Benjamin Franklin*, ed. W. B. Willcox. New Haven : Yale University Press.

Gilain, Christian. 1993. "Condorcet, les mathématiques et le *Supplément à l'Encyclopédie*", Lekton, vol. 3, n° 1, pp. 79–92.

———. 1996. "Sur la correspondance de Condorcet avec Euler et ses disciples de Pétersbourg", dans Catherine Brice (dir.), *France et Italie au miroir des Lumières : Condorcet 1794–1994 : Mélanges de l'École française de Rome*, t. 108, 517–531.

Hamy, Ernest-Antoine. 1905. *Joseph Dombey. Médecin, naturaliste, archéologue. Explorateur du Pérou, du Chili et du Brésil. (1778–1785). Sa vie, son œuvre, sa correspondance* [. . .], Paris : E. Guilmoto.

Histoire de l'Académie Royale Des Sciences [...] *AVEC LES MÉMOIRES DE MATHÉMATIQUE&DE PHYSIQUE* [...] *TIRÉS DES REGISTRES DE CETTEACADÉMIE*, Paris : Imprimerie royale, 1699–1790.

Jefferson, Thomas. 1950–. *The Papers of Thomas Jefferson*, éd. Barbara B. Oberg *et alii*. Princeton : Princeton University Press.

Jovy, Ernest. 1922. *Une illustration scientifique vitryate : le P. François Jacquier et ses correspondants – 116 lettres inédites conservées pour la plupart à la bibliothèque de la ville de Vitry-le-François*, ed. Vitry-le-François, 50–51. Saint Dizier : A. Brulliard.

Lacroix, Alfred. 1938. Lettre de Joseph Le Pante d'Agelet. *La Géographie* LXIX : 112–114.

Lagrange, Joseph-Louis. *Œuvres de Lagrange publiées par les soins de M. J.-A. Serret* [*et G. Darboux*] ; [*précédé d'une notice sur la vie et les ouvrages de J.-L. Lagrange, par M. Delambre*], 14 vol. Paris : Gauthier-Villars, 1867–1892.

Lalanne, Ludovic. 1880. "Lettre de Séguier, de Nîmes", *Revue des sociétés savantes des départements, publiée sous les auspices du Ministère de l'instruction publique et des beaux-arts*, 7e série, t. II, 145–148. Paris : Imprimerie nationale.

Laplace, Pierre-Simon. 2013. *Correspondance de Pierre-Simon Laplace (1749–1827)*, éd. Roger Hahn, 2 t. Turnhout : Brepols.

Launay, Françoise. 2019. "L'identification de quelques mystérieux correspondants de Condorcet à l'Académie des sciences", dans Nicolas Rieucau (dir.) (2019a).

Lavoisier Antoine Laurent de. 1955–2012. *Œuvres de Lavoisier. Correspondance*, éds. René Fric, Michelle Goupil et Patrice Bret, 8 vol. Paris : Albert Blanchard puis Hermann.

Magnan, André. 2014. "La navette Condorcet-Ruault, archive centrale du "Voltaire" de Kehl : outils, pilotis, datations, documents pour l'histoire littéraire", dans Nicolas Rieucau (dir.), 139–202.

Mairé, Béatrice. 2003. La lettre de Paul de Lamanon à Condorcet sur l'intensité magnétique, ou l'énigme d'une lettre perdue. *Revue de la Bibliothèque nationale de France* 14 : 29–39.

MÉMOIRES DE MATHÉMATIQUE ET DE PHYSIQUE PRÉSENTÉS À L'ACADÉMIE ROYALE DES SCIENCES PAR DIVERS SAVANTS [dits "SAVANTS ÉTRANGERS"] ET LUS DANS SES ASSEMBLÉES. 1750–1786, 11 vol. Paris : Imprimerie royale.

Ondo-Grečenkovà, Martina. 2007. La noblesse de la monarchie des Habsbourg au cœur de la République des Lettres européenne : le cas de la relation entre Windischgrätz et Condorcet. *Revue des études slaves* LXXVIII : 451–467.

———. 2009. Windischgrätz and Condorcet : A Tale of One Enlightenment Project. *The Czech Historical Review* 107 (3) : 569–598.

Passeron, Irène (dir.). 2009. avec la collaboration de Anne-Marie Chouillet et Jean-Daniel Candaux, "Introduction générale", *Jean Le Rond D'Alembert. Inventaire analytique de la correspondance : 1741–1783*, xiii–xlvii. Paris : CNRS Éditions.

Pepe, Luigi. 1994. "Condorcet et l'Italie : Lagrange, Frisi et Canterzani", dans Anne-Marie Chouillet et Pierre Crépel (dir.), *Condorcet, Homme des Lumières et de la Révolution*, 73–92. Fontenay-aux-Roses : ENS Éditions.

Piva, Franco. 1985. *Anton Maria Lorgna e la Francia*, Accademia di agricultura scienze e lettere di Verona.

PLUMITIF DES SÉANCES DE L'ACADÉMIE ROYALE DES SCIENCES DEPARIS, AAdS.

Price, Richard. 1994. *Correspondence of Richard Price*, William Bernard Peach & Oswald Thomas David (1994), Duke University Press, Durham, vol. III (fév. 1786–fév. 1791).

Rieucau, Nicolas. 1997. *Nature et diffusion du savoir dans la pensée économique de Condorcet*, Thèse pour le Doctorat de Sciences économiques, Université Paris I.

———. 2005. "Condorcet' Science Obscured : Shadows Cast by the Enlightenment", *Proceedings of the Western Society for French History* (Lubbock, 30 sept.–2 oct. 2004), 82–106, vol. 32. University of Michigan Library.

———. 2009. "Éliza O'Connor propose, François Arago dispose : l'écriture de la *Biographie de Condorcet* (1841, 1849) et l'édition de ses *Œuvres* (1847–1849)", *Les Arago, acteurs de leur temps*, Archives départementales des Pyrénées-Orientales, 367–379.

———. 2013. Comment dater un manuscrit sans le comprendre ? Le cas des archives Condorcet. *Dix-huitième siècle* 45 : 681–718.

——— (dir.). 2014. avec la collaboration d'Annie Chassagne et Christian Gilain, analyse matérielle : Claire Bustarret, *La correspondance de Condorcet, documents inédits nouveaux éclairages. Engagements politiques (1775–1791)*, 39–62. Ferney-Voltaire : Centre international d'étude du XVIIIe siècle.

——— (dir.). 2019a (à paraître). avec la collaboration de Françoise Launay et Jean-Daniel Candaux. Analyse matérielle sous la responsabilité de Claire Bustarret. Préface de Daniel Roche, *La correspondance de Condorcet. Documents inédits, nouveaux éclairages. Relations scientifiques avec les provinces françaises (1772–1791)*, Ferney-Voltaire, Centre international d'étude du XVIIIe siècle.

——— (dir.). 2019b (à paraître). "Présentation générale", dans Nicolas Rieucau (dir.) (2019a).

Sieffert, Charles-Henry. 1955. Deux lettres de Condorcet à Sabbathier, secrétaire de l'Académie des sciences, arts et belles lettres de Châlons (1777). *Mémoires de la Société de la Marne* XXIX : 89–91.

Szulman, Éric. 2014. "Aux sources d'une activité administrative : la correspondance entre Condorcet et Trudaine au sujet de la navigation intérieure sous le ministère Turgot", dans Nicolas Rieucau (dir.) (2014), 39–62.

———. 2019 (à paraître). "La correspondance de Condorcet avec les États de Bretagne dans les années 1780 : les enjeux d'une contre-expertise", dans Nicolas Rieucau (dir.) (2019a).

Taton, René. 1947. Une correspondance mathématique inédite de Monge. *Revue scientifique*, 85e année, 963–969.

———. 1959. Condorcet et Sylvestre-François Lacroix. *Revue d'histoire des sciences*, XII : 128–158, 243–262.

Thuillier, Guy. 1997. *Le premier actuaire de France : Duvillard (1755–1832)*. Paris : Comité d'histoire de la Sécurité sociale.

Voltaire François-Marie Arouet dit. *Correspondence and related documents*, definitive edition by Theodore Besterman, Genève, puis Banbury, puis Oxford, 1968–1977, *The Complete works of Voltaire*, vol. 85–135.

Publication of the Complete Works of Lagrange in the Digital Age

Luigi Pepe

Abstract
This work is divided into three parts, the first of which will examine the nineteenth century edition of the *Oeuvres de Lagrange* published in Paris from 1867 to 1892 in 14 volumes, edited by the mathematicians Alfred Serret, for Volumes I–X, XIII, and, after Serret's death, by Gaston Darboux for Volumes XI and XII. Volumes XIII and XIV, containing the correspondence, were edited by the historian and librarian, Ludovic Lalanne. The second part concerns the question of the manuscripts and correspondence not contained in the *Oeuvres*, while the third part takes stock of some editions currently being published, and presents some possibilities for the collection of the complete works of Lagrange.

1 Publishing the Works of the Great Mathematicians

The publishing of complete or collected works of the great classical authors, like Homer, Virgil, Plato and Aristotle, etc., has posed a problem since printing began. Widespread dissemination was closely linked to editorial choices; for this reason, publications of Greek works in the original language were preceded by translations in Latin, which was the language of the Church and places of learning like the University. Limiting ourselves to works by mathematicians, we may observe that Euclid's *Elements* were first printed in Latin (Venice 1482), then in Greek (Basel 1533), the *Opera* by Archimedes was similarly printed in Latin (Venice 1503), then in Greek with a Latin translation (Basel 1544), the *Almagestum* by Ptolemy first came to light in Latin (Venice 1517), but was subsequently printed in Greek (Basel 1538), and the *Arithmetica* by Diophantus appeared in Latin (Basel 1575), with

L. Pepe (✉)
Department of Mathematics and Computer Science, Ferrara University, Ferrara, Italy
e-mail: pep@unife.it

M. T. Borgato et al. (eds.), *Mathematical Correspondences and Critical Editions*, Trends in the History of Science, https://doi.org/10.1007/978-3-319-73577-1_7

its Greek printing not coming about until the next century (Paris 1621). Almost two centuries passed between the publication in Latin of the first four books of the *Conics* by Apollonius of Perga (Venice 1537) and their first edition in Greek (Oxford 1710). The *Mathematicae Collectiones* by Pappus were published in Latin in the sixteenth century (Pesaro 1588), but not until the nineteenth century was the original Greek published (Berlin 1876–1878).

The Jesuit Christophorus Clavius (1538–1612) was possibly the first mathematician to print his own work, *Opera mathematica*, during his lifetime (Mainz 1612). A series of works by Galileo Galilei (1564–1642), excluding the *Dialogo dei massimi sistemi*, condemned by the Church, was printed in Bologna shortly after his death, between 1655 and 1656. The works by Descartes had been published several times as early as the seventeenth century, and those by Leibniz and Newton in the eighteenth century. In the latter century, Maupertuis (Amsterdam 1744, Dresden 1752, Lyon 1756) and, in the field of natural sciences, Charles Bonnet and Buffon served as both authors and editors of their own works. The most prolific century, as far as the publishing of works by great mathematicians was concerned, was the nineteenth century, a time when the combination of two needs came together; firstly, the urgency of making readily available works that had been disseminated in the academic journals for decades, but had become difficult to find, and secondly, the political will to contribute to the glory of the nation through the publication of works by their most illustrious authors (*Edizioni* 1986; *History* 2004).

This work is divided into three parts, the first of which will examine the eighteenth century edition of the *Oeuvres de Lagrange*; the second concerns the question of the manuscripts and correspondence not contained in the *Oeuvres*, while the third takes stock of some editions currently being published, and presents some possibilities for the collection of the complete works of Lagrange.

2 The Edition of the *Oeuvres de Lagrange*

Publication of the *Oeuvres* of Joseph Louis Lagrange took place in Paris from 1867 to 1892 in 14 volumes, edited by the mathematicians Alfred Serret (1819–1885), for Volumes I–X, XIII, and Gaston Darboux (1842–1917), for Volumes XI and XII, after the death of Serret. Volumes XIII and XIV, containing the correspondence, were edited by the historian and librarian Ludovic Lalanne (1815–1898). A facsimile edition of the *Oeuvres* was made by Georg Olms Verlag in 1973. Serret first collected the memoirs printed in Turin, Vols. I–II, then those published in the Acts of the *Académie Royale des Sciences et de Belles-lettres de Berlin*, Vols. II–III–IV–V, followed by the memoirs printed in Paris, Vol. VI, and those not included in the academic acts together with the *Additions aux Elements d'Algèbre* by Euler and the *Leçons élémentaires sur les mathématiques*, Vol. VII. Volume VIII is entirely devoted to the *Traité de la Résolution des Equations Numériques* (third edition, 1826); Volume IX to the *Théorie des Fonctions Analytiques* (third edition edited by Serret, 1847); Volume X to the *Leçons sur le Calcul des Fonctions* (second edition, 1806); Vols. XI–XII to the *Mécanique Analytique* (fourth edition, 1853);

Vol. XIII collects the correspondence, unedited at that time, between Lagrange and D'Alembert; and Vol. XIV, the last one, collects the correspondence between Lagrange and Condorcet, Laplace, Euler and other scientists. A *Table alphabetique* of Vols. XIII and XIV completes Volume XIV. Volume I is preceded by the republication of the *Notice sur la vie et les ouvrages de M. le comte J.-L. Lagrange, par M. Delambre*, whom he presented as a "mathématicien français né a Turin." Today, in the digital era, we can read the *Oeuvres de Lagrange* on the *Gallica* site of the *Bibliothèque Nationale de France*:

Oeuvres de Lagrange, tome I, Paris, Gauthier-Villars, 1867, pp. LI, 733
http://gallica.bnf.fr/ark:/12148/bpt6k2155691.r=Oeuvres%20de%20Lagrange?rk=236052;4

Oeuvres de Lagrange, tome II, Paris, Gauthier-Villars, 1868, p. 727
http://gallica.bnf.fr/ark:/12148/bpt6k2292245.r=Oeuvres%20de%20Lagrange?rk=278971;2

Oeuvres de Lagrange, tome III, Paris, Gauthier-Villars, 1869, p. 797
http://gallica.bnf.fr/ark:/12148/bpt6k229222d.r=Oeuvres%20de%20Lagrange?rk=300430;4

Oeuvres de Lagrange, tome IV, Paris, Gauthier-Villars, 1869, p. 750
http://gallica.bnf.fr/ark:/12148/bpt6k229223s.r=Oeuvres%20de%20Lagrange?rk=214593;2

Oeuvres de Lagrange, tome V, Paris, Gauthier-Villars, 1870, p. 720
http://gallica.bnf.fr/ark:/12148/bpt6k2292245.r=Oeuvres%20de%20Lagrange?rk=278971;2

Oeuvres de Lagrange, tome VI, Paris, Gauthier-Villars, 1873, p. 818
http://gallica.bnf.fr/ark:/12148/bpt6k229225j.r=Oeuvres%20de%20Lagrange?rk=193134;0

Oeuvres de Lagrange, tome VII, Paris, Gauthier-Villars, 1877, p. 626
http://gallica.bnf.fr/ark:/12148/bpt6k2299428.r=Oeuvres%20de%20Lagrange?rk=128756;0

Oeuvres de Lagrange, tome VIII, Paris, Gauthier-Villars, 1879, p. 370
http://gallica.bnf.fr/ark:/12148/bpt6k229943n.r=Oeuvres%20de%20Lagrange?rk=42918;4

Oeuvres de Lagrange, tome IX, Paris, Gauthier-Villars, 1881, p. 427
http://gallica.bnf.fr/ark:/12148/bpt6k2299441.r=Oeuvres%20de%20Lagrange?rk=171674;4

Oeuvres de Lagrange, tome X, Paris, Gauthier-Villars, 1884, p. 455
http://gallica.bnf.fr/ark:/12148/bpt6k229945d.r=Oeuvres%20de%20Lagrange?rk=64378;0

Oeuvres de Lagrange, tome XI, Paris, Gauthier-Villars, 1888, p. XXII, 502
http://gallica.bnf.fr/ark:/12148/bpt6k229946s.r=Oeuvres%20de%20Lagrange?rk=107296;4

Oeuvres de Lagrange, tome XII, Paris, Gauthier-Villars, 1889, p. VIII, 391
http://gallica.bnf.fr/ark:/12148/bpt6k2299475.r=Oeuvres%20de%20Lagrange?rk=21459;2

Oeuvres de Lagrange, tome XIII, Paris, Gauthier-Villars, 1882, pp. (8), 401
http://gallica.bnf.fr/ark:/12148/bpt6k229948j.r=Oeuvres%20de%20Lagrange?rk=150215;2
Oeuvres de Lagrange, tome XIV, Paris, Gauthier-Villars, 1892, pp. XIV, 346, (2)
http://gallica.bnf.fr/ark:/12148/bpt6k229949x.r=Oeuvres%20de%20Lagrange?rk=85837;2

The main phases of this edition may be followed through the correspondence between the publisher, Gauthier-Villars, and the librarian of the *Institut*, Ludovic Lalanne. The editor was able to substitute the *Imprimerie Imperiale*, which had been edited a few years previously, for the works of Lavoisier (1862) and Fresnel (1866). For the edition of the *Oeuvres de Lagrange*, Gauthier-Villars had obtained a public subscription of 300 copies on the part of French public institutions. Alfred Serret (1819–1885), Professor of celestial mechanics at the *Collège de France* and of differential and integral calculus at the Faculty of Science in Paris, was scientific advisor for the edition up to his death, after which, on March 29th, 1885, the position was taken over by Gaston Darboux (1842–1917), who had recently been nominated as a member of the section of geometry of the Académie des sciences (Verdier 2011).[1]

This edition was harshly criticized by the historian of science, George Sarton (1884–1956), in a famous essay:

> The editor followed the path of least resistance, putting together first the Turin memoirs, then the Berlin ones, then the Paris ones, then the rest. Corrections were almost exclusively restricted to mathematical ones; there are hardly any historical notes. The editor has no historical training, no idea whatsoever of the responsibilities that the publication of ancient texts implies. Such an undertaking should always begin with the printing of the correspondence, in order that the letters might be conveniently used and referred to for the elucidation of the memoirs and books. This was done admirably in the case of Huygens, the edition of whose works is a paragon. For the printed or unpublished memoirs and books, one of two methods might be followed. They might be published either in strict chronological sequence, or divided into groups by subjects, the chronological sequence being adhered to in each group. In any case, it is not permissible to reprint a memoir first printed a century earlier without adding a number of notes, explaining the circumstances of the original publication and its repercussions. Of course, this requires considerable labor, which Serret and Darboux were neither willing nor capable of doing (Sarton 1944).

The defects pointed out by Sarton as to the eighteenth edition of the *Oeuvres* can now be rectified, in part, by using the original editions of the seventeenth century academic acts easily available on the net. For example, the following sites can be consulted:

Memorie dell'Accademia delle Scienze di Torino:

[1] A dozen letters addressed to Lalanne were found in the Lagrange fund of the library of the *Institut de France*. The economic arrangements between the editor and the public institutions and the roles of the editors may be reconstructed by means of the dossier: Paris, *Archives nationales*, F/17/3247.

http://www.accademiadellescienze.it/attivita/editoria/periodici-e-collane/memorie
Nouveaux Mémoires de l'Académie de Berlin
http://bibliothek.bbaw.de/bibliothek-digital/digitalequellen/schriften/#A3
Histoire de l'Académie royale des sciences avec les mémoires de mathématique et de physique
http://gallica.bnf.fr/ark:/12148/cb32786820s/date

Publications of the didactic works of Lagrange, reviewed by Lagrange himself, are available on the net:
Méchanique analitique, Paris, Veuve Desaint, 1788
http://gallica.bnf.fr/ark:/12148/bpt6k29071 2/f20.image.r=lagrange%20mecanique %20analytique
Mecanique analytique, Paris, Courcier, Vol. 2, 1811–1815
https://books.google.it/books?id=Q8MKAAAAYAAJ&pg=PA74&dq=lagrange+Mecanique+analytique&hl=it&sa=X&redir_esc=y#v=onepage&q=lagrange%20Mecanique%20analytique&f=false
Théorie des fonctions analytiques, Paris, Imprimerie de la République, an V
http://gallica.bnf.fr/ark:/12148/bpt6k10512862/f332.image.r=joseph-louis%20Lagrange%20th%C3%A9orie%20des%20fonctions%20analytiques
Théorie des fonctions analytiques, nouvelle edition, Paris, Courcier, 1808
https://books.google.it/books?id=XLW4dEEpp1IC&printsec=frontcover&dq=lagrange+th%C3%A9orie+des+fonctions+analytiques&hl=it&sa=X&redir_esc=y#v=onepage&q=lagrange%20th%C3%A9orie%20des%20fonctions%20analytiques&f=false
Elemens d'algebre par Léonard Euler, Lyon, Bruyset, 1774, Vol. 2
http://mathematica.sns.it/opere/13/
Leçons sur le calcul des fonctions, nouvelle édition revue, corrigée et augmentée par l'Auteur, Paris, Courcier, 1806
http://mathematica.sns.it/media/volumi/478/calculfoncions.pdf
https://archive.org/details/leonssurlecalcu01lagrgoog
Traité de la résolution des équations numériques de tous les degrés, Paris, Courcier, 1808
http://gallica.bnf.fr/ark:/12148/bpt6k1042793z

3 The Manuscripts and Correspondence of Lagrange

Lagrange has left us a large part of his autographed manuscripts, now preserved in Paris in the Bibliothèque de l'Institut, Ms 901–916, and summarised in:

> Catalogue général des manuscrits des bibliothèques publiques de France, Paris, Bibliothèque de l'Institut, Anciens et nouveau fonds, Paris, 1928, p. 215.

These manuscripts are now presented in a little more detail in:

http://www.calames.abes.fr/pub/institut.aspx#details?id=IF2B10539

Lagrange's manuscripts from the *Institut* were purchased from his widow by Lazare Carnot, Minister of Home Affairs at the time (1815), and examined with a view towards possible publication by a commission proposed by the class of science of mathematics and physics of the *Institut*, set up by Carnot himself, and composed of Legendre, Prony, Poisson and Lacroix. This commission, which co-opted Frédéric Maurice, decided not to print the manuscripts, except for a few pages, but arranged for them to be classified and collected into 16 volumes. The final report, endorsed by the other members of the commission, was made public on November 3rd, 1817, and then printed in Vol. XII of the *Oeuvres*, pp. 387–388. Serret and Darboux essentially confirmed this choice for the publication of the *Oeuvres*, but they inserted Lagrange's correspondence with D'Alembert and Euler into Vols. XIII and XIV of the *Oeuvres*. The severity of the commission's judgment was not universally shared; Giuseppe Peano saw the manuscript in 1913 and, besides extrapolating an identity from it, which he inserted into the *Formulario*, he found confirmation of his opinion that Lagrange had a great interest in the history of mathematics. He also revealed preparatory material for an edition of Diophantus and studies on Euclid's fifth postulate. The notes on Diophantus were published by Rashed (1988), the studies on the fifth postulate by Borgato and Pepe (1988). The historical notes on mechanics, also containing the figures omitted in the printed edition, were published by Borgato and Pepe (1990). According to René Taton, about 200 pages can be extrapolated from Lagrange's manuscripts for a supplementary volume of his works (Pepe 1986c). The most important publications of Lagrange's manuscripts from the Institut were carried out by Borgato (2013): the work includes almost all the Ms 916 manuscript.

The Saluzzo 736 manuscript, entitled *Principi di analisi sublime dettati da Lagrange nelle Reggie scuole di Artiglieria di Torino* is preserved in the *Biblioteca Reale* in Turin. It was published, with a presentation by Borgato and Pepe (1987), by M. T. Borgato. Another important manuscript ascribable to Lagrange's lessons at the *Ecole politechnique* is preserved in the Library of the *Ecole des ponts et chaussées* Ms 1323 (Pepe 1986b). Neither of these manuscripts is autographed.

After the publication of Vols. XIII and XIV of the *Oeuvres de Lagrange* containing the correspondence, many of Lagrange's other letters were published, and still more remain unedited in various libraries. Several of these are to be found in Italy; among those published, we may recall those addressed to Paolo Frisi (Favaro 1895); Pietro Paoli (Riccardi 1897); Tommaso Valperga di Caluso (Pittarelli 1908); Fagnano (Volterra et al. 1912); Daniel Bernoulli (Delsedime 1971); and Leonhard Euler (Juskevic and Taton 1980). All of these, and others, can be found in Pepe's bibliography (1986c). The bulk of Lagrange's unedited letters, concerning family matters, preserved either in the Library of the *Ecole polytechnique* in Paris or inherited within private collections, have now been printed in Borgato and Pepe (1989).

The catalogue of Lagrange's private library was published shortly after his death for sale at auction. It has been examined in Borgato (1989).

Some unedited letters from Lagrange can still be found in Genève (Lesage), at the Bernoullis (Basel), in Berlin, etc.

In conclusion, there remains sufficient material for a further volume of the *Oeuvres de Lagrange* of similar size as the others, if one were to follow the example of the *Oeuvres* by Cauchy, to which another volume was added in 1974.

4 Further Considerations

Sarton's criticisms of the *Oeuvres de Lagrange* are all justifiable, and the methodology is to be recommended: first should come the correspondence, then the memoirs and the treatises, further contextualized through the correspondence, a method that was, in fact, followed in the admirable edition of the *Oeuvres complètes* of Christian Huygens (1629–1695), printed in Holland in 22 volumes from 1888 to 1950. The last volume also included the catalogue of the sales of Huygens' books, preserved in Amsterdam and reproduced in facsimile. Two mathematicians, Diederik J. Korteweg (1848–1942) and his pupil, G. De Vries, contributed to this splendid edition: their names are also linked to a famous partial differential equation.

The edition of the works of Huygens was, however, published within a different historical context than that of the *Oeuvres de Lagrange*. It may be compared to the edition of the *Opere di Galileo Galilei* (Firenze, Barbera, 1890–1907, Vol. 20), edited by Antonio Favaro, in which the letters were printed at the end, but the contextualization of the works was guaranteed by the many years of work that Favaro devoted to the study of Galileo's friends, correspondents and adversaries. Two historians of mathematics, Paul Tannery and Charles Henry, are to be thanked for the long-awaited critical edition of the works of Pierre de Fermat: *Oeuvres de Fermat*, Paris, Gauthiers-Villars, 1891–1896, Vol. 4, Supplément 1922:

http://www.biodiversitylibrary.org/item/62832#page/27/mode/1up

Since the publication of the *Oeuvres de Lagrange* was seen as a useful tool for the study of mathematics, when Serret and Darboux edited Lagrange's didactic works, they did so not from the first editions, but from the last, so they could be sure of making as many corrections as possible. Serret also edited the memoirs by Lagrange scattered throughout the academic acts, modifying and updating the mathematical notes. This choice, which makes these volumes of the *Oeuvres* virtually unusable for historical study, instead presented them in such a way as to be very familiar to the mathematicians of the second half of the nineteenth century.

From a historical point of view of the publications, the Lagrange edition is to be compared with the two editions of the *Oeuvres* by Laplace, first edition, Paris 1843–1847 in 7 volumes, second edition, Paris 1878–1912 in 14 volumes. Similar criteria, i.e., priority given to a full understanding of the mathematical content, was used for the publication of the *Oeuvres* of Cauchy, Paris 1882–1974, in 27 volumes (Vol. XV of the II series, 1974). In praise of the *Oeuvres* of Lagrange, it must be said that the editions of the *Oeuvres* of Laplace and Cauchy do not contain any correspondence. This fact turned out to be particularly damaging in the case of Laplace, given that his scientific correspondence, preserved in the hands of his heirs, was almost completely lost in a fire (Hahn 2013).

For nineteenth century nations, publication of the works of great mathematicians was not merely a question of their scientific value, as they also served to contribute to the glory of the nation. The publication of the *Oeuvres complètes de François Arago* (Paris Baudry, 1854–1862, 17 Vol.) was in line with this aim. Arago's role was both scientific and political, and he had been an excellent promoter of science (1786–1853). This nationalistic aspect was not a French prerogative: it is much more evident in the publication of the works of Gauss. On the other hand, nationalism connected to publication of complete works of great scientists of the past made more sense then than it would in these present times of globalization. The publication of works of Euler and the Bernoullis, for some years entrusted to private funds in Switzerland and only recently newly funded by Swiss institutions, is being dragged out ad infinitum, with various changes in editorial criteria, which make them more like a collection of volumes rather than a harmonious edition (Kleinert and Mattmüller 2007).

In Italy, national editions are underfunded, and so, without continuity of funding, various solutions have had to be found for publications. The *Accademia Nazionale dei Lincei* has promoted the printed publication of *Opere matematiche. Memorie e note di Guido Castelnuovo* (Vol 4, 2002–2007), while the *Lettere e i quaderni delle lezioni dell'archivio di Guido Castelnuovo*, edited by P. Gario, are available on the net:
http://operedigitali.lincei.it/Castelnuovo/Lettere_E_Quaderni/menu.htm

Four national editions regarding the mathematical sciences are currently in progress:
National Edition of the works of Federigo Enriques:
http://enriques.mat.uniroma2.it/
National Edition of *Mathematica italiana*[2]:
http://mathematica.sns.it/
National Edition of the works and correspondence of Ruggiero Giuseppe Boscovich:
http://www.edizionenazionaleboscovich.it/
National Edition of the mathematical work of Francesco Maurolico:
http://www.maurolico.it/Maurolico/index.html

As far as the publication of the complete works of Lagrange is concerned today, a minimal programme could be given space on an institutional site, in Turin, Berlin or Paris, the three cities where his mathematical work was developed, to upload:

[2]The digital library includes about four hundred works of Italian mathematicians, rare or not easily accessible, printed from the Renaissance to the beginning of the twentieth century.

a. The links to the edition of the *Oeuvres de Lagrange* and to all of the printed editions that appeared in his lifetime
b. All of Lagrange's unedited works
c. The scanning of the 16 volumes of Lagrange's manuscripts in collaboration with the *Institut de France*
d. A Lagrange bibliography that includes the historic and scientific works truly related to Lagrange.[3]

The *Principi di analisi sublime*, the letters and unedited works printed after the publication of the *Oeuvres*, could be collected into a single volume. In this case, the works of Cauchy could be taken as an example, giving historians the opportunity to study texts, at times difficult to find, which are essential for the publication of a sadly lacking scientific biography of Lagrange. It could be a sort of continuation of the *Oeuvres*, but, based on philological editing and editorial criteria, quite unlike those of the nineteenth century edition.

As historians of mathematics, we can only hope for a better solution: a new edition of the complete works of Lagrange, starting from his edited and unedited correspondence, as Sarton wished. It would be a very demanding job, even for a team of scholars, but very useful for reviving studies on the science and culture of the eighteenth century. Something similar to what is being carried out in France with the publication of the works of d'Alembert:

http://dalembert.academie-sciences.fr/

Such a work, however, is not possible without continuous and considerable intervention on the part of public institutions, which only France would seem to be able to guarantee, but for how much longer, we do not know. As far as the publication of complete works is concerned, as Plato pointed out with regard to solid geometry in the Greece of his day, the question of public intervention once again reappears:

> There is no state that sets any value on it, and so, being difficult, it is not pursued with energy; and research is not likely to progress without a director, who is difficult to find and, even if found, is unlikely to be obeyed in the present intolerant mood of those who study the subject. But under the general direction of a state that set a value on it, their obedience would be assured, and research pressed forward continuously and energetically until the problems are solved. Even now, with all the neglect and inadequate treatment it has suffered from the students who do not understand its real uses, the subject is so attractive that it makes progress in spite of all handicaps, and it would not be surprising if its problems were solved.[4]

[3]Something similar was carried out for the works of Lavoisier: http://www.lavoisier.cnrs.fr/index.html

[4]Plato, *The Republic*, VII, 526–527 (translated by H.D.P. Lee made from the Oxford text, Adam's edition, The Republic of Plato; C.U.P. 1926, The Penguin Classics).

Bibliography

Borgato, M.T. 1989. *La storia delle Matematiche nell'opera di Lagrange*, in *Pietro Riccardi (1828–1898) e la storiografia delle matematiche in Italia*. Atti del Convegno, a cura di F Barbieri, F Cattelani Degani, Modena, Università degli Studi, pp. 107–131.

———. 2013. Lagrange et les fonds de pension pour les veuves. *Bollettino di Storia delle Scienze. Matematiche* 23(1): 39–109 (Lagrange J L, *Solution d'un probleme sur les rentes viageres*, Ib. 111–149).

Borgato, M.T., and L. Pepe. 1987. Lagrange a Torino (1750–1759) e le sue lezioni inedite nelle R. Scuole di Artiglieria. *Bollettino* di Storia *delle Scienze. Matematiche* 7(2): 3–43 (Giuseppe Luigi Lagrange, *Principi di Analisi Sublime*. edizione a cura di Maria Teresa Borgato, ivi, 45–200).

———. 1988. Una memoria inedita di Lagrange sulla teoria delle parallele. *Bollettino di Storia delle Scienze. Matematiche* 8 (2): 307–335.

———. 1989. Sulle lettere familiari di Giuseppe Luigi Lagrange. *Bollettino di Storia delle Scienze. Matematiche* 9: 193–318.

———. 1990. L'inventaire des manuscrits de Lagrange et la mécanique avec l'édition du manuscrit de Lagrange "Differentes notes sur les ouvrages de mecanique". *Atti della Accademia delle Scienze di. Torino. Classe di Scienze Fisiche, Matematiche e Naturali* 124 (suppl): 25–49.

Delsedime, P. 1971. La disputa sulle corde vibranti e una lettera inedita di Lagrange a Daniel Bernoulli. *Physics* 13: 117–146.

Edizioni. 1986. *Edizioni critiche e storia delle matematiche*, a cura di E Giusti e L. Pepe, Atti Convegno CIRM Trento 1985, Pisa, ETS.

Favaro, A. 1895. Sette lettere di Lagrange al P. Paolo Frisi. *Atti della Accademia delle Scienze di. Torino* 31: 182–194.

Hahn, R. 2013. *Correspondence de Pierre Simon Laplace (1849–1827)*, publiée et annotée, Brepols, Turnhout, vol. 2.

History. 2004. *History of science, history of text*. ed. K. Chemla. Dordrecht: Springer.

Juskevic, A.P., and R. Taton. 1980. *Leonhard Euler, Opera Omnia*, series IV, A, vol. 5, Basel, 359–518.

Kleinert, A., and Mattmüller, M. 2007. *Leonhardi Euleri Opera Omnia: a centenary project*, EMS Newsletter, September, 25–31.

Pepe, L. 1986a. Lagrange e la trattatistica dell'Analisi matematica. *Symposia Mathematica* 27: 69–99.

———. 1986b. Tre "prime edizioni" ed un'introduzione inedita della Théorie des fonctions analytiques di Lagrange. *Bollettino di Storia delle Scienze Matematiche* 6: 17–44.

———. 1986c. *Sull'edizione delle opere di Lagrange*, in *Edizioni* (1986), pp. 109–122.

———. 1991. *Angelo Genocchi e l'edizione della corrispondenza di Lagrange*, in *Angelo Genocchi e i suoi interlocutori scientifici* a cura di A. Conte e L. Giacardi, Torino, Dep. Subalpina di Storia Patria, 221–240.

———. 1992. Supplemento alla bibliografia di Lagrange: I "Rapports" alla prima classe dell'Institut. *Bollettino di Storia delle Scienze Matematiche* 12 (2): 279–301.

———. 1993. *Lagrange e i suoi biografi*, prefazione a F. Burzio, *Lagrange*, II edizione, Utet Libreria, XI–XLIII.

———. 2008. *Il giovane Lagrange e i fondamenti dell'analisi*, in *Sfogliando la Méchanique analitique*, a cura di Giannantonio Sacchi Landriani e Antonio Giorgilli, Milano, Edizioni Universitarie di Lettere Economia Diritto, 37–49.

———. 2012. *Lagrange, citoyen ou "sans papiers"?*, in *Aventures de l'analyse de Fermat à Borel. Mélanges en l'honneur de Christian Gilain*, ouvrage dirigé par Suzanne Féry, Nancy, Pun Editions Universitaires de Lorraine, 473–481.

———. 2015. *Lagrange tra meccanica e analisi*, in *Il sapere scientifico in Italia nel secolo dei Lumi*, a cura di Gianpiero Sironi, Alberto Conte, Gian Antonio Danieli, Venezia, Istituto Veneto di Scienze Lettere ed Arti, 59–70.

Pittarelli, G. 1908. *Due lettere inedite di Lagrange all'abate di Caluso*. Atti del IV Congresso Internazionale dei Matematici, Roma, vol. III, 554–556.

Rashed, R. 1988. *Lagrange historien de Diophante*, in *Sciences à l'époque de la Révolution française. Recherches historiques*. Travaux de l'équipe REHSEIS, édités, Paris, Blanchard.

Riccardi, P. 1897. Alcune lettere di Lagrange, di Laplace e di Lacroix dirette al matematico Pietro Paoli. *Mem Accad. Modena*, sez. 3, 1 sci, 105–129.

Sarton, G. 1944. Lagrange's Personality (1736–1813). *Proceedings of the American Philosophical Society* 88 (1944): 457–496.

Taton, R. 1974. Inventaire chronologique de l'oeuvre de Lagrange. *Revue d'Histoire des Sciences* 27: 1–36.

Verdier, N. 2011. *Editer des oeuvres complètes avec Gauthier-Villars au XIXème siècle*. Images des mathématiques, CNRS. http://images.math.cnrs.fr/Editer-des-OEuvres-completes-avec.html

Volterra, V., Loria, G., and D. Gambioli. 1912. *Opere matematiche del marchese Giulio Carlo de' Toschi di Fagnano*, Milano, vol. III, 179–213.

A Critical Survey and Inventory of the Edited Works of Carl Friedrich Gauss

Karin Reich and Elena Roussanova

Abstract
Gauss's works were edited by the Academy of Sciences of Göttingen; this work took 70 years, from 1863 to 1933. The main editors were Ernst Schering (1833–1897) and Felix Klein (1849–1925). Klein's collaborators finally finished the edition, consisting of 12 volumes. This paper describes the genesis of these works. Gauss's works are exceptional, because they include not only the published work, but also manuscripts, letters, commentaries, reviews, and so on. The situation is quite confusing and it is not easy to obtain a general overview. The reason for this is that Felix Klein did not begin where Schering ended. While Schering had a clear concept, which he followed, Klein and his followers chose to publish what they thought to be of interest. The result is an edition that cannot claim to be critical.

1 Introduction

If we consider, at the present time, a directory of the works of Carl Friedrich Gauss, we have Uta C. Merzbach to thank for the best and most comprehensive bibliography available (Merzbach 1984, pp. 1–53). Excluding translations, excerpts

This article is a revised, supplemented and abridged version of (Reich and Roussanova 2013).

K. Reich (✉)
Faculty of Mathematics, University of Hamburg, Hamburg, Germany
e-mail: reich@math.uni-hamburg.de

E. Roussanova
Saxonian Academy of Sciences, Leipzig, Germany
e-mail: Elena.Roussanova@medizin.uni-leipzig.de

M. T. Borgato et al. (eds.), *Mathematical Correspondences and Critical Editions*,
Trends in the History of Science, https://doi.org/10.1007/978-3-319-73577-1_8

and later publications, some 286 titles of separate publications by Gauss are listed there. Yet even the aforementioned bibliography does not encompass all of his published works.[1] If one takes a closer look inside the 12-volume edition of the works of Carl Friedrich Gauss (hereafter Gauss's works), approximately 1600 writings can be found listed there. What a striking difference! Both the works, as well as excerpts from works or other documents and materials, have been included, and yet not all of Gauss's printed works have been recorded. This clearly shows that we should first analyze the present situation and genesis of Gauss's works, in the manner in which they are presented to the reader and researcher, before criticism is raised.

2 The Present Situation on Gauss's Works

Gauss's works were published from 1863 to 1933 and cover 12 volumes, although volumes 10 and 11 each consist of 2 sub-volumes.

The edited works of well-known scholars, such as Leonhard Euler (1707–1783), Augustin Louis Cauchy (1789–1857), Wilhelm Weber (1804–1891), and so forth, largely contain the printed works of the author in question, arranged chronologically or systematically, sometimes both at the same time. Occasionally, an exchange of letters was also integrated into the edition, either complete exchanges of letters or simply parts. Carl Friedrich Gauss's edited works (1777–1855) differ greatly from this tried and tested formula.

Gauss's printed works consist of monographs or individual writings, essays, prefaces to monographs by other authors, tables and charts, which sometimes appeared alone or accompanied other essays, indexes of his own writings or that of others, i.e., reviews of works by other authors and not always clearly identified inserts into the works of other authors. It further includes short reports on observations made by himself or his colleagues in Göttingen, or measurements, as well as charts and illustrations (mostly lithographs), which were attached to many of his works, but sometimes also appeared additionally as "tables," and much more besides this. At the very least, one would expect to discover all of these contents in an edition of the collected works. However, this is not the case. Due to the complexity of the situation, it is extremely difficult to state the exact number of publications of works by Gauss. However, it is clear that not all printed writings have been incorporated into this edition.

Besides printed writings by Gauss, the following materials were additionally incorporated into Gauss's works:

a. posthumous material;
b. particular letters and excerpts from letters that Gauss wrote or received, but also including letters and excerpts from letters between third parties;

[1] Kenneth O. May considers Gauss to have published 323 works (May 1972, p. 300).

c. comments made by publishers, that is to say, third parties;
d. material for a scientific biography;
e. writings by third parties.

Generally speaking, the posthumous materials and letters are documents taken from Gauss's *Nachlass,* which are housed at the State and University Library of Göttingen. These are very extensive. Unfortunately, readers of Gauss's works are not able to ascertain the part of the *Nachlass* in which the original document is to be found. Only imprecise information, if any at all, is given about the location within Gauss's works. In addition, Gauss's *Nachlass* has, in the meantime, been re-organized and given new shelf-marks.

Also, the manner in which the document is presented in the editions makes it unclear whether or not it was reproduced in a shortened form. Furthermore, there are no indications as to the original context of the materials added to the edited works. The same can also be said of the letters and the excerpts from letters. In this case, readers of Gauss's works are once again not informed of the location of the respective letter, nor often whether or not the letter was reproduced in its complete form.

In addition, Gauss's works contain explanations by other authors, for example, observations, additions and comments, which the publishers of specific works have attached to or inserted into Gauss's text, as well as comments marked with [ ] within Gauss's text and related writings by third parties. There are also a further ten longer essays, the so-called "Materials for a Scientific Biography," which were added as a second part to Volumes 10 and 11, although the pagination was not continued. Each of these essays retains the original pagination, i.e., this material was only attached to the volumes and not integrated.

The volumes of Gauss's works do not contain an index. This will be addressed below (see Sect. 3.5).

3 The Genesis of the Edited Works

Carl Friedrich Gauss died on February 23rd, 1855, in Göttingen, where he was also buried. One of his last students, Ernst Schering, was given the task of initiating the publication of Gauss's works.

3.1 Gauss's Works Under the Aegis of Ernst Julius Schering

Ernst Christian Julius Schering (1833–1897), who was born in the Forsthaus Sandbergen near Bleckede on the Elbe (in the district of Lüneburg), had initially begun by studying construction and mechanical engineering at the polytechnic in

Hannover in 1850 before passing his university entrance examination at the lyceum in Hannover in 1853. Schering was a pupil of Gauss as early as 1852 (Patterson 2001; Wittmann 2009, p. 83).

In the winter term of 1852/1853, Gauss gave a lecture entitled "The Method of Least Squares," which was attended by 14 students. That is the highest recorded number of students to attend a Gauss lecture. One of the attendees was Ernst Julius Schering (Folkerts 2002, p. 91), who later, on November 17th, 1854, enrolled at Göttingen University to study mathematics (Ebel 1974, p. 194). Simultaneously, Schering also studied astronomy intensively.

In 1857, Schering received his Ph.D. from Göttingen University. His thesis "On the mathematical theory of electric currents: proof of general theorems of electrodynamics, in particular, the theory of induction from basic laws of electricity" (Schering 1857) won a prize from the Faculty of Philosophy at the Georgia Augusta (Göttingen University). From February 1857, Schering lived at the observatory in Göttingen (Wittmann 2009, p. 85). On June 26th, 1858, he qualified as a professor at Göttingen, with his thesis entitled "Concerning the Conformal Mapping of an Ellipsoid on a Plane" (Schering 1858). This thesis was also awarded a prize by the Faculty of Philosophy at the Georgia Augusta. In 1860, after Schering had rejected an offer from Gießen University, he became Associate Professor of mathematics and mathematical physics at Göttingen. On June 6th, 1868, he was appointed Full Professor and acted as one of the two directors of the Göttingen Observatory. From that time on, he lived in the west wing of the observatory in the same apartment where Carl Friedrich Gauss had once lived (Wittmann 2009, pp. 85–86). In 1869, Schering also became head of the Magnetic Observatory, which Gauss had had built on the grounds of the Astronomic Observatory (Patterson 2001).

Not only was Schering a mathematician, he was also a physicist, an astronomer and a geodesist, and thus met all the desired requirements for a publisher of Gauss's works. From 1859, having been commissioned by the Royal Society of Sciences in Göttingen, he focused on the publication of Gauss's works. In 1860, he became assessor of the mathematics class, and in 1862, he became a full member of the Society (cf. Fritsch and Schmeidler 2005; Wittmann 2009).

Under the aegis of Schering, a systematically and chronologically arranged edition consisting of seven volumes was produced. Volumes 1–6 were published in Göttingen between 1863 and 1874; Volume 7 was published in Gotha in 1871 (see Fig. 1). In this seven volume edition, Schering "made the real achievements of Gauss available to the whole scientific world, and in so doing, decisively influenced the wider development of mathematics" (Patterson 2001). These volumes brought Schering "international recognition as a mathematician and a mathematics historian" (Fritsch and Schmeidler 2005, p. 695). The edition produced by Ernst Julius Schering was structured as follows:

Volume	Year of publication	Volume title	Number of pages
1	1863	Disquisitiones arithmeticae	478
2	1863	Number Theory	504
3	1866	Analysis	499
4	1873	Probability Calculation and Geometry	492
5	1867	Mathematical Physics	642
6	1874	Essays on Astronomy	664
7	1871	Theoria Motus	291

Originally, a "complete edition of the works by Gauss" with a "complete index of all the works by the great geometrician" was to be produced within 5 years. In a report from January 1862, which appeared in the "Archiv der Mathematik und Physik" published by Johann August Grunert (1797–1872), it was announced that "printing has begun and will be accelerated such that the first six volumes[2] will be published within, at most, 5 years" (Grunert 1862, p. 188, 198). The tables of contents of the six volumes were also presented.

The seventh volume (Fig. 1), which was published in Gotha, is exceptional in that the Friedrich Andreas Perthes Publishing House owned the publication rights

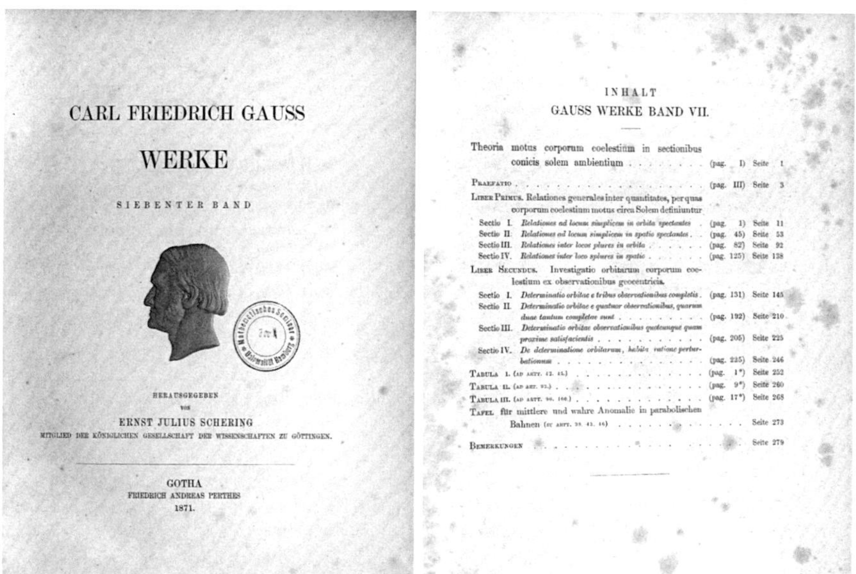

CARL FRIEDRICH GAUSS
WERKE
SIEBENTER BAND
HERAUSGEGEBEN VON
ERNST JULIUS SCHERING
MITGLIED DER KÖNIGLICHEN GESELLSCHAFT DER WISSENSCHAFTEN ZU GÖTTINGEN.
GOTHA
FRIEDRICH ANDREAS PERTHES
1871.

INHALT
GAUSS WERKE BAND VII.

Fig. 1 The title page and table of contents of the seventh volume of "The Works of Carl Friedrich Gauss," published by Ernst Julius Schering (Gotha 1871). Copy from the Library of Mathematics and History of Natural Sciences, Hamburg University

[2]The seventh volume, a reprint of "Theoria Motus" was not included here.

for Gauss's work "Theoria motus corporum coelestium in sectionibus conicis solem ambientium" printed in this volume. This work was, in fact, published in 1809 by Friedrich Perthes (1772–1843), whose publishing house at that time was still located in Hamburg.[3]

Schering admitted that he had omitted a few tables and charts from Gauss's works. In June 1871, in his comments on Volume seven, he wrote the following: "This copy of Theoria motus [. . .], in combination with the six volumes of Gauss's works, which I am publishing on behalf of the Royal Society of Sciences in Göttingen, constitutes a complete edition of the works by Gauss. With the exception of a few sets of tables and charts on geomagnetism, it contains all of Gauss's published works, plus all previously unpublished written work from the *Nachlass* [. . .] that seemed to me to be of scientific interest" (Gauss's works: 7: 1871, pp. 279–290, here p. 279). Volume 5, "Mathematical Physics," is particularly badly affected. For example, it contains Gauss's "General Theory of the Earth's Magnetism" (Gauß 1839), an essay that, in the original publication, was accompanied by seven charts, of which only one was reproduced in the works.

Between 1870 and 1880, a more or less expanded or respectively improved second edition of the first five volumes of Gauss's works was published, once more under the aegis of Schering.

Volume	Year of publication of the second edition	Number of pages
1	1870	478
2	1876	528
3	1876	499
4	1880	492
5	1877	642

Although the second edition was only to have corrections or minimal changes added to it (cf. Scriba 1977, p. 46), the second volume in particular contained significant additions. While the first edition of 1863 comprised a total of 504 pages,[4] the second edition of 1876 contained 528 pages. In the second edition, Schering added the following seven papers, which can be found on pages 497–518:

[3]It is sometimes said that Schering was *only* directly responsible for the publication of Volumes 1–6. In fact, Schering was also responsible for the publication of Volume 7 and the comments written by him (Gauß-Werke: 7: 1871, pp. 279–290). Although Felix Klein was to remark, in 1898, "that Ernst Schering, who was entrusted with the publication, released the first six volumes to the public," he then commented in 1906 in the second edition of Volume 7 on page 643: "It is well-known that in 1871, the [...] seventh volume [...] was published anew by Schering with the publisher F. A. Perthes in a form compatible with volumes I–VI and with the addition of several notes from the *Nachlass* [. . .]." On the negotiations with the publisher Perthes, see also Grunert (1862, p. 189).

[4]Pages 496–504 contained "comments" written by Ernst Schering.

Titles of the papers added to Volume 2 in the 2nd edition	Page references
Circuli quadratura nova	497–500
On the calculation of common logarithms	501–503
Quadratorum myrias prima	504–505
Indices of prime numbers in number theory	506
Table for the solution of the indeterminate equation $A = fxx + gyy$ using the exclusion method	507–509
Sectio octava. Quarumdam disquisitionum ad circuli sectionem pertinentium uberior consideratio	510–514
Letters from Gauss to Dirichlet[5]	514–518

None of the above seven titles had previously been published, but instead came from Gauss's *Nachlass* or from an exchange of letters found therein.

Schering had already incorporated several previously unpublished pieces from Gauss's *Nachlass* into Gauss's works, such as writings on elliptic functions in Volume 3. He had also integrated many of Gauss's reviews into this edition.[6]

On April 20th, 1877, the centenary of Gauss's birth, Schering gave a much acclaimed commemorative speech at a public meeting of the Göttingen Academy of Sciences entitled "On the Centenary Anniversary of Gauss's Birth." On November 2nd, 1897, Schering died in Göttingen, where he was also buried (cp. Wittmann 2009, pp. 86–88). As Felix Klein remarked in 1898, the editing of Gauss's *Nachlass* and the publication of his works initially "proceeded rather quickly [...], but then, however, adverse circumstances hampered the completion of the undertaking, so that today, when Schering is no longer with us, to all outward appearances, we have not come any further than before. This is all the more regrettable, since the thread of personal tradition, which connected us to Gauss, has been broken" (Klein 1899–1922 [1899:1], p. 128).

In 1897, in his obituary for Ernst Schering, Felix Klein wrote: "Schering's name will forever be associated with the collected works of Gauss [...]. The great care and expertise that Schering invested in his task are universally recognized; which makes it even more regrettable that he himself was not able to bring the undertaking to its final fruition. Apart from the formal conclusion, important parts of the *Nachlass* still remain unincorporated, above all, Gauss's theory on planetary disturbance and his correspondence relating to non-Euclidean geometry. The Göttingen Society of Sciences will ensure that, with the assistance of younger colleagues, this undertaking is actively promoted" (Klein 1897, p. 25). Regarding Schering's own research activities, Klein commented: "Schering continues, sometimes from this perspective, sometimes from another, to unravel questions that can be traced back to Gauss. This therefore meant that he had, from the outset, an extraordinary plethora of questions to grapple with [...]" (ibid. p. 26).

[5]This refers to two letters dated '13.9.1826' and '30.5.1828.'

[6]See Reich (1996).

3.2 Gauss's Works Under the Aegis of Felix Klein and in the Period Following Him

Felix Klein (1849–1925) was born in Düsseldorf and studied at Bonn University, where he earned his PhD in 1868. He continued his studies in Berlin and Paris and, in 1871, qualified as a professor at Göttingen University. In 1872, he was appointed Professor of Mathematics at Erlangen University. Then, in 1875, he moved to the Technical University of Munich, whereupon in 1880, he changed to Leipzig University, and finally, in 1886, took up the post of Professor at Göttingen University, where he remained for the rest of his life. Klein was not only an outstanding mathematician; he also pursued, to a large degree, historical and didactic studies. He began his famous "Lectures on the development of mathematics in the 19th century" with an extensive chapter on Carl Friedrich Gauss, which ended with the following quotation: "Gauss appears to me to be like the Zugspitze in the overall vista of our Bavarian mountains, as it appears to an observer from the North. The summits gradually rise from the East, peaking in the one giant colossus, which falls away steeply into the lowlands in a new formation, in which the foothills extend for miles and from which flowing water creates new life" (Klein 1926: 1, p. 62).

In 1898, after Schering's death, Felix Klein took charge of the publication of Gauss's works, which had been in abeyance for almost 20 years. Under Klein's supervision, great progress was made and he rapidly became the "heart and soul of the enterprise" (Born 1924). Klein had recruited some excellent collaborators for the new edition of Gauss's works, in particular, Martin Brendel (1862–1939) and Ludwig Schlesinger (1864–1933). Brendel was an astronomer, who, from 1898, was Associate Professor of theoretical astronomy at Göttingen University, where he lectured on actuarial theory and geodesy. In 1907, he moved to Frankfurt am Main. Schlesinger was a mathematician and a historian of mathematics. After working for a short time in Bonn, he moved to Klausenburg in 1897 and later, in 1911, to Gießen. In 1902, during his time in Klausenburg (today Cluj), he organized the centenary celebrations in honor of János Bolyai (1802–1860).

Due to ill health, Felix Klein was forced to relinquish his position on the publication of Gauss's works in 1922. Continuation of the work was subsequently assigned to the stewardship of Max Born (1882–1970), a later winner of the Nobel Prize, who was a professor at Göttingen University between 1921 and 1933. Born stressed that he, "as a physicist, was only to a minor extent an expert on the topics covered by Gaussian research," and had only accepted the honorable duty of continuing work on the edition because "Klein had promised him his advice and help in the future" (Born 1924, p. 25). Klein died on June 22nd, 1925, in Göttingen. He did not live to see the publication of the last volumes of the publication, nor thus the completed edition in 1933. From around 1928, the work was led by the mathematician Richard Courant (1888–1972), who was appointed the successor to Felix Klein as professor and later as head of the mathematics institute in Göttingen. In 1933, along with Born, Courant left Germany.

Brendel and Schlesinger immediately took up the responsibility for the continuation and completion of the publication of Gauss's works. At the end of Volume 10.2, which was the final volume published in 1933, the two scholars noted: "Only now, with these two volumes of essays, has the publication of GAUSS's collected works reached the satisfactory conclusion that KLEIN, [. . .] had essentially referred to as the desired objective in the sense that it concerns a task that was set a long time ago by the Society of Sciences in Göttingen. In particular, these essays provide further confirmation that GAUSS's work, although it dates back over half a century, does not belong to a period of science in the past, but is directly related to present-day concerns" (Gauss's works: 10.2: 1922–1933, in the unpaginated part; cf. Scriba 1977, p. 50).

Under the aegis of Felix Klein and his colleagues and successors, the following volumes of the edited works were published:

Volume	Year	Volume title	Number of pages
6	1907–1910	Anastatic reprint (Fig. 2)	664
7	1906	Theoretical astronomy	650
8	1900	Additions to Volumes I–III	458
9	1903	Geodesy. Continuation of Volume IV	528
10.1	1917	Additions to pure mathematics	586
10.2	1922–1933	Seven essays on Gauss's scientific work in the fields of pure mathematics and mechanics	75 + 222 + 18 + 123 + 95 + 76 + 61 [=672]
11.1	1927	Additions to physics, chronology and astronomy	518
11.2	1924–1929	Three essays on Gauss's scientific work in the fields of geodesy, physics and astronomy	165 + 217 + 259 [=642]
12	1929	Miscellaneous, including the "Atlas of the Earth's Magnetism"	415

The second edition of Volume Seven (1906)—"Theoretical Astronomy"—now contained not only the "Theoria motus," but also further relevant parts of the *Nachlass*, as well as numerous letters and excerpts from letters concerning astronomy. In Schering's edition of 1871, Volume Seven comprised only 290 pages, whereas Klein's second edition, published in 1906, contained 650 pages!

After 1907, a new edition of Volume Six was required, as it was out of print. A revised edition of the volume was not planned, as this would have taken a long period of time. Thus, from 1907 to 1910, an anastatic, or relief print,[7] was published

[7] A printing process whereby an original document is used.

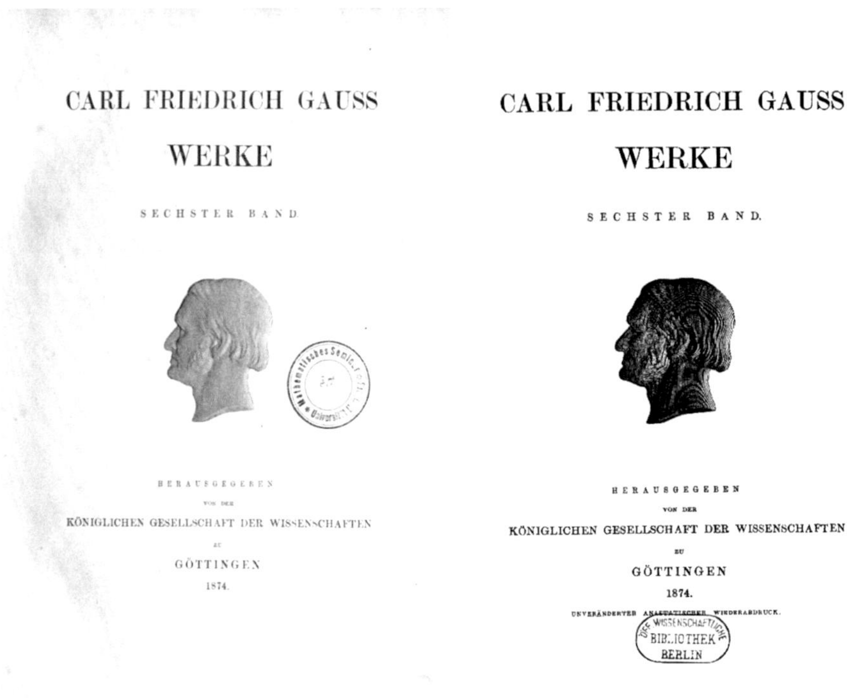

Fig. 2 The title page of Volume Six of "The Works of Carl Friedrich Gauss" (1874) and its unaltered anastatic reprint (1907–1910). 1. Copy belonging to the Library of Mathematics and History of Natural Science, Hamburg University. 2. Copy belonging to the Berlin State Library—Prussian Cultural Heritage

(Fig. 2). It was, however, inevitable that there would be a number of reproduction errors (Klein 1899–1922 [1910:8], p. 444). The anastatically printed edition did not, however, include a corrigenda.

Primarily under Klein and after Klein, numerous documents from Gauss's *Nachlass*, as well as letters and excerpts from letters, were added to the collected works of Gauss. This is, in fact, to be evaluated positively, as only then did it become clear how important the *Nachlass* and the exchange of letters are for Gauss's work.[8] It is worth considering that there are approximately 1600 independent titles in Gauss's works, of which only about 286 are articles published by Gauss. However, with the volumes published by Klein and his colleagues, Schering's hard-earned and

[8]"Using extracts from Gauss's correspondence, Klein [. . .] also tried for the first time to shed light on the genesis of Gaussian thinking" (Scriba 1977, p. 47). It should be noted here that, under Schering, several letters had already been added to the edition.

desired systematic and chronological order was irrevocably lost. There is no longer an identifiable classification system for all of the volumes or, to put it more clearly, confusion reigns. To make matters worse, there is no comprehensive index for the twelve volumes, making it extremely difficult to find particular essays, letters, etc., in the complete edition.

3.3 Change of Publisher

As already discussed, the first six volumes of the first edition were published by the Royal Society of Sciences in Göttingen; the seventh volume was published by the publishing house Friedrich Andreas Perthes in Gotha. On taking over Gauss's edited works, Felix Klein immediately announced a change of publisher to B. G. Teubner Verlag in Leipzig: "Meanwhile, we hope to revitalize sales of the six volumes already published by handing them over to a capable and effective publisher." He further commented: "Negotiations with Teubner, the bookseller's, have, in the meantime, reached a successful conclusion, and from now on, the first six volumes of the collected works can be obtained from Teubner's and no longer from the sales office of Göttingen University" (Klein 1899–1922 [1899:1], p. 132).

In the promotional prospectus signed by Felix Klein, which appeared on October 1st, 1900, entitled "Continuation of the collected edition of works by Gauss managed by the Royal Society of Sciences in Göttingen" (Fig. 3), Klein declared that the aim was "to revitalize the undertaking and to promote it in such a way that a continual development and a determinate conclusion of the same can, with certainty, be anticipated."[9]

Volumes eight to ten (first section, Volume 10.1), as well as the reprint of the seventh volume, were published on commission by B. G. Teubner Verlag in Leipzig.

Under Max Born, there was an adjustment to the operation of the business; the publisher on commission was changed once more. In the annual report of 1923/24, Born disclosed that the publisher Julius Springer had agreed "to take over the commission publishing under terms that were considerably more favorable than the old contract with Teubner" (Born 1924, pp. 25–26). Volumes 10.2–12 were published on commission by Julius Springer Verlag in Berlin. At the same time, in 1936, Julius Springer Verlag was prepared to take over the production of the planned index volume "under the same terms as the earlier volumes of Gauss's works" (Kienle 1936–1944 [1936], p. 45). However, this volume never materialized.

[9]On the topic of C. F. Gauss and B. G. Teubner, see Wußing (2011).

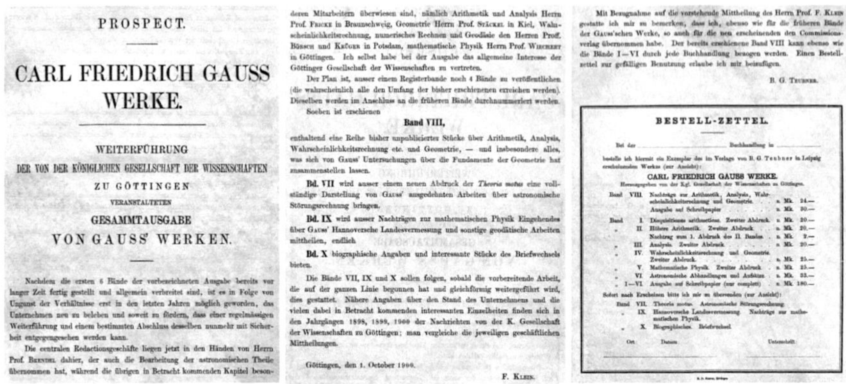

PROSPECT.

CARL FRIEDRICH GAUSS WERKE.

WEITERFÜHRUNG DER VON DER KÖNIGLICHEN GESELLSCHAFT DER WISSENSCHAFTEN ZU GÖTTINGEN VERANSTALTETEN GESAMMTAUSGABE VON GAUSS' WERKEN.

Nachdem die ersten 6 Bände der vorbezeichneten Ausgabe bereits vor langer Zeit fertig gestellt und allgemein verbreitet sind, ist es in Folge von Ungunst der Verhältnisse erst in den letzten Jahren möglich geworden, das Unternehmen neu zu beleben und soweit zu fördern, dass einer regelmässigen Weiterführung und einem bestimmten Abschluss desselben nunmehr mit Sicherheit entgegengesehen werden kann.

Die centralen Redactionsgeschäfte liegen jetzt in den Händen von Herrn Prof. Brendel dahier, der auch die Bearbeitung der astronomischen Theile übernommen hat, während die übrigen in Betracht kommenden Kapitel besonderen Mitarbeitern überwiesen sind, nämlich Arithmetik und Analysis Herrn Prof. Fricke in Braunschweig, Geometrie Herrn Prof. Stäckel in Kiel, Wahrscheinlichkeitsrechnung, numerisches Rechnen und Geodäsie den Herren Proff. Börsch und Krüger in Potsdam, mathematische Physik Herrn Prof. Wiechert in Göttingen. Ich selbst habe bei der Ausgabe das allgemeine Interesse der Göttinger Gesellschaft der Wissenschaften zu vertreten.

Der Plan ist, ausser einem Registerbande noch 4 Bände zu veröffentlichen (die wahrscheinlich alle den Umfang der bisher erschienenen erreichen werden). Dieselben werden im Anschluss an die früheren Bände durchnummeriert werden.

Soeben ist erschienen

Band VIII,

enthaltend eine Reihe bisher unpublicierter Stücke über Arithmetik, Analysis, Wahrscheinlichkeitsrechnung etc. und Geometrie, — und insbesondere alles, was sich von Gauss' Untersuchungen über die Fundamente der Geometrie hat zusammenstellen lassen.

Bd. VII wird ausser einem neuen Abdruck der *Theoria motus* eine vollständige Darstellung von Gauss' ausgedehnten Arbeiten über astronomische Störungsrechnung bringen.

Bd. IX wird ausser Nachträgen zur mathematischen Physik Eingehendes über Gauss' Hannoversche Landesvermessung und sonstige geodätische Arbeiten mittheilen, endlich

Bd. X biographische Angaben und interessante Stücke des Briefwechsels bieten.

Die Bände VII, IX und X sollen folgen, sobald die vorbereitende Arbeit, die auf der ganzen Linie begonnen hat und gleichförmig weitergeführt wird, dies gestattet. Nähere Angaben über den Stand des Unternehmens und die vielen dabei in Betracht kommenden interessanten Einzelheiten finden sich in den Jahrgängen 1898, 1899, 1900 der Nachrichten von der K. Gesellschaft der Wissenschaften zu Göttingen; man vergleiche die jeweiligen geschäftlichen Mittheilungen.

Göttingen, den 1. October 1900.

F. Klein.

Mit Bezugnahme auf die vorstehende Mittheilung des Herrn Prof. F. Klein gestatte ich mir zu bemerken, dass ich, ebenso wie für die früheren Bände der Gauss'schen Werke, so auch für die neu erscheinenden den Commissionsverlag übernommen habe. Der bereits erschienene Band VIII kann ebenso wie die Bände I—VI durch jede Buchhandlung bezogen werden. Einen Bestellzettel zur gefälligen Benutzung erlaube ich mir beizufügen.

B. G. Teubner.

BESTELL-ZETTEL.

Bei der .... Buchhandlung in ....

bestelle ich hiermit ein Exemplar des im Verlage von B. G. Teubner in Leipzig erscheinenden Werkes (zur Ansicht):

CARL FRIEDRICH GAUSS WERKE.

Herausgegeben von der Kgl. Gesellschaft der Wissenschaften zu Göttingen.

Band	VIII.	Nachträge zur Arithmetik, Analysis, Wahrscheinlichkeitsrechnung und Geometrie	à Mk. 24.—
		Ausgabe auf Schreibpapier	à Mk. 30.—
Band	I.	Disquisitiones arithmeticae. Zweiter Abdruck	à Mk. 20.—
"	II.	Höhere Arithmetik. Zweiter Abdruck	à Mk. 20.—
		Nachtrag zum 1. Abdruck des II. Bandes	à Mk. 2.—
"	III.	Analysis. Zweiter Abdruck	à Mk. 20.—
"	IV.	Wahrscheinlichkeitsrechnung und Geometrie. Zweiter Abdruck	à Mk. 25.—
"	V.	Mathematische Physik. Zweiter Abdruck	à Mk. 25.—
"	VI.	Astronomische Abhandlungen und Aufsätze	à Mk. 33.—
"	I—VI.	Ausgabe auf Schreibpapier (nur complett)	à Mk. 180.—

Sofort nach Erscheinen bitte ich mir zu übersenden (zur Ansicht):

Band VII. Theoria motus. Astronomische Störungsrechnung.

" IX. Hannoversche Landesvermessung. Nachträge zur mathematischen Physik.

" X. Biographisches. Briefwechsel.

Ort. Datum. Unterschrift.

Fig. 3 Promotional prospectus by B. G. Teubner for Gauss's works, signed by Felix Klein on October 1st, 1900. Printed in: Wußing (2011, pp. 224–226). By kind permission of the 'Edition am Gutenbergplatz', Leipzig (Sincere thanks to Herrn Jürgen Weiß (Edition am Gutenbergplatz Leipzig) for sending the prospectus)

3.4 Changes to the Concept of the Edited Works: Reports on the Present Situation Regarding the Publication of Gauss's Works

From the very beginning of his work with the "Gauss Project," Felix Klein publicly presented detailed reports on the progress of the publication of Gauss's works. These presentations, which were given at public sessions of the Royal Society of Sciences in Göttingen, were entitled "Reports on the present situation regarding the publication of Gauss's works" and published in "Communication from the Royal Society of Sciences in Göttingen: Business announcements" (Nachrichten von der (Königlichen) Gesellschaft der Wissenschaften zu Göttingen: Geschäftliche Mitteilungen), and then also printed in the "Mathematical Annals" Mathematische Annalen), published by Felix Klein, amongst others. This publication, one of the highest ranked mathematics journals in the world, was at that time published by B. G. Teubner. Under the aegis of Klein, a total of 14 comprehensive reports appeared (Klein 1899–1922). Klein presented the first report to the Society on April 30th, 1898; this was printed in the "Mathematical Annals" in 1899 (Klein 1899–1922 [1899:1], pp. 128–133). The fourteenth and final report by Klein appeared in the "Mathematical Annals" in 1922 (Klein 1899–1922 [1922:14], pp. 326–328). The next, technically the series' fifteenth, which consists of just two pages, was written by Max Born (Born 1924). Between 1928 and 1933, only very short "Gauss Commission Reports" were published, signed by the mathematician Richard Courant (Courant 1928–1933).

The comprehensive reports by Felix Klein afford a deep insight into the implementation, alteration and developments in the concept of the complete edition of Gauss's work. We also learn of several findings in Gaussian research, as well as details on the acquisition of Gauss-related historical documents, letters, manuscripts, Gauss's diary,[10] etc. Felix Klein harbored great plans to broaden the edition to include material from Gauss's scientific *Nachlass*, as well as a "scientific biography," which will be examined more closely below.

Here is an overview of the evolving concept of Gauss's works based on reports presented by Felix Klein:

Concepts for the Structure of Gauss's Edited Works

1899	In: Klein 1899–1922 [1899:1]
Volume 7 (=new edition)	Astronomy
Volume 8	Scientific additions
Volume 9	Biographical material
Supplementary volume	Comprehensive indexes
1902	In: Klein 1899–1922 [1902:3]
Volume 7 (=new edition)	Astronomy
Volume 8	published 1900
Volume 9	Geodesy, Physics
Volume 10	Biographical material
Supplementary volume	Comprehensive Indexes
1907	In: Klein 1899–1922 [1907:7]
Volume 7 (=new edition)	published 1906
Volume 9	published 1903
Volume 10	Final volume (!), Papers from the *Nachlass* on objects of practical and stellar astronomy, chronology, theoretical physics and optics, messages from exchanges of letters, biographical material, report on the publication of Gauss's works, report on the Gauss archive
Supplementary volume	Comprehensive general index
1910	In: Klein 1899–1922 [1910:8]

(continued)

[10]Gauss's 19-page mathematical diary appeared as a printed facsimile in Volume 10.1 of Gauss's works published by B. G. Teubner in 1917. This print was prefixed by introductory remarks from Klein and Schlesinger: "Reproduction of the diary (journals of notes) by C. F. Gauss 1796 Mar. 30–1814 Jul. 9" (Gauss's Works: 10.1: 1917, after p. 482, unpaginated) also "Print of the diary (journals of notes) with comments" (Gauss's Works: 10.1: 1917, pp. 483–574). The comments stem from Klein, Schlesinger, Bachmann, Stäckel, Loewy, Dedekind, Brendel and Galle. For the edition of the diary of the great mathematician discovered by Paul Stäckel in the *Nachlass* of a grandson of Gauss in Hameln, see, e.g., Klein (1899–1922 [1900:2], pp. 46–48 and also [1918:12], pp. 417–419).

Volume 6 (=reprint)	Anastatic reprint (published 1907–1910)
Volume 10	Final Volume (!), in preparation
Supplementary volume	Comprehensive general index
1912	In: Klein 1899–1922 [1912:9]
Volume 10	Four sections 1. biographical material of a general nature 2. monographic illustrations in the following order: arithmetic, analysis, geometry, geodesy, physics, astronomy 3. print of the scientific diary with extensive comments 4. interesting points from correspondence
Supplementary volume	Comprehensive general index

In 1913, in his tenth "Report on the present situation regarding the publication of Gauss's works," Klein confided that a few of the essays that had been scheduled for printing in the tenth volume would initially be published in separate booklets under the collective title "Material for a scientific biography of Gauss, collected by F. Klein and M. Brendel" by the publishing house B. G. Teubner in Leipzig (Klein 1899–1922 [1913:10], pp. 410–412).

But as early as 1916, Klein reported: "The wealth of material has resulted in an eleventh volume being planned in addition to the tenth. The tenth volume should comprise two sections, the first of which will be a series of writings from the *Nachlass* on arithmetic and algebra, on analysis, geometry and physics, as well as a copy of the diary with detailed comments, whilst essays on Gauss's scientific work are destined for the second section. The eleventh volume will then contain a broad biography of Gauss, with a copy of relevant excerpts from letters, a description and catalogue of the complete handwritten *Nachlass,* a history of the publication of the works and an index" (Klein 1899–1922 [1916:11], pp. 303–304). In addition, Klein reported that the esteemed mathematician and historian of mathematics Paul Stäckel (1862–1919) had agreed to write an additional essay on "Gauss as a member of Göttingen University" (ibid. p. 305). This essay was to appear in the eleventh volume of the works as a discourse included among the "material" and was to consist of the following sections[11]:

[11]The death of Paul Stäckel precluded these plans. Not until much later was this theme revived in a paper by Menso Folkerts "Carl Friedrich Gauss's activities at Göttingen University (Folkerts 2002).

1.	Gauss as a student at Göttingen
2.	Gauss's position on teaching in academia
3.	His appointment to Professor at Göttingen
4.	Academic career at Göttingen
5.	Gauss's lectures arranged according to contents
6.	Gauss's students
7.	Dissertations supervised by Gauss
8.	Prize Competitions
9.	Gauss in the Faculty of Philosophy and in the Society of Sciences
10.	Gauss's position on academic questions

After certain delays due to the First World War, preparations for printing the tenth volume continued, and in 1918, in his twelfth report, Klein announced that Volume eleven would also comprise two sections (Klein 1899–1922 [1918:12], p. 16). Numerous details regarding the concept of the first section of Volume eleven can be found in Klein's two final reports (Klein 1899–1922 [1921:13], as well as [1922:14]). In 1922, there was no mention of a twelfth volume, and Volume eleven was still referred to as the final one. The first mention of a twelfth volume appeared in a short communication from Richard Courant in the annual report of 1929/1930 (Courant 1928–1933 [1930], p. 29). In his penultimate communication in the annual report of 1930/1931, Courant recorded: "As concerns scientific work by Gauss, work on finalizing the edition of Gauss's complete works is finished. What is lacking is, in the main, a few essays on the scientific biography of Gauss, as well as an index" (Courant 1928–1933 [1931], p. 37).

A few more details will help to elaborate upon Felix Klein's intention, as quoted above, to develop a so-called "scientific biography" within the framework of the collected works. Academic specialists were commissioned by Klein to prepare historical articles on Gauss's various fields of academic work. These individual essays were initially published externally between 1911 and 1920 in the series "Material for a scientific biography of Gauss. Collected by F. Klein, L. Schlesinger und M. Brendel." Many of these articles appeared in "Communication from the Society of Sciences in Göttingen" (Nachrichten der Gesellschaft der Wissenschaften zu Göttingen) or at least in their supplementary brochures (Hefte) (see appendix). However, not all of the articles published there were later assimilated into Gauss's works, and several articles that were not previously published in the "Nachrichten" or supplementary brochures (Hefte) can be found in Gauss's works.

The problem regarding an index, that is, a comprehensive general index, has, to this day, not been solved.

3.5 The Index

None of the volumes of Gauss's works contains an index.[12] Felix Klein had originally planned a supplementary volume, the tenth volume, to be the index. Hence, in his first report on the contemporary situation regarding the publication of Gauss's works in 1898, he announced: "The *supplementary volume* will finally contain a comprehensive index.—Provided there is no unforeseen disruption, the entire undertaking should be completed in around 3 years."[13] On completion of the Edited Works in 1933, the idea of an index, that is, a complete index, had still not been abandoned.

The idea of an "index" was raised again and again in the "Reports from the Gauss Commission," which, after 1936, were published in "Communication from the Society of Sciences in Göttingen" (in the annual report of 1935/1936) and which were signed by Hans Kienle, the Chairman of the Gauss Commission. The astronomer and astrophysicist Hans (Johann Georg) Kienle (1895–1975) was Professor of Astrophysics and Astronomy at Göttingen University in 1924 and, at the same time, became Director of the Göttingen Observatory in 1925; he moved to Potsdam in 1939. In Kienle's first "Gauss Commission Report," it is stated: "On December 6th, 1935, a meeting was held at the observatory, where the guidelines for the planned index-volume to conclude the edition of Gauss's works were established. [...] Resulting from this meeting and after clarification of several questions in writing, the following resolutions were adopted: [...] Volume 13, being the definitive index volume, will complete the collected works. The question of Gauss's scientific bibliography is not to be linked with this task"[14] (Kienle 1936–1944 [1936], p. 45).

Progress on the different indexes was described in the annual report for the business year 1936/1937 as follows: "Preparatory work, which began last year, on the Index-Volume XIII of Gauss's works was continued by Messrs Bessel-Hagen,[15]

[12]The index of proper names and the index on subject matter accompanying the paper by Ludwig Schlesinger (Gauss's Works: 10.2: 1922–1933, the second paper, pp. 211–216), as well as the index accompanying the paper by Clemens Schaefer (Gauss's Works: 11.2: 1924–1929) second paper, pp. 212–217), are exceptions and can in no way be described as an index or a partial index of Gauss's Works.

[13]Statement made by Felix Klein at the public meeting of the Göttingen Academy of Sciences on April 30, 1898 (cited according to: Klein 1899–1922 [1899:1], p. 132).

[14]At this meeting, it was also decided: "In order to satisfy the demand for a conclusive biographical account, Sartorius von Waltershausen's biography on Gauss's death will be reprinted with supplementary notes." The plan to publish a supplemented edition of Sartorius von Waltershausen's biography "Gauss zum Gedächtniss" ("To the memory of Gauss") has only recently materialized (Reich 2012).

[15]Erich Bessel-Hagen (1898–1946) earned his doctorate under Constantin Carathéodory in Berlin in 1920, was private assistant to Felix Klein in Göttingen from 1921 to 1924, assistant to Helmut Hasse at Halle University in 1927, and in 1931, became associate professor and, later, in 1939, adjunct professor at Bonn University.

Geppert[16] and Kopff,[17] along with their colleagues; the ultimate completion of the index can be expected this year. Work is under way on three indexes: (1) a detailed index of Gauss's writings printed in the "Works," which contains all necessary bibliographic data, listed in order of subject area and chronology; (2) an index of names in Gauss's works, with a reference to the profession and chronology of the relevant persons; (3) a detailed guide to Gauss's works, which enables the reader not acquainted with the Gauss edition to find any desired subject matter in old or new terminology. It will be broken down into four single indexes; mathematics including geodesy, astronomy, physics and biography" (Kienle 1936–1944 [1937], p. 45).

This task has, in fact, to this very day still not been completed. A complete index would, however, be very difficult, if not impossible, to create, as Gauss's works comprise a complex variety of material, and the lack of a consistent system of page numbering makes it virtually unthinkable. The difficulty of the task is evident from Kienle's reports:

1939 "The preliminary work for the index-volume, as undertaken by Mr. Geppert und Mr. v. Schelling, has been completed. A part of the material to be delivered by Herrn Bessel-Hagen is still missing" (Kienle 1936–1944 [1939], p. 101).

1940 "Work on the index-volume could not be continued further, as Herr Bessel-Hagen was not in a position to complete the part he had undertaken to an extent that a definitive structural order could have been put in place" (Kienle 1936–1944 [1940], p. 69).[18]

1942 "Resumption of work on the index-volume must be postponed until after the war" (Kienle 1936–1944 [1942], p. 117).

1943 "The present situation as regards the publication of the index-volume remains unchanged from last year's report" (Kienle 1936–1944 [1943], p. 164).

1944 "Work on the index-volume could not be continued" (Kienle 1936–1944 [1944], p. 109).

Christoph Scriba mentions that Harald Geppert and Erich Bessel-Hagen had managed to write various notes for an index, which fell victim, however, to the Second World War (Scriba 1977, p. 54).

[16]Harald Geppert (1902–1945) worked in the mathematics department at the University of Gießen from 1935 to 1940, and was, from 1940 to 1945, full professor at Berlin University.

[17]August Kopff (1882–1960) was Director of the Astronomical Calculation Institute (ARI) between 1924 and 1954, initially in Berlin and later in Heidelberg.

[18]In 1940/1941, the commission responsible for the publication of Gauss's works consisted of: Kienle (Chairman), David Hilbert (1862–1943), Gustav Herglotz (1881–1953) and Helmut Hasse (1898–1979).

4 Further Editions of Gauss's works

4.1 The Reprint Editions by the Publisher Georg Olms

In 1973 (1st reprint), and again in 1981 (2nd reprint), photo mechanical reprints of Gauss's works, which had been produced by the publisher Georg Olms (Hildesheim, New York), were published. Unfortunately, these reprints were only based on volumes of the first edition of Gauss's works. One exception was the seventh volume, which was a reprint of the second edition of 1906. However, the significantly expanded second edition of Volume two was unfortunately disregarded in both the first and second reprints (see Sect. 3.1). Volumes 11 and 12, which originally appeared as 2 half volumes, were printed in 1 volume. Both reprinted editions by Georg Olms were published without amendments and supplements, i.e., also without an index.[19]

4.2 Digital Editions of Gauss's Works

Digitized editions of works by Carl Friedrich Gauss are available to today's reader. In this context, mention should be made primarily of the digital copies at the Göttingen State and University Library (GDZ Göttinger Digitalisierungszentrum)[20] and the National Library of France (Gallica).[21] The volumes produced by the Göttingen Digitization Centre (GDZ) are also displayed on the DFG-Viewer.[22] No details have been given on the editions upon which the volumes available on the web are based. The digitized texts are normally presented in the internet as graphic, i.e., PDF files. In other words, research or searching within Gauss's works is not possible.

Volume	Digital Source GDZ or DFG-Viewer	Digital Source Gallica
1	1st edition of 1863	1st edition of 1863
2	1st edition of 1863	1st edition of 1863
2*	Appendix to first print of Volume 2 of 1876, pp. 495–528	Missing
3	1st edition of 1866	1st edition of 1866
4	1st edition of 1873	1st edition of 1873

(continued)

[19]http://www.olms.de/artikel_2669.ahtml?T=1330207733513

[20]http://resolver.sub.uni-goettingen.de/purl?PPN235957348

[21]http://gallica.bnf.fr

[22]http://dfg-viewer.de

5	1st edition of 1867	1st edition of 1867
6	1st edition of 1874	1st edition of 1874
7	2nd edition of 1906	1st edition of 1871
8	1900 edition	1900 edition
9	1903 edition	1903 edition
10.1	1917 edition	Missing
10.2	1922–1933 edition	Missing
11.1	1927 edition	Missing
11.2	1924–1929 edition	Missing
12	1929 edition	Missing

5 The Project "Gauss Translated into German"

Carl Friedrich Gauss published many of his early works in Latin, and the edition of Gauss's works adopted the original publications. Owing to the fact that knowledge of Latin is no longer widespread, the publishing house Olms decided to publish three supplementary volumes entitled "Gauss translated into German," which contain a German translation of all of the works Gauss published in Latin. They are as follows:

Supplementary Volume 1: Carl Friedrich Gauss: Investigations into Higher Arithmetic. Edited and published in German by Hermann Maser. With an introduction, bibliography and index edited by Karin Reich. Hildesheim, Zürich, New York 2015, XVI+720 p.

Supplementary Volume 2: Carl Friedrich Gauss: Theory of the motion of the heavenly bodies moving around the sun in conic sections. Translated into German by Carl Haase. With an introduction, bibliography and index edited by Karin Reich. Hildesheim, Zürich, New York 2015, p. 5*–39*, pp. 1–279, p. A–F, pp. 1–96.

Supplementary Volume 3: Carl Friedrich Gauss: Miscellaneous. This volume comprises 15 essays by Gauss dating from 1799 to 1841. This volume also contains an introduction, bibliography and index. However, it will no longer be published by Olms, but instead by the Göttingen Academy of Sciences and Humanities and, in fact, in electronic form under "res doctae" (http://rep.adw-goe.de). The volume will comprise approx. 700 pages and will appear in 2019.

Hence, for the first time, these three supplementary volumes come with an index of proper names, a subject matter index and detailed bibliographies. This is, of course, only a first step.

It should be mentioned here that it is the aim of the Göttingen Academy of Sciences to make all three supplementary volumes available electronically.

6 Résumé

It must be said, for the record, that the classic edition of Gauss's works does not meet present-day criteria. It is a critical edition that can only be deemed unsatisfactory, and it is particularly unfortunate that only a selection of Gauss's extensive maps and tables is included in the works. Also, it has, to date, not been possible to search for Gauss's writings on particular topics, owing to the fact that, following Schering's departure, the editors did not keep to any systematic organisation of the material. This can be illustrated here by a particularly stark example. In 1837, on the occasion of the centenary celebration of the founding of Göttingen University, Gauss presented his bifilar magnetometer. Gauss's introduction to this speech is in Volume 12, whilst his explanation of the bifilar magnetometer appears in Volume 5.[23] Undoubtedly, the publishers of Gauss's works, Ernst Schering and Felix Klein, along with their colleagues and successors, did their best. At the time of the publication of the works, some things could, perhaps, not have been produced in any better form.

It is unrealistic to hope that, in the future, a new edition of Gauss's works will appear that does not have the faults of the old edition. We will continue to have to live with the old edition of the works.

However, in the twenty first century, several things can now be solved electronically for which there was no practical solution in the past. Problems that can be solved electronically include a complete table of contents of Gauss's works, as well as an index of proper names and subject matter, that is to say, a keyword index for all available volumes. Insufficient bibliographic details should also be supplemented and improved upon. All of these details could be held in an online database, and thus would help to solve the most pressing problems of the edited works. This is still up in the air, but nevertheless a task that may well be resolved in the foreseeable future.

Appendix: "Materialien für eine wissenschaftliche Biographie von Gauß"

The brochures (Hefte) I bis VIII "Materialien für eine wissenschaftliche Biographie von Gauß, gesammelt von F. Klein, M. Brendel und L. Schlesinger" were published by B. G. Teubner in Leipzig in the years 1911 to 1920 (Wußing 2011, p. 244).

Heft I

Bachmann, Paul: Über Gauss' zahlentheoretische Arbeiten. In: Nachrichten von der Königlichen Gesellschaft der Wissenschaften zu Göttingen aus dem Jahre 1911, Mathematisch-physikalische Klasse, Berlin 1911, pp. 455–508.

Reprinted in: Gauß-Werke: 10.2: 1922–1933 [1922], 75 p.

[23]Gauss's Works 12, pp. 109–110; Gauss's Works 5, pp. 357–373.

Heft II

Schlesinger, Ludwig: Fragmente zur Theorie des arithmetisch-geometrischen Mittels aus den Jahren 1797–1799. In: Nachrichten von der Gesellschaft der Wissenschaften aus dem Jahre 1912, Mathematisch-physikalische Klasse, Berlin 1912, pp. 513–546.

Not in: Gauß-Werke.

Heft III

Schlesinger, Ludwig: Über Gauss' Arbeiten zur Funktionentheorie. In: Nachrichten von der Königlichen Gesellschaft der Wissenschaften zu Göttingen aus dem Jahre 1912, Mathematisch-physikalische Klasse, Beiheft, Berlin 1912, 143 p.

Reprinted in: Gauß-Werke: 10.2: 1922–1933 [1933] (revised edition; from p. 1 to p. 118; Appendix (pp. 119–143) not in: Gauß-Werke.

Heft IV

Galle, Andreas: C. F. Gauß als Zahlenrechner (= ein Ausschnitt aus ,Über die geodätischen Arbeiten von Gauß). In: Nachrichten von der Königlichen Gesellschaft der Wissenschaften zu Göttingen aus dem Jahre 1912, Mathematisch-physikalische Klasse, Beiheft, Berlin 1917, pp. 1–24.

Not in: Gauß-Werke.

Heft V

Stäckel, Paul: Gauss als Geometer. In: Nachrichten von der Königlichen Gesellschaft der Wissenschaften zu Göttingen aus dem Jahre 1917, Mathematisch-physikalische Klasse, Beiheft, Berlin 1917, pp. 25–142.

Reprinted in: Gauß-Werke: 10.2: 1922–1933 [1923], 123 p.

Heft VI

Maennchen, Philipp: Die Wechselwirkung zwischen Zahlenrechnen und Zahlentheorie bei C. F. Gauss. In: Nachrichten von der Königlichen Gesellschaft der Wissenschaften zu Göttingen aus dem Jahre 1918, Mathematisch-physikalische Klasse, Beiheft, Berlin 1918, 46 p.

Gauss als Zahlenrechner. (Revised reprint of the treatise *Die Wechselwirkung zwischen Zahlenrechnen und Zahlentheorie bei C. F. Gauss*).

Reprinted in: Gauß-Werke: 10.2: 1922–1933 [1930], 76 p.

Heft VII

Brendel, Martin: Über die astronomischen Arbeiten von Gauß. 1. Abschnitt: Theoretische Astronomie. In: Nachrichten von der Gesellschaft der Wissenschaften aus dem Jahre 1919, Mathematisch-physikalische Klasse, Beiheft, Berlin 1919, 104 p.

Reprinted in: Gauß-Werke: 11.2: 1924–1929 [1929], pp. 145–254 as second part of: Brendel, Martin: Über die astronomischen Arbeiten von Gauss. In: Gauß-Werke: 11.2: 1924–1929 [1929], 258 p.

Heft VIII

Fraenkel, Adolf Abraham Halevi: Zahlbegriff und Algebra bei Gauß. Mit einem Anhang von A. Ostrowski in Göttingen. Zum ersten und vierten Gaußschen Beweise des Fundamentalsatzes der Algebra (pp. 50–59). In: Nachrichten von der Gesellschaft der Wissenschaften aus dem Jahre 1920, Mathematisch-physikalische Klasse, Beiheft, Berlin 1920, 59 p.

The treatise by Fraenkel was not reprinted in Gauß-Werke.

Revised edition: Ostrowski, Alexander: Über den ersten und vierten Gaussschen Beweis des Fundamentalsatzes der Algebra. **In: Gauß-Werke: 10.2: 1922–1933, 18 p.**

Bibliography

Born, Max. 1924. *Bericht über den Stand der Herausgabe von Gauss' Werken* (Fünfzehnter Bericht). Nachrichten von der Gesellschaft der Wissenschaften zu Göttingen. Gesellschaftliche Mitteilungen aus dem Berichtsjahr 1923/24. Berlin, 25–26.

Courant, Richard. 1928–1933. *Bericht der Gauss-Kommission.* Nachrichten von der Gesellschaft der Wissenschaften zu Göttingen. Gesellschaftliche Mitteilungen aus dem Berichtsjahr 1927/28. Berlin 1928, 33; aus dem Berichtsjahr 1928/29. Berlin 1929, 36; aus dem Berichtsjahr 1929/30. Berlin 1930, 29; aus dem Berichtsjahr 1930/31. Berlin 1931, 37; aus dem Berichtsjahr 1932/33. Berlin 1933, 37.

Ebel, Wilhelm, ed. 1974. *Die Matrikel der Georg-August-Universität zu Göttingen. Vol. 2: 1837–1900.* (= Veröffentlichungen der Historischen Kommission für Niedersachsen und Bremen; 9,3). Hildesheim.

Folkerts, Menso. 2002. Carl Friedrich Gauß' Aktivitäten an der Universität Göttingen. *Nachrichten der Akademie der Wissenschaften zu Göttingen; Mathematisch-Physikalische Klasse* 2: 23–131.

Fritsch, Rudolf, and Felix Schmeidler. 2005. Schering, Ernst Christian Julius. *Neue Deutsche Biographie* 22: 695–696.

Gauß, Carl Friedrich. 1839. Allgemeine Theorie des Erdmagnetismus. Resultate aus den Beobachtungen des magnetischen Vereins im Jahre 1838, Leipzig, 1–57 (Gauß-Werke 5, 119–175).

Grunert, Johann August. 1862. Carl Friedrich Gauss Werke. *Archiv der Mathematik und Physik mit besonderer Rücksicht auf die Bedürfnisse der Lehrer an höheren Unterrichtsanstalten* 38: 188–198.

Kienle, Hans. 1936–1944. *Herausgabe der Werke von Gauß.* Nachrichten von der Gesellschaft der Wissenschaften zu Göttingen. Jahresbericht über das Geschäftsjahr 1935/36 [1936], 45; über das Geschäftsjahr 1936/37 [1937], 45; über das Geschäftsjahr 1938/39 [1939], 101; weiter in: Jahrbuch der Akademie der Wissenschaften in Göttingen (Societät der Reichsakademie) für das Geschäftsjahr 1939/40 [1940], 69; für das Geschäftsjahr 1941/42 [1942], 117; für das Geschäftsjahr 1942/43 [1943], 164; für das Geschäftsjahr 1943/44 [1944], 109.

Klein, Felix. 1897. Ernst Schering. [Obituary]. *Jahresbericht der Deutschen Mathematiker-Vereinigung* 6: 25–27.

———. 1899–1922: Ueber den Stand der Herausgabe von Gauss' Werken. *Nachrichten von der Königlichen Gesellschaft der Wissenschaften zu Göttingen*, Geschäftliche Mitteilungen. *Mathematische Annalen* 51, 1899, 128–133 [1]; 53, 1900, 45–48 [2]; 55, 1902, 136–138 [3]; 55, 1902, 139–142 [4]; 57, 1903, 35–43 [5]; 61, 1905, 72–76 [6]; 63, 1907, 333–336 [7]; 69, 1910, 444–445 [8]; 71, 1912, 251–256 [9]; 74, 1913, 410–412 [10]; 77, 1916, 303–306 [11]; 78, 1918, 416–419 [12]; 80, 1921, 82–84 [13]; 85, 1922, 326–328 [14].

———. 1926. *Vorlesungen über die Entwicklung der Mathematik im 19. Jahrhundert.* 2 Bde, Berlin, 1927. Reprint Berlin, Heidelberg New York 1979 in one volume. Development of mathematics in the 19th century. Brookline. Mass. 1979.

May, Kenneth O. 1972. Gauss, Carl Friedrich. *Dictionary of Scientific Biography, New York* 5: 298–315.

Merzbach, Uta C. 1984. *Carl Friedrich Gauss, A Bibliography*. Wilmington: Scholarly Resources.

Patterson, Samuel James. 2001. Ernst Schering, 1833–1897. In *Göttinger Gelehrte. Die Akademie der Wissenschaften zu Göttingen in Bildnissen und Würdigungen 1751–2001*. Hrsg. von Karl Arndt u.a. Göttingen, 190.

Reich, Karin. 1996. Die Rezensionen von Gauß in den Göttingischen Gelehrten Anzeigen. *Mitteilungen der Gauß-Gesellschaft* 33: 21–41.

———, ed. 2012. *Wolfgang Sartorius von Waltershausen, Carl Friedrich Gauß zum Gedächtniss. Biographie Carl Friedrich Gauß, Leipzig 1856. Mit dem von Karin Reich verfassten Essay "Wolfgang Sartorius von Waltershausen (1809–1876)"*. Leipzig: Edition am Gutenbergplatz.

Reich, Karin, and Elena Roussanova. 2013. Eine kritische Bestandsaufnahme der Werkausgabe von Carl Friedrich Gauß. *Mathematische Semesterberichte* 60: 217–247.

Schering, Ernst. 1857. *Zur mathematischen Theorie electrischer Ströme: Beweis der allgemeinen Lehrsätze der Electrodynamik insbesondere der Inductionslehre aus dem electrischen Grundgesetze. Eine von der philosophischen Facultät der Georgia Augusta am 13. Juni 1857 gekrönte Preisschrift*. Göttingen: Druck der Dieterichschen universitäts-Buchdruckerei (W. Fr. Kaestner), 35 p.

———. 1858. Ueber die conforme Abbildung des Ellipsoids auf der Ebene. Eine von der philosophischen Facultät der Georgia Augusta am 4. Juni 1858 gekrönte Preisschrift. Göttingen: Niedersächsische Staats- und Universitätsbibliothek, 34 p.

Scriba, Christoph. 1977. Carl Friedrich Gauß in der Wissenschaftsgeschichte. *Abhandlungen der Braunschweigischen Wissenschaftlichen Gesellschaft* 27: 39–56 (Festschrift der Braunschweigischen Wissenschaftlichen Gesellschaft und der Technischen Universität Carolo Wilhelmina zu Braunschweig zur 200. Wiederkehr des Geburtstages von Carl Friedrich Gauss.).

Wittmann, Axel. 2009. Ernst Christian Julius Schering (1833–1897). Ein Göttinger Sternwartendirektor aus Bleckede. *Mitteilungen der Gauß-Gesellschaft* 46: 81–89.

Wußing, Hans. 2011. *Carl Friedrich Gauß: Biographie und Dokumente*. 6., revised and enlarged edition (=Leipziger Manuskripte zur Verlags-, Buchhandels-, Firmen- und Kulturgeschichte). Leipzig.

On the Correspondence of Sophie Germain

Andrea Del Centina and Alessandra Fiocca

Abstract

The aim of this paper is to give a thorough account of the presently known correspondence of Sophie Germain, as well as the history of its discovery and editing. In particular, we will focus on the correspondence with Gauss and Guglielmo Libri.

1 Introduction

Sophie Germain is known for her studies in Mathematical Physics and Number Theory. Although she earned fame as a recipient of the French Academy prize for her research on elasticity in 1816 (Bucciarelli and Dworsky 1980), she produced her best results in the theory of numbers (Del Centina 2008; Laubenbacher and Pengelley 2010).

In her lifetime, she corresponded with several mathematicians and scientists, including Legendre, Lagrange, Fourier, Poinsot, Cauchy, Gauss and her friend and fellow mathematician, Guglielmo Libri.

The letters she wrote have a twofold importance, as they provide a better understanding of not only her scientific interests and mathematical achievements, but also her personality and life. Those she addressed to Gauss are probably the most relevant of the whole correspondence as regards the former, whereas those written to Libri are the most important as far as the latter aspect is concerned. A large part of Sophie Germain's correspondence has been edited over the years in Stupuy (1879, 1896). Other documents have only recently been published, as were the letters she

A. Del Centina (✉) · A. Fiocca
Dipartimento di Matematica e Informatica, Università di Ferrara, Ferrara, Italy
e-mail: a.delcentina@unife.it; a.fiocca@unife.it

M. T. Borgato et al. (eds.), *Mathematical Correspondences and Critical Editions*,
Trends in the History of Science, https://doi.org/10.1007/978-3-319-73577-1_9

wrote to Libri (Del Centina 2005) and the mathematical notes included in her first five letters to Gauss (Del Centina and Fiocca 2012).

Our aim here is to give a thorough account of the presently known correspondence of Sophie Germain, as well as the history of its discovery and editing, focusing, in particular, on that with Gauss and Guglielmo Libri.

The second section is devoted to a presentation of Sophie Germain's life and work. In the third section, we present Sophie Germain's correspondence, and we discuss its dispersion after Sophie's death and its re-discovery from the late 1870s until recently. In section four we discuss the editorial project of Boncompagni. Section 5 is devoted to a description of the correspondence with Libri, while in Sect. 6, we present the correspondence with Gauss, with a brief comment on it. Finally, in the appendix, we list all known letters (and drafts), written or received by her, giving their storage location, the year in which they were edited for the first time, and who published them.

2 Sophie Germain's Life and Works

Marie-Sophie Germain was born in Paris on April 1st, 1776, the middle daughter of Ambroise-François, a rich silk-merchant—who was later elected a deputy in the National Assembly—and Marie-Madeline Gruguelu. In the years of the Great Revolution, she was in her teens, and being forced to stay at home when the Terror began in 1793, she found diversion from her fear in reading. Libri, who wrote Sophie's first biography, tells us that, in her father's library, she found the *Histoire des Mathématiques* by J.E. Montucla, which she read with great interest (Libri 1832).[1] Fascinated by the story about Archimedes' fate, she decided to devote herself to mathematics and science. She studied with great passion and indomitable will, in opposition to her parents, who considered her desire as being alien to her gender and social class.

Sophie began to learn mathematics by studying the *Cours des mathématiques* by Étienne Bézout, and then the *Calcul differentiel et calcul integral* by Joseph Cousin. With the establishment of the École Polytecnique in 1794, to which women were not admitted, Sophie obtained the lecture notes of various professors for her own use. The students were typically invited to present the professors with written works pertaining to the course topics, and she used the name "Antoine-August Le Blanc" in order to submit her own. Joseph Louis Lagrange, who was one of the professors, praised Le Blanc's essays highly, and wanted to meet him. He was really astonished to discover that Le Blanc was a young lady, but went to her home to demonstrate to her all of his appreciation and good will.

[1] Much of what we know about her youth, personality and attitude comes from the biographical obituary written by Libri, shortly after her death. Other information comes from Stupuy (1879, 1896).

She immediately caused curiosity among men of science, many of whom were eager to help her in her education, and although she did not wait long before receiving visits from great mathematicians, not all of those encounters were pleasant, comfortable and encouraging.[2]

Sophie Germain became known in the Parisian intellectual milieu more as a sort of prodigy over which to marvel, rather than as a student to be taught. Nevertheless, at those times, regular studies were available only to men, and this prevented Germain from having a scientific career.

She became interested in the number theory by reading Legendre's *Essai sur la théorie des nombres* (1798). We do not know when or how Sophie learned of Gauss's *Disquisitiones Arithmeticae*, which was published in 1801, but she was certainly amazed by the originality of this work. In it, she found new stimulus toward number theory. She studied Gauss's treatise for a couple of years, solving exercises and trying to give her own proofs of many of the theorems therein. Then, eager for encouragement, on November 21st, 1804, she wrote her first letter to Gauss, which she signed "Mr. Le Blanc," fearing rejection if she had used her true name. Gauss replied after a 7-month interlude, but in his letter, he showed appreciation for the work of the young Parisian geometer.

Sophie Germain wrote three letters to Gauss under the name of Le Blanc, and Gauss welcomed Le Blanc among his correspondents. Although Gauss's praises were sincere, as is clearly demonstrated in some letters he wrote to his friend, Olbers, he scarcely commented on her work.

In February 1807, as a consequence of the Franco–Prussian War, Sophie had to reveal her true identity to Gauss. This time, he responded promptly and at length, praising her work more than before.

Gauss stopped writing to Germain in January 1808, even though she wrote three other letters to him, to which he probably never replied. Germain stopped writing to him in 1809.

In 1808, after a series of spectacular experiments performed in Paris by the German physicist E.F. Chladni on vibrating thin plates, the Paris Academy of Sciences announced a contest for the best memoir on the mathematical theory of vibrating elastic surfaces supported by experimental evidence.

Sophie Germain set aside her beloved number theory and began to study this problem intensely. In 1811, assisted by Legendre, she presented, as the only competitor, her first contribution to the Academy. The memoir she offered, based on a generalization of Euler's theory of vibrating beams, was marked by a serious error. Since the contest was extended for 2 years, in 1813, helped by her mentors Lagrange and Legendre, she submitted an amended version of her first memoir to the Academy, again as sole competitor. This time, although the way she had arrived

[2]Sometimes, she felt intellectually diminished by them in certain ways, as was the case with the astronomer Joseph Lalande, who gallantly proposed that she read his Astronomie des dames. Irritated, she responded that she had already read Laplace's Système du monde (Bucciarelli and Dworsky 1980, pp. 12–13).

at her fundamental equation was still judged incorrect, the memoir was awarded an "honorable mention" for its experimental part.

The competition was once more prolonged, this time until October 15th, 1815. Sophie Germain's third memoir, presented again as the only competitor, was still considered deficient, despite the fact that many aspects had been revised. Nevertheless, the commission, consisting of Poisson, Laplace, Legendre, Poinsot and Biot, decided to award her the prize with reservation, declaring that her (correct) fundamental equation was not clearly deduced from the hypothesis, but that the comparison made with the results observed by Chladni, and the new experiments carried out, appeared to merit the prize.[3]

In late December 1815, the Academy of Paris established a new contest and prize, which was extended in 1818, aimed at the proof of Fermat's Last Theorem (FLT), i.e., the impossibility of solving in nonzero integers the equation $x^n + y^n = z^n$ when $n \geq 3$. It seems that only after the second call did Sophie Germain forcefully return to work on this challenging question, which she had been considering since 1804 when she mentioned it in her first letter to Gauss. She was convinced that the theory of congruences and power residues, developed by Gauss in the *Disquisitiones*, was the right tool for solving that old problem, but, in 1818, she received new stimulus in tackling the question when Louis Poinsot announced publication of the memoir *Sur l'application de l'algébre à la théorie des nombres* (Poinsot 1820) to the Academy, subsequently giving her a manuscript copy of it (Del Centina 2005).

In May 1819, she received a visit from H.C. Schumacher, a friend of Gauss. This created occasion for her to write to Gauss again after 10 years of silence. In her letter, she put forth her ideas and her progress toward the resolution of Fermat's problem. For a long time, her only contribution to FLT had been entirely described by the so-called "Sophie Germain theorem" (see Dickson 1971, II, p. 734; Edwards 1977, chap. 3; Ribenboim 1999, p. 110), which Legendre attributed to her in a footnote of his paper (Legendre 1827), but very recently—through in-depth studies of some of her unpublished manuscripts—it has been recognized that her results related to FLT went far beyond the content of that theorem (Del Centina 2008; Laubenbacher and Pengelley 2010).

In the early 1820s, through the work of Poisson, Fourier, Navier, and Cauchy, a new theory of elastic surfaces emerged. Unfortunately, it was impossible for Sophie Germain to take part in its development; this was not only because of her lack of knowledge of mathematical analysis, but also because of the difficulties she encountered in getting information on the works of others, and the arrogance with which she was treated, since she was never included in serious scientific discussions. She felt herself to be in a very uncomfortable position (Bucciarelli and Dworsky 1980, ch. 9).

---

[3]For a detailed analysis of Sophie Germain's work in elasticity, see Bucciarelli and Dworsky (1980), Dahan-Dalmédico (1987), and Truesdell (1991).

In May 1825, during one of the Thursday parties given by François Arago at the Observatory, Sophie Germain met Guglielmo Libri.[4]

Germain and Libri had many interests in common, most importantly a real passion for the theory of numbers and FLT, and—in spite of the fact that he was 26 years her junior—a strong friendship immediately grew between them, and they met several times during Libri's stay in Paris. In the summer, Libri went back to Florence, and he started to correspond with her (Del Centina 2005).

In 1826, Sophie Germain wrote a new memoir on elasticity, which she considered a clearer version of her third entry, and presented it to the Academy. But Cauchy, who had been appointed reviewer, suggested she should publish it elsewhere, in order to relieve the Academy of the embarrassment of having to deal with her memoir. She published her work in *Annales de Chimie* (Germain 1828).

In 1829, Sophie Germain wrote her last letter to Gauss. The occasion arose from the visit she had received from Mr. Bader, a pupil of Gauss's, who delivered a copy of Gauss's *Theoria residuorum biquadraticorum* (published a year earlier) to her.[5] That year, Sophie Germain got breast cancer, and was unable to do any more real work. She died on June 17th 1831, at one o'clock in the morning.

As previously mentioned, Gauss scarcely commented on Germain's work on number theory, although he regarded it as being not unworthy of consideration. It was not out of mere respect for her sex that, in 1837, for the Centennial Jubilee of the University of Göttingen, Gauss was to recommend that, had she lived, the faculty should have awarded her an honorary doctorate.[6]

In addition to mathematics, Sophie Germain studied natural sciences and philosophy. Her philosophical writings, *Considérations générales sur l'état des sciences et des lettres aux différentes époques de leur culture* and *Pensées diverses*, were published posthumously. The first volume was edited by her nephew Jacques-Amant

[4]Guglielmo Libri, count of Bagnano (1802–1869), was born in Florence. He enrolled at the University of Pisa in 1816, and graduated in Mathematics in 1820. The same year, he wrote his first paper, *Memoria sulla teoria dei numeri* [Memoir on Number Theory], which he sent to Legendre and Cauchy. In 1823, he became Professor of Mathematical Physics in Pisa. In the winter of 1824, Libri, already known as a talented young mathematician to Legendre and Cauchy as a result of his correspondence with them, went to Paris to present his memoirs in person to the Academy. Libri remained in Paris until August 1825, when he returned to Florence. He returned to Paris in 1830, and he took part in the fighting of the July Revolution. In 1833, Libri became a French citizen and was nominated to become a member of the Academy. In 1843, he became a professor at the Sorbonne. In 1848, accused of being involved in thefts from several French public libraries, he fled to London, where he continued his trade in books and manuscripts. He returned to Florence in December 1868. He died the following year in Fiesole, a small town in the hills surrounding Florence (Maccioni Ruju and Monstert 1995; Del Centina and Fiocca 2010).

[5]A survey of the relations between Gauss and Sophie Germain based on their correspondence can be also found in Leibrock (2001), in which some excerpts of the letters are translated into German.

[6]According to Dunnington (1955, p. 68), on this occasion, Gauss declared, "She (Sophie Germain) proved to the world that even a woman can accomplish something worthwhile in the most rigorous and abstract of the sciences, and for that reason would have well deserved an honorary degree." Among the documents for the centenary celebration preserved at the Historical Archives of the University of Göttingen, no record of Gauss's words has been found.

Lherbette, 2 years after her death (Germain 1833; republished in Stupuy 1879, 1896). This work was highly appreciated by the philosopher Auguste Comte (Comte 1864, II, p. 415). The second volume, which is a list of reflections on the history of science and mathematics, was edited, together with a few of her letters, by Hyppolite Stupuy (1879, 1896).

After Sophie Germain's death, her nephew Jacques-Amant Lherbette—son of Sophie's eldest sister Marie-Madeleine—donated some of her manuscripts and autographs to Guglielmo Libri. Among them were the letters she had received from Gauss. Libri, in his short biography of her (Libri 1832), was the first to make mention of an exchange of letters between Germain and Gauss.

3 The Fate of Germain's Correspondence and Archive

When, in 1848, Libri fled to London, he took with him most of the content of his vast library and archive, but not all of it. Many papers were confiscated by the French authorities in his apartment at the Sorbonne, and among them, there were writings by Sophie Germain and letters addressed to her. All of these documents are presently housed at the Bibliothèque Nationale, in the funds Ms. Français 9114–9116; 9118; NAF 4073. Among what remained in Libri's hands, there was the most important of the letters she had received from Gauss, dated April 30th, 1807 (see below), other correspondence, drafts of her letters, and manuscripts of her mathematical works, two of which were later sold by Libri at the auction sale he held in London in March 1859 (Libri 1859).

That same year, another part of Germain's archive was donated by her sister, Angelique-Ambroise, Madam Dutrochet and her nephew Lherbette, to the French Academy (Bertrand 1859).

When Libri died in 1869, the portion of Germain's archive that had remained in his possession suffered the fate of the dispersal of Libri's own archive (Del Centina and Fiocca 2004). We can estimate that half of Libri's archive is presently preserved at the Biblioteca Moreniana in Florence[7] (Del Centina and Fiocca 2004, pp. 267–270), while the other half has probably been lost forever. Gauss's letter to Germain, dated April 30th, 1807, was acquired by Prince Boncompagni in 1878 (Del Centina and Fiocca 2004, p. 53 e p. 182). This letter, which is quoted in Boncompagni (1892, art. 605, n. 40),[8] but not in Boncompagni (1895), is presently in the Niedersächsische Staats- und Universitätsbibliothek Göttingen (SUB).[9]

[7]The material is distributed in three fonds: Fondo Palagi-Libri (FPL), acquired in 1878; Nuovo fondo Libri (NFL), acquired in 1959; and Fondo carte Libri (FCL), acquired in 1999.

[8]In Del Centina and Fiocca (2012), it was erroneously quoted as being (Boncompagni 1895).

[9]Cod. Ms. Gauss Briefe B: Germain. In Del Centina and Fiocca (2012), we stated that this letter is dispersed, and that the Niedersächsische Staats und Universitätsbibliothek in Göttingen holds the photolithographic reproduction made by Boncompagni in 1879. Thanks to Karin Reich (Zbl 1268.01029), we learned that the above-mentioned Library holds the original letter. The librarian of the Niedersächsische Staats- und Universitätsbibliothek Göttingen, Ms. Dietlind Willer, confirmed

The existence of letters from Germain to Libri was, for the first time, indicated in FPL by the historian Giacomo Candido (Candido 1941). Among Libri's papers in NFL, more than two hundred sheets written in Germain's hand can be found: scientific works, reports of experiments, drafts of letters to Gauss, Legendre, Lagrange and Poinsot, remarks on papers by Cauchy and Navier, etc. (Del Centina and Fiocca 2004, pp. 103–105; 234–235). During the celebrations for the first centenary of Gauss's birthday, Ernst Schering, one of the editors of Gauss's works, gave an official speech to the Academy of Science of Göttingen. This was published, together with some letters from Gauss's archives (Schering 1877). Among these was Sophie Germain's letter to Gauss dated February 20th, 1807. The same year, Schering informed the Academy of Sciences of Paris that the library and the archive of Gauss had been acquired by the Academy of Science of Göttingen. He sent copies to Paris of some letters by Lagrange, Laplace, Delambre and Germain addressed to Gauss,[10] with a request for copies of letters, manuscripts, and documents concerning Gauss owned by Parisian scientists, collectors and institutions, in order to include them in the new volumes of Gauss's works that he was preparing. Schering's request was presented to the Parisian Academy of Science by Joseph Bertrand (Bertrand 1877).

In 1879, Hyppolite Stupuy, in his reevaluation of Sophie Germain's philosophical works (Stupuy 1879), included some letters from her correspondence that he had found at the Bibliothèque Nationale (Ms. François 9118). Among these, there were three letters from Gauss and two undated drafts signed "Le Blanc." In Stupuy's work, Gauss's first letter was erroneously dated June 16th, 1806 and, moreover, Stupuy did not mention the whereabouts of these documents. Charles Henry expressed severe criticism of Stupuy's work (Henry 1879) and published other letters from Germain's correspondence, found at the Bibliothèque Nationale Ms. NAF 4073.

That same year, Baldassarre Boncompagni published the photolithographic reproduction of Gauss's letter to Germain of April 30th, 1807, which was in his hands (Boncompagni 1879a).[11] In November of that year, Ernst Schering presented Boncompagni's work to the Academy of Science of Göttingen. Schering stressed the importance of that letter, not only because it was the first that Gauss had addressed to Germain after having been informed of her true identity, but mainly because it allowed him to date Gauss's studies on biquadratic residues (Schering 1879).

At the end of 1879, only seven letters of the Germain–Gauss correspondence were known to the public: (1) the two undated drafts, from Germain, published by Stupuy, (2) three letters from Gauss, dated June 16th, August 20th, 1805 and

this, and added that the original letter came to the Library after 1920, but the circumstances of the acquisition are unknown.

[10]These copies, verified by Schering, are now at the Bibliothèque de l'Institut de France, ms. 2031 (Bucciarelli and Dworsky 1980, pp. 126–127, nota 4).

[11]This letter, which came from Libri's archives, was acquired by Boncompagni in December 1878 (Del Centina and Fiocca 2004).

January 19th, 1808 published by Stupuy, (3) Germain's letter dated February 20th, 1807, published by Schering, (4) Gauss's reply, dated April 30th, 1807, published by Boncompagni. In particular, at that time, nobody, except Schering, knew of the existence of the mathematical notes that Sophie Germain had enclosed in the first five letters.

The existence of a mathematical note enclosed in Germain's first letter, as published by Stupuy, was suggested to Angelo Genocchi by a phrase Germain wrote in it. Genocchi then asked Schering about the existence of mathematical notes written by her among the correspondence with Gauss. At the same time, searches were commissioned by Boncompagni in order to determine the whereabouts of the letters between Germain and Gauss published by Stupuy. The three letters from Gauss and the two drafts from Sophie Germain were finally rediscovered at the Bibliothèque Nationale in Paris (Ms. François 9118).

On March 3rd, 1880, Schering informed Genocchi that, among Gauss's papers, there were ten letters from Sophie Germain, and that to the first five, there were attached special mathematical notes filling up many sheets, while the other five contained extended mathematical research.

Boncompagni convinced Schering to allow a photolithographic reproduction to be made of Germain's letters, which had already been edited. The first five letters were published both in facsimile and in printed versions, but without the additional mathematical notes (Boncompagni 1880).

Thanks to this publication, it became clear that the two undated letters signed "Le Blanc," published by Stupuy, were drafts of the actual letters dated November 21st, 1804, and November 16th, 1805, housed in Göttingen. Genocchi also corrected the misprint of the date of the first letter from Gauss (June 16th, 1806, instead of June 16th, 1805) in Stupuy's edition. The misprint had induced Schering to believe that Gauss's reply to Germain's first letter was still to be found (Schering 1880). To thank Genocchi, he sent some excerpts from Germain's mathematical notes.

On May 30th, 1880, Genocchi, while presenting Boncompagni's new editorial work to the Academy of Science of Turin, announced his own yet-to-be-published writing on the correspondence between Germain and Gauss. The note appeared that same year in the memoirs of the Academy of Turin, under the title *Il carteggio di Sofia Germain e Carlo Federico Gauss* (Genocchi 1880). With it, Genocchi revealed to the community of mathematicians and historians that the (known) correspondence consisted of ten letters from Germain—five with extended mathematical notes—and four from Gauss. He also informed them of the exact location of the letters, and commented briefly on the letters published up to that time. He remarked on the importance of Gauss's letter of April 30th, 1807, for the two theorems on cubic and biquadratic residues that Gauss had stated therein.

4 The Editorial Project of Boncompagni

Boncompagni asked Schering for a copy of all of Germain's letters and notes. From Schering's reply on March 3rd, 1880, it appears that Schering had not read all of Germain's mathematical notes, and those he had were read with no great attention. In fact, Schering made comments only on the contents of the first letter and its mathematical note, and judged them to be uninteresting, as far as both the mathematics and the history were concerned: some results had already been published in 1831 in Crelle's Journal, and the others regarding the proof of the first case of the FLT were wrong. So, on the same day that Schering wrote to Genocchi about his plan to publish all of the letters and notes by Germain, he changed his mind. The reason for such a rapid retraction is not clear.

On July 13th, Boncompagni informed Genocchi that he had finally received a copy of the five unedited letters and all of the mathematical notes by Germain. In his letter, Boncompagni declared his intention to publish the correspondence in its entirety, mathematical notes included—in the form of a little book to be entitled *Carteggio tra Sofia Germain e Gauss*—and asked Genocchi's opinion of the project.

In contrast with Schering, Genocchi was convinced that the notes were worthy of publication. He also suggested that Boncompagni publish them in printed form and not in facsimile. Boncompagni agreed, and hoped to publish the volume in the first half of 1881.

In the middle of November 1880, the proofs were ready. They consisted of 69 printed pages in all, 26 for Germain's letters, 7 for those from Gauss, and 36 for the mathematical notes.[12] Boncompagni invited Genocchi to cooperate in the correction of the proofs and, especially, being well aware of his own lack of knowledge in number theory, to help him with the mathematics. The correction of the proofs lasted a couple of months, and during this time, Genocchi visited Boncompagni in Rome.

In 1880, two reviews of Boncompagni (1880) appeared, Henry (1880) and Mansion (1880). In the latter, a note signed by Eugène Catalan at the end of Mansion's article announced Boncompagni's plan for the publication of the Germain–Gauss correspondence in its entirety.

One more review appeared in 1881 (Günther 1881). In it, the author claimed that the letters already published constituted the entire correspondence between Germain and Gauss, and that the mathematical notes had been lost.

It is clear that the existence of still unpublished letters and mathematical notes by Sophie Germain was not known to everyone who wrote about the Germain–Gauss correspondence. In the Italian translation of Günther's article, published in the *Bullettino* (Günther 1882), the editor (in a footnote signed "B.B."), rectified Günther's claim:

[12]We do not know whether these proofs still exist or not.

> L'illustre Società Reale di Göttingen possiede dieci lettere di Sofia Germain, cinque delle quali sono qui menzionate dal sig. Günther, ed altre cinque finora inedite saranno da me in breve date in luce. Unitamente a queste cinque lettere inedite pubblicherò cinque note matematiche di Sofia Germain, di ciascuna delle quali la medesima Società delle Scienze di Güttingen possiede un esemplare autografo. Tali note sono annesse alle prime 5 delle dieci suddette lettere di Sofia Germain [The Royal Society of Göttingen possesses ten letters by Sophie Germain, five of which are mentioned here by Mr. Günther, and the other five, still unedited, I will publish shortly. Together with these five letters, I will publish five mathematical notes by Sophie Germain, the originals of which are also held by the same Society. These notes are attached to the first five letters of the above mentioned letters by Sophie Germain].

Boncompagni had himself planned to write an introduction to the *Carteggio*, but because of the duties involved in managing his *Bullettino,* and the printing of the *Regula Abaci* by Adelard of Bath, which took more time than expected (for one reason or another), he was unable to find the time to complete it. From what appears in Boncompagni (1879b–1880), we may deduce that he presented a note entitled *Intorno al carteggio tra Sofia Germain e Carlo Federico Gauss* to the Pontificia Accademia dei Nuovi Lincei. This was probably a draft version of the planned introduction that he withdrew shortly afterwards, with the aim of perfecting it.

A few years later, Boncompagni published the paper (Boncompagni 1884),[13] in which he commented on and annotated at great length (as was his style) Gauss's letter to Olbers of September 3rd, 1805 [this letter had already appeared in facsimile (Boncompagni 1883)]. In this letter, as in others to Olbers, Gauss again mentioned his correspondent "Le Blanc" with great appreciation, so Boncompagni had the opportunity to comment on the first two letters from Germain to Gauss and their *addenda*. In particular, Boncompagni stressed the fact that, in the second *addendum*, she had given new proof that 2 is a quadratic residue for primes of the form 8k + 1 and 8k + 7 and a nonresidue for those of the form $8k + 3$ and $8k + 5$.

Genocchi, who had read the letters exchanged between Germain and Gauss with great attention and interest, wrote, in 1884, two other notes on the subject (Genocchi 1884a, b). In his papers, he credited Germain with certain results contained in her mathematical notes (see Del Centina and Fiocca 2012, section 8).

However, Boncompagni never completed his project, and a few years after the initial interest aroused by the first incomplete edition of the correspondence, the still-unpublished letters from Germain to Gauss, along with her mathematical notes, were completely forgotten, and—what is even worse—are now believed to be lost forever.

---

[13]Here, Boncompagni also gave detailed information about the contents of Gauss's correspondence housed at the Academy of Sciences of Göttingen. The letters addressed to Gauss by 164 authors were distributed in alphabetical order in 180 portfolios contained in 19 cardboard folders. The letters by Sophie Germain were contained in the fourth folder, numbered 98, covered in dark green paper, and inside the second portfolio, numbered 38. On the spine of the folder was written, in golden letters and digits, "98 Gauss Nachlass 98 Briefe Fuss—Harding," and on the portfolio, in Schering's handwriting, was written "38, Sophie German Leblanc 1804—1829." The drafts of the letters written by Gauss were in alphabetical order, and distributed in portfolios contained in 19 folders numbered 95b–113b.

5 The Correspondence with Libri

In a short note on Libri's archives in Florence, Grattan-Guinness (1984) remarked, with some regret, that the letters Sophie Germain addressed to Libri, which he considered the most interesting writings therein, were not taken into account in Bucciarelli and Dworsky (1980).[14] It is presumable that Bucciarelli and Dworski were not aware of this correspondence, and, in fact, in their work, they only referred to the letter she wrote to Libri in March 1831, which is preserved at the Bibliothèque Nationale in Paris and published in Henry (1879, p. 631) (see Bucciarelli and Dworsky 1980, pp. 121–122). The letters held in Florence, which actually cast new light on some episodes of Sophie's life and her personality, as well as on her friendship with Libri, have been published in their entirety in Del Centina (2005).

One of these letters, which is the most interesting from a mathematical point of view, was actually not addressed to Libri, as supposed by Grattan-Guinness, but rather to Louis Poinsot. This is a draft of the letter she wrote to him in 1819 to tell him of the great help she had received from his memoir on the use of complex numbers in algebra (Poinsot 1820). In this memoir, by simultaneously developing the congruence $x^n \equiv 1 \pmod{p}$ and the equation $x^{n-1} = 0$, Poinsot gave an analytical representation of power residues by means of imaginary roots of unity.

Germain found in Poinsot's method a justification of her method toward a proof of FLT (see also Del Centina 2008).

Libri started to be interested in the theory of numbers in his youth. He was seventeen when, hearing of the prize established by the Paris Academy for the proof of FLT, he began his research on the subject, eagerly studying the works of Euler, Legendre and Gauss. In January 1821, he presented to the Academy a memoir on the theory of numbers, which was praised by Cauchy. The work of Libri was also read and carefully studied by Sophie Germain. This is attested to by a three page-long comment in her hand, entitled *Notes sur Memoria sopra la teoria dei numeri* and dated June 22nd 1822, preserved at the Moreniana Library (Del Centina 2005, pp. 62–63).

When Libri went to Paris, late in December 1824, he was welcomed by the scientific community as a talented young mathematician. He met Sophie Germain for the first time on May 13th at one of the Thursday evening parties given by the astronomer François Arago at the Observatory. Having enjoyed each other's company, the two met again several times, and she also invited him for lunch at her home. From the letters she wrote to him in that period, it clearly appears that their relationship quickly grew beyond the common interest in the theory of numbers (Del Centina 2005, p. 65).

Libri stayed in Paris until August 1825. On the way home, he stopped in Geneva for about a month. Once back in Florence, Libri immediately wrote to Germain. In his letter, dated November 17th (Henry 1879), he asked her to send him her news from time to time. She probably replied promptly, but of the letters she sent to him

[14]BMF, FPL 432 (12).

in Florence over the following 4 years, only two have survived. Both letters, dated November 15th 1826 and February 8th 1830, have been published in Del Centina (2005).

In the first, she told him about the new memoir on elasticity she had written that year, and of the difficulties she had encountered in preparing it. In the second, she told him that she had been seriously ill and has not been able to work. She also informed him about Legendre's and Fourier's health, of Jacobi's stay in Paris in August 1829, and of Abel's death.[15]

Libri returned to Paris in June 1830, and he did not fail to visit Sophie Germain. In Paris, Libri knew August L. Crelle, who, charged by the Prussian Government, was on an official tour to study the methods of teaching mathematics in France. One day, most likely in July, Germain invited Libri for dinner at her home, asking him to extend the invitation to Crelle.[16]

At the end of July, the revolution against King Charles X began. Libri took an active part in it, and the newspapers praised his conduct on the barricade. With this exploit, Libri laid the foundation for his future career in France. In November, Libri left Paris. Travelling back to Florence, he stopped in Geneva for about a month. Here, he met Cauchy, who had followed his King into exile. Libri reached Florence around the middle of January 1831. A suffering Sophie Germain wrote to him on February 2nd. She informed him that she had received from Crelle, together with the invitation to submit other papers to his journal, the reprints of her work (Germain 1831). In this paper, she published a generalization of certain results on the equation

$$\frac{4\,(x^p - 1)}{x - 1} = Y^2 \pm Z^2$$

with the Y, Z polynomial in x with integral coefficients and p a prime $\neq 2$ (Gauss 1801, art. 357), that she had announced 25 years earlier in her first letter to Gauss. It should be noted that this is the only paper she had ever published on the theory of numbers.

This letter is also interesting for the presence of other information. For instance, we learn from it that, while in Geneva, Cauchy, whose correspondence was then being intercepted by the police, entrusted Libri to forward a letter to his wife. Libri sent the letter to Sophie Germain, who, in turn, forwarded it to Cauchy's wife.

On his arrival in Florence, imbued with the spirit of the July revolution, Libri took part in the plot to coerce the Grand Duke into accepting a constitution. But the plot failed, and Libri was forced into exile by the end of February. He reached Marseilles on March 21st, 1831, and there, he wrote to Germain of his difficult situation. She answered on April 18th. Her letter is published in Henry (1879, pp. 631–632), and,

[15]Legendre officially informed the Academy of Abel's death in the meeting of June 22nd, 1829 (Del Centina 2005, p. 68 nota 23).

[16]Undated card addressed to G. Libri in Germain's handwriting at Biblioteca Moreniana of Florence (BMF), FPL 431 (95) (Del Centina 2005).

translated into English, cane also be found in (Bucciarelli and Dworsky 1980, pp. 121–122).

The last words we have from Germain's pen are probably those she wrote to Libri on May 17th, one month before her death (Del Centina 2005, p. 71). These words and her handwriting clearly reveal that the letter was written by a person in the greatest of pain. Looking at herself, she saw only illness and suffering; she had lost all hope of getting better. Nevertheless, she wanted help her friend, and offered to send him books for his studies.

One year later, Libri wrote a short, but impassioned, biography of Sophie Germain (Libri 1832).

6 The Correspondence with Gauss

The known correspondence between Germain and Gauss consists of fourteen letters: ten from Germain[17] and four from Gauss.

She addressed eight letters to Gauss in the period 1804–1809, and the remaining two in 1819 and 1829, respectively. The first five were published (in facsimile) in Boncompagni (1880). This work has long led those interested to think that there were only five letters of hers housed at Göttingen. Bucciarelli and Dworsky, despite acknowledging the existence of Germain's unpublished letters (1980, pp. 114–115, p. 143), believed the mathematical notes to have been lost (1980, p. 21) (see also Sampson 1990; Ribenboim 1999, p. 203).[18] Excerpts from these, and the one dated 1829, have been published in English in Bucciarelli and Dworsky (1980). The remaining letters by Sophie, except for the one dated 1819, which was edited in Del Centina (2008), were published for the first time in Del Centina and Fiocca (2012), together with the mathematical notes she had enclosed with the first five letters. The publication of these notes has contributed to completing the re-evaluation of Sophie Germain as a number theorist, as initiated by Del Centina (2008) and Laubenbacher and Pengelley (2010).

Gauss's letters are preserved in Paris at the Bibliothèque Nationale (Fonds Français n. 9118), except for the one dated April 30th, 1807. Of this letter, listed as n. 40 art. 605 in the *Catalogo dei manoscritti ora posseduti da D. Baldassarre Boncompagn* (Boncompagni 1892), the Library of the University of Göttingen has only a photolithographic reproduction. This copy was a gift from Prince Boncompagni.

The first three letters by Germain—dated November 21st, 1804, July 21st, and November 16th, 1805, respectively—are signed "Le Blanc." In the first letter, she

[17]These were preserved by Gauss himself, and reached the Academy of Göttingen after his death. Presently, they are housed at the Niedersächsische Staats- und Universitätsbibliothek Göttingen (SUB).

[18]They are preserved, together with Germain's letters to Gauss, at Göttingen (SUB) (Cod. ms. Gauss Briefe A: Germain).

asked Gauss to send his reply to the address of Sylvestre de Sacy, who would have forwarded it to her.

Gauss responded to the first letter after almost 7 months, on June 16th, 1805, while to the second, he replied more promptly, on August 20th, 1805. Both times, Gauss praised "Mr. Le Blanc" for his results and progress in the difficult field of higher arithmetic. Gauss's reply to Sophie's third letter, if it ever existed, has not been found.

As we have already stated, it was only with the fourth letter that Sophie Germain disclosed her true identity to Gauss.[19] Gauss responded promptly with a long letter containing many mathematical suggestions and even more praise for his correspondent than before.

The epistolary exchange between the two continued with a letter from Germain dated June 27th, 1807, the reply from Gauss dated January 19th, 1808, and a new letter from Germain dated March 19th, 1808.

The letter of January 1808 is likely the last Gauss addressed to Sophie Germain. In this letter, he seems to be saying that, due to various circumstances, he would no longer have much time to correspond with her, although, as is clear from his words, he maintained great esteem for her talent.

In the first month (probably in the spring) of 1809, Gauss forwarded to Sophie Germain, through Legendre, a copy of his memoir *Summatio quarundam serierum singularium*, a work that he had presented to the Academy of Göttingen the year before. In the letter dated May 22nd, after having thanked Gauss for this gift, she professed admiration for Gauss's memoir *Theorematis arithmetici demonstratio nova*,[20] a copy of which Gauss had sent to her on January 19th, 1808. She then put forth her results concerning biquadratic residues.

As we have seen, Sophie Germain stopped writing to Gauss in 1809. In the following years, Sophie Germain became more and more involved in her studies

[19]As is well known, the circumstances were as follows: in October 1806, the success of the Napoleonic Army at Jena opened the way for the invasion of a large part of Prussia, and French troops occupied Brunswick, Gauss's hometown. Sophie, fearing for Gauss's safety, asked General Pernety (a family friend and commander of the French artillery in the Prussian campaign, responsible for the siege of Breslau) the favour of determining Gauss's whereabouts and ensuring that he was not mistreated. The General ordered Captain Chantal to ride two hundred miles west to Brunswick and carry out this mission. The meeting between Chantal and Gauss is described in the letter that the former sent to his General once he had fulfilled his duty; this letter is published in Stupuy (1879). When Chantal revealed to Gauss the name of his protector, "Mademoiselle Sophie Germain," Gauss replied that he had not had the honour of knowing her. Informed of these events by General Pernety (Stupuy 1879, 1896) on February 20th, 1807, Sophie wrote her fourth letter to Gauss, finally revealing her true identity. The letter from Chantal to Pernety and that from Pernety to Germain, are housed at the Bibliothèque Nationale of Paris, MS Fr. 9118, p. 266 and p. 269. A copy of the first, probably requested by E. Schering, is in Göttingen (SUB) with the Gauss–Germain correspondence. For an English translation of these letters, see (Bucciarelli and Dworsky 1980, pp. 23–24).

[20]In this paper, Gauss gave a new and shorter proof of his "Fundamental Theorem," as he called it in the *Disquisitiones*, which is the theorem today known as the "Quadratic Reciprocity Law."

on elasticity and vibrating surfaces. After an interlude of 10 years, she wrote to him again on May 12th, 1819. This letter is very important, because in it, she explains at some length her ideas and progress toward a proof of FLT. The last letter Germain addressed to Gauss is dated May 28th, 1829. It is likely that Gauss never responded to these two letters.

A reading of the letters and mathematical notes that Sophie Germain wrote to Gauss shows that she was quick to grasp—long before any others—the content of the *Disquisitiones*. She was not only able to learn the techniques and theorems of Gauss's work, but also—a testament to her merit—to achieve, on her own initiative, new results that were by no means trivial and develop ideas and conjectures that demonstrate a strong taste for generalization.

Unfortunately, due to her gender, she was prevented from attaining adequate university instruction. Still worse, she was even denied the possibility of working in the academic world, as would have been possible for a man of her ability. Thus, apart from the benevolence of Lagrange and Legendre—largely as a result of the respect due to a woman—she worked in almost total isolation, often without being guaranteed access to the scientific information and debates within the Academy.

She saw in her correspondence with Gauss the possibility of escaping from the bell jar under which she felt herself eternally banished rather than protected. Gauss greatly appreciated her intelligence and admired the wisdom with which she knew how to confront and demonstrate theorems that he himself had deemed to be among the most difficult and enigmatic. However, she would certainly have needed more support and interaction with Gauss to develop all of her potential for number theory.

Nothing can better express the sentiment that forever hung over her than the words with which she closed her last letter to Gauss: "I regret that I am deprived of the advantage that I would find in enjoying your learned conversation, as Mr. Bader does. What he told me does not astonish me, but it is an object of my envy. Apart from what I could learn from you, I regret again that I can't submit for your judgement so many ideas that I have not published, and that would be too long to explain in letter form."

We believe that the analysis found within the correspondences with Gauss reinforces the conviction—already extolled, on the basis of the study of other unedited writings by her, in Del Centina (2008) and Laubenbacher and Pengelley (2010)—that Sophie Germain has a rightful place among the greatest of professional number theorists.

Appendix: The Published Correspondence of S. Germain

Funds Where the Correspondence is Preserved:

In the Bibliothèque Nationale de France, the known correspondence of Sophie Germain is contained in four archive folders:

Ms. Français 9118: Correspondence of Sophie Germain with the mathematicians Cauchy, Delambre, Fourier, ..., edited (Stupuy 1879)
NAF 5166: Papers of the mathematician J.-L. Lagrange (1813). This manuscript is listed in L. Delisle, *Catalogue des manuscripts des fonds Libri et Barrois* (Paris, Champion, 1888) p. 177.
NAF 9544: Collection of autograph letters of d'Alembert, Bossut, Sophie Germain, C. Huyghens, Mersenne. This manuscript belonged to Libri (on page 27, there is a *Notice des papiers de Roberval par G. Libri*).
NAF 4073: Correspondence of Sophie Germain, edited (Henry 1879)

In the Biblioteca Moreniana of Florence (BMF), the known correspondence of Sophie Germain is contained in two archive folders:

Palagi-Libri (FPL) filza 432 ins. 12: eight letters from Sophie Germain to Guglielmo Libri (1819–1831)
Nuovo fondo Libri (NFL) cass. 10 ins. 214; cass. 11 ins. 271

In the Niedersächsische Staats- und Universitätsbibliothek Göttingen (SUB), there are ten letters from Sophie Germain to Carl Friedrich Gauss, five mathematical notes attached to the first five letters (Cod. Ms. Gauss Briefe A: Germain) and the letter from Gauss to Sophie German, dated April 30th, 1807 (Cod. Ms. Gauss Briefe B: Germain).

Letters Addressed to Sophie Germain from:

Bernard libraire, Paris November 4th (Ms. Français 9118; Stupuy 1879, pp. 291–292).

Cauchy Augustin Louis, Paris July 24th, 1821 (Ms. Français 9118; Stupuy 1879, p. 355); Sceaux-Penthièvre July 23rd, 1826 (Ms. Français 9118; Stupuy 1879, pp. 367–368, wrongly dated 1823; Bucciarelli e Dwrosky, p. 108).

Choron Alexandre-Étienne, (NAF 4073; Henry 1879, p. 628).

D. (?) March 13th, 1814 (NAF 4073; Henry 1879, pp. 637–638).

D'Ansse de Villoison Jean-Baptiste Gaspard, one undated letter; July 12th and 14th, 1802 (Ms. Français 9118; Stupuy 1879, pp. 293–297).

Delambre Jean-Baptiste, one undated letter (NAF 4073; Henry 1879, pp. 627–628).

Fourier Jean-Baptiste Joseph, June 1st, 1820 (Ms. Français 9118; Stupuy 1879, pp. 350–351); one undated letter (but 1822) (Ms. Français 9118; Stupuy 1879, pp. 357–359); one undated letter (Ms. Français 9118; Stupuy 1879, pp. 360–363); Paris, September 19th, 1820 (NAF 4073; Henry 1879, pp. 628–629); Paris January 30th, 1827 (NAF 4073; Henry 1879, p. 629); one undated letter (NAF 4073; Henry 1879, pp. 629–630); one undated letter (but 1827) (NAF 4073; Henry 1879, p. 630); May 2nd (NAF 4073; Henry 1879, p. 630); one undated letter (NAF 4073; Henry 1879, p. 631); one undated letter (NAF 4073; Henry 1879, p. 631). One letter from

Fourier to Germain, dated May 4th, 1821, is listed in the *Catalogo dei manoscritti ora posseduti da D. Baldassarre Boncompagni*, (Roma, Tipografia delle Scienze Matematiche e Fisiche 1892, art. 605, n. 37), but now, it is lost.

Gauss Carl Friedrich, Brunswick June 16th and August 20th, 1805 (Ms. Français 9118; Stupuy 1879, pp. 302–307); Brunswick April 30th, 1807 (Göttingen (SUB), Cod. Ms. Gauss Briefe B: Germain; Boncompagni 1879a; Stupuy 1896, pp. 274–282; Gauss 1917, pp. 70–74); Göttingue January 19th, 1808 (Ms. Français 9118; Stupuy 1879, pp. 318–320).

Institut de France, Classe des sciences physiques et mathématiques (Jean-Baptiste Delambre), Paris January 1816 (Ms. Français 9118; Stupuy 1879, pp. 343–344).

Lagrange Joseph Louis, Paris 17 Germinal (Wednesday) (NAF 4073; Henry 1879, p. 633).

Lalande Jérôme, November 4th, 1797 (NAF 4073; Henry 1879, p. 635).

Le secrétaire perpétuel de l'Académie (Jean-Baptiste Delambre), Paris July 23rd, 1821 (Ms. Français 9118; Stupuy 1879, p. 354); (Fourier), Paris May 30th, 1823; June 1st, 1823; June 3rd, 1823; July 24th, 1823 (Ms. Français 9118; Stupuy 1879, pp. 363–369).

Legendre Adrien-Marie, one undated letter; Paris January 19th, 1811; January 28th, 1811; Paris October 22nd, 1811; Paris November 10th, 1811; December 4th, 1811; December 4th, 1813; Paris December 31st, 1819; Paris June 23rd, 1821 (Ms. Français 9118; Stupuy 1879, pp. 321–342; 348–349; 351–353); one undated letter (The New York Public Library, box 7 of the Samuel Ward papers, see Laubenbacher and Pengelley 2010).

Libri Guglielmo, Firenze November 17th, 1825 (NAF 4073; Henry 1879, pp. 636–637).

Navier Claude-Louis, Paris August 2nd, 1821 (Ms. Français 9118; Stupuy 1879, p. 356).

Pernety Le Général, Cotel pres Breslau, December 23rd, 1806 (Ms. Français 9118; Stupuy 1879, pp. 316–317).

Poisson Siméon-Denis, Paris January 15th, 1816 (Ms. Français 9118; Stupuy 1879, pp. 347–348).

Without sender and undated letter (but from Victor Cousin, between 1790 and 1800) (NAF 4073; Henry 1879, pp. 623–625).

Tessier Alexandre-Henry, Pluviôse 17th (Ms. Français 9118; Stupuy 1879, pp. 289–291).

Université Impériale. Le trésorier Delambre, Paris May 14th, 1810 (NAF 4073; Henry 1879, pp. 626–627).

Letters and Drafts of Sophie Germain Addressed to:

Gauss Carl Friedrich, one undated draft (but November 21st, 1804); one undated draft (but November 16th, 1805) (Ms. Français 9118; Stupuy 1879, pp. 298–302; 308–311); November 21st, 1804; July 21st, 1805; November 16th, 1805; February

20th, 1807; June 27th, 1808; March 19th, 1808; May 22nd, 1809; May 26th, 1809; May 12th, 1819 (Del Centina 2008, pp. 356–362); March 28th, 1829 [partially published in English in Bucciarelli and Dworsky (1980, pp. 114–115)]. All of the letters and the mathematical notes attached to the first five letters are housed in Göttingen (SUB), Cod. Ms. Gauss Briefe A: Germain and are published in Del Centina and Fiocca (2012).

Lagrange Joseph Louis, one undated draft (but autumn 1808) (NFL cass. 10 ins. 214, Del Centina 2005, p. 74).

Legendre Adrien-Marie, December 1811 (NAF 5166; Bucciarelli, Dworsky in English, pp. 58–59); one undated draft (but January 1811) (NFL cass. 11 ins. 271 cc 1–4, Del Centina 2005, p. 72).

Libri Guglielmo, two undated letters (FPL, filza 432 ins. 12, Del Centina 2005, pp. 64–65); September 17th, 1826 (FPL, filza 432 ins.12, Del Centina 2005, p. 66); February 8th, 1830 (FPL, filza 432 ins. 12, Del Centina 2005, p. 68); February 2nd, 1831 (FPL, filza 432 ins.12 Del Centina 2005, p. 69); May 17th, 1831 (FPL, filza 432 ins.12, Del Centina 2005, p. 71); Paris April 18th, 1831 (NAF 4073; Henry 1879; Bucciarelli e Dworsky p. 121–122 in English); two undated letters (FPL, filza 432 ins.12, Del Centina 2005, pp. 64–65); September 17th, 1826 (FPL, filza 432 ins.12, Del Centina 2005, p. 66); February 8th, 1830 (FPL, filza 432 ins.12, Del Centina 2005, p. 68); February 2nd, 1831 (FPL, filza 432 ins.12, Del Centina 2005, p. 69); May 17th, 1831 (FPL, filza 432 ins.12, Del Centina 2005, p. 71).

Poisson Siméon-Denis, one undated letter (but 1816) (Ms. Français 9118; Stupuy 1879, pp. 344–346).

- draft without recipient and year, July 18th (but addressed to Augustin Louis Cauchy and 1826) (Ms. Français 9118; Stupuy 1879, pp. 369–373; Bucciarelli e Dworsky in English, p. 106).
- draft without recipient, about October 1813, (NAF 5166, Bucciarelli, Dworsky in English, p. 61);
- draft without recipient, July 2nd, 1819 (but addressed to Louis Poinsot) (FPL, filza 432 ins.12, Del Centina 2005, p. 63).
- draft without recipient, Paris April 18th, 1831 (but addressed to Guglielmo Libri) (NAF 4073; Henry 1879, pp. 631–632).

References

Bertrand, J. 1859. *Comptes Rendus hebdomadaires des Séances de l'Académie des Sciences* 49: 45.

———. 1877. *Comptes Rendus hebdomadaires des Séances de l'Académie des Sciences* 84: 1017.

Boncompagni, B. 1879a. *Lettera inedita di Carlo Federico Gauss a Sofia Germain pubblicata da B. Boncompagni*, Calcografia e autografia Achille Paris, Firenze (Printed version in: *Archiv der Mathematik und Physik* 63 (1880), Litterarischer Bericht 259: 27–31).

———. 1879b–1880. Presentazione di opere, Atti dell'Accademia Pontificia de' Nuovi Lincei, 33: 427.

———. 1880. *Cinq lettres de Sophie Germain à Charles Frédéric Gauss publiées par B. Boncompagni d'après les originaux possédés par la Société Royale des Sciences de Göttingen*, Berlin, Institut de photolithographie des frères Burchard, imprimerie de Gustave Schade (Printed version in: *Archiv der Mathematik und Physik* 65, Litterarischer Bericht 257: 5–9 and *Archiv der Mathematik und Physik* 66, Litterarischer Bericht 256: 4–10).

———. 1883. *Lettre de Charles-Frédéric Gauss au D.r Henri-Guillaume-Mathias Olbers en date de Braunschweig den 3 September 1805, publiée par B. Boncompagni d'après l'original possédés par la Société Royale des Sciences de Göttingen*, Berlin, Institut de photolithographie des frères Burchard, imprimerie de Gustave Schade (the letter and its Italian translation by A. Sparagna, were printed in *Bullettino di Bibliografia e Storia delle Scienze Matematiche e Fisiche* 16: 215–220).

———. 1884. Intorno ad una lettera di Carlo Federico Gauss al Dr. Enrico Guglielmo Mattia Olbers. *Atti dell'Accademia Pontificia de' Nuovi Lincei* 36: 201–295.

———. 1892. *Catalogo dei manoscritti ora posseduti da D. Baldassarre Boncompagni, by E. Narducci*. Roma: Tipografia delle Scienze Matematiche e Fisiche.

———. 1895. *Catalogo della insigne biblioteca appartenuta alla chiara memoria del Principe Baldassarre Boncompagni, Parte I*. Roma: Cecchini.

Bucciarelli, L.L., and N. Dworsky. 1980. *Sophie Germain: An Essay in the History of the Theory of Elasticity*, Studies in the History of Modern Science. Vol. 6. Dordrect: Reidel.

Candido, G. 1941. Il "fondo Palagi-Libri" della Biblioteca "Moreniana" di Firenze. Atti del secondo Congresso dell'Unione Matematica Italiana, 841–885. Roma: Edizioni Cremonese.

Comte, A. 1864. *Cours de philosophie positive*. Paris: Bachelier.

Dahan-Dalmédico, A. 1987. Mécanique et théorie des surfaces: les travaux de Sophie Germain. *Historia Mathematica* 14: 347–365.

Del Centina, A. 2005. Letters of Sophie Germain Preserved in Florence. *Historia Mathematica* 32: 60–75.

———. 2008. Unpublished manuscripts of Sophie Germain and a revaluation of her work on Fermat's last theorem. *Archive for History of Exact Sciences* 62: 349–392.

Del Centina, A., and A. Fiocca. 2004. *L'archivio di Guglielmo Libri dalla sua dispersione ai fondi della Biblioteca Moreniana* [The Archive of Guglielmo Libri from its Dispersal to the Collections at the Biblioteca Moreniana]. Firenze: Olschki.

———. 2010. *Guglielmo Libri matematico e storico della matematica*. Firenze: Olschki.

———. 2012. The Correspondence of Sophie Germain with Carl Friedrich Gauss. *Archive for History of Exact Sciences* 66: 585–700.

Dickson, L.E. 1971. *History of the Theory of Numbers*, 3 vols. New York: Chelsea.

Dunnington, G.W. 1955. *Gauss, Titan of Science* (Reprinted by The Mathematical Association of America in 2004. New-York: Hafner).

Edwards, H.M. 1977. *Fermat's Last Theorem: A Genetic Introduction to Algebraic Number Theory*. New York: Springer.

Gauss, C.F. 1801. *Disquisitiones Arithmeticae*, Lipsia apud Gerh. Fleischer (Also in: *Werke*, erster Band, herausgegeben von der Königlichen Gesellschaft der Wissenschaften zu Göttingen, 1863: 1–474).

———. 1917. *Werke*, zehnten Bandes, Erste Abteilung, herausgegeben von der Königlichen Gesellschaft der Wissenschaften zu Göttingen.

Genocchi, A. 1880. Il carteggio di Sofia Germain e Carlo Federico Gauss. *Atti della Accademia delle scienze di Torino* 15: 795–808.

———. 1884a. Alcune asserzioni di C.F. Gauss circa le forme quadratiche YY±nZZ. *Bullettino di Bibliografia e di Storia delle Scienze Matematiche e Fisiche* 17: 245–247.

———. 1884b. Teoremi di Sofia Germain intorno ai residui biquadratici. *Bullettino di Bibliografia e di Storia delle Scienze Matematiche e Fisiche* 17: 248–251.

Germain, S. 1828. Examen des principes qui peuvent conduire à la connaissance des lois de l'équilibre et du mouvement des solides élastiques. *Annales de Chimie* 38: 123–131.

———. 1831. Note sur la manière dont se composent les valeurs de y et z dans l'équation $\frac{4(x^p-1)}{x-1} = y^2 \pm z^2$ et celles de Y' et de Z' dans l'équation $\frac{4\left(x^{p^2}-1\right)}{x-1} = Y'^2 \pm Z'^2$. *Journal für die reine und angewandte Mathematik* 7: 201–204.
———. 1833. *Considérations générales sur l'état des sciences et des letters*, éd. M. Lherbette. Paris: Lachevardière.
Grattan-Guinness, I. 1984. Note sur les manuscrits de Libri conservés à Florence. *Revue d'Histoire des Sciences* 37: 75–76.
Günther, S. 1881. Die Briefwechsel zwischen Gauss and Sophie Germain. *Historisch-literarische Abteilung der Zeitschrift für Mathematik und Physik* 26: 19–25.
———. 1882. Il carteggio tra Gauss e Sofia Germain (Italian translation by Alfonso Sparagna). *Bullettino di Bibliografia e di Storia delle Scienze Matematiche e Fisiche* 15: 174–179.
Henry, C. 1879. Les manuscrits de Sophie Germain et leur récent éditeur, Documents nouveaux. *Review of Philosophy* 8: 619–641.
———. 1880. Comptes rendus et analyses. *Bulletin des Sciences Mathématiques et Astronomiques* 4 (2): 273–276 [Review of the five letters by Sophie Germain to Gauss edited by Boncompagni].
Laubenbacher, R., and D. Pengelley. 2010. Voici ce que j'ai trouvé: Sophie Germain's Grand Plan to Prove Fermat's Last Theorem. *Historia Mathematica* 37: 641–692.
Legendre, A.-M. 1827. Recherches sur quelques objets d'analyse indéterminée et particulièrement sur le théorème de Fermat. *Mémoires de l'Académie des sciences de l'Institut de France* 6 (2): 1–60.
Leibrock, G. 2001. Meine Freundin Sophie. Carl Friedrich Gauß' Brieffreudschaft mit Sophie Geermain. *Mitteilungen Gauß-Gesellschaft Göttingen* 38: 17–28.
Libri, G. 1832. Notice sur M.lle Sophie Germain. *Journal des Débats*, 18 mai 1832 (Also in: *Considérations générales sur l'état des sciences et des letters*, M. Lherbette éd. Paris: Lachevardière).
———. 1859. *Catalogue of the Extraordinary Collection of Splendid Manuscripts ... Formed by M. Guglielmo Libri*. Londra: Sotheby & Wilkinson.
Maccioni Ruju, A., and M. Monstert. 1995. *The Life and Time of Guglielmo Libri (1802–1869): Scientist, Patriot, Scholar, Journalist and Thief: a Nineteenth-Century Story*. Hilversum: Verloren.
Mansion, P. 1880. *Nouvelle Correspondance Mathématique* 6: 315–317 (Review of the five letters by Sophie Germain to Gauss edited by Boncompagni).
Poinsot, L. 1820. Mémoire sur l'application de l'algèbre à la théorie des nombres. *Journal de l'École Polytechnique* 11: 342–410 (Also in: *Mémoires de l'Académie des sciences de l'Institut de France* 4(2) (1824): 99–184).
Ribenboim, P. 1999. *Fermat's Last Theorem for Amateurs*. New York: Springer.
Sampson, J.H. 1990. Sophie Germain and the Theory of Numbers. *Archive for History of Exact Sciences* 41: 157–161.
Schering, E. 1877. *Carl Friedrich Gauss' Geburtstag nach Hundertjahriger Wiederkehr Festrede*, Abhandlungen der Mathematischen Classe der K. Gesellschaft der Wissenschaften in Göttingen, Band 22. Italian translation in: *Annali di Matematica pura e applicata* 9: 210–239.
———. 1879. Bemerkungen über Gauss' Brief vom 30 April 1807 an Sophie Germain. *Nachrichten von der K. Gesellschaft der Wissenschaften und der G.A. Universität zu Göttingen* 1879: 381–384.
———. 1880. Briefe der Sophie Germain an Gauss in Photographie veröffentlicht von B. Boncompagni. *Nachrichten von der K. Gesellschaft der Wissenschaften und der G. A. Universität zu Göttingen* 1880: 367–369.
Stupuy, H. 1879. *Oeuvres Philosophiques de Sophie Germain*. Paris: Ritti.
———. 1896. *Oeuvres Philosophiques de Sophie Germain*. Nouv. ed. Paris: Ritti.
Truesdell, C. 1991. Sophie Germain: Fame Earned by a Stubborn Error. *Bollettino di Storia delle Scienze Matematiche* 11 (2): 3–24.

Hermite and Lipschitz: A Correspondence and Its Echoes

Catherine Goldstein

Abstract
In the second half of the nineteenth century, the French mathematician Charles Hermite wrote thousands of letters to dozens of correspondents. Mixing personal, political, academic and mathematical matters, as well as views on mathematics and its development, these letters offer a vivid picture of the mathematical landscape of the time. Particularly interesting is the fact that many themes appear repetitively among several correspondents, while some others, contrastingly, are specific to only one. Such echoes and contradictions are, of course, evocative, but also constitute a challenge to a potential editor: neither strict chronology nor restriction to one correspondent allow us to take them into account. We discuss here these problems and some solutions, while focussing on the exchanges between Hermite and the German mathematician Rudolf Lipschitz.

1 Charles Hermite, Rudolf Lipschitz and Their Correspondence

1.1 Parallel Lives

Charles Hermite (1822–1901) and Rudolf Lipschitz (1832–1903) were separated by a decade and a frontier, but their professional lives evolved in intertwined patterns. In 1842, Hermite succeeded in entering the École polytechnique, the main incubator of French mathematicians at the time. But barred from the standard careers open to

C. Goldstein (✉)
CNRS, Institut de mathématiques de Jussieu-Paris Rive Gauche, UMR 7586, Sorbonne Université, Univ Paris Diderot, Paris, France
e-mail: catherine.goldstein@imj-prg.fr

M. T. Borgato et al. (eds.), *Mathematical Correspondences and Critical Editions*, Trends in the History of Science, https://doi.org/10.1007/978-3-319-73577-1_10

Polytechnique graduates by a lame foot and possessed by a much greater enthusiasm for pure mathematics than for engineering, Hermite left the school after a year, thus cutting short the French royal road to mathematical excellence. He briefly tried an alternative path, passing examinations as a prerequisite to a career as a high school teacher, but this did not appeal much to him either. Giving lectures at the Collège de France, then holding positions as *répétiteur* and *examinateur* at the Polytechnique, but, above all, obtaining important results on quadratic forms and elliptic functions paved Hermite's unorthodox and slightly chaotic way to institutional recognition. When he was elected to the French Academy of Sciences in 1856, he still did not have any professorship.

At first glance, Lipschitz's path seems more straightforward, if not more prestigious: from 1847 on, he studied in Königsberg, then in Berlin. After his 1853 *Doktorarbeit*, he became a high school teacher for a few years, during which time he published papers on quadratic forms and series, and prepared his *Habilitation*. From 1857 on, he was Privatdozent in Bonn, but left five years later for an extraordinary professorship at the University of Breslau. That very year, 1862, an equivalent position was at last created for Hermite, as *maître de conférences* at the École normale supérieure that the recently appointed director of studies, Louis Pasteur, wanted to reorganize. But two years later, Lipschitz returned to Bonn, this time as a full professor; Hermite would not obtain such a position before the end of the decade, first at Polytechnique, then—and for a while simultaneously—at the Sorbonne.

In the 1870s, both mathematicians had reached the forefront of their professions in their respective countries. Lipschitz's choice to stay at Bonn, despite other attractive proposals, may seem from our perspective today to have put him in a backwater, compared to Hermite, who, in Paris, was at the center of all things mathematical. On the other hand, Lipschitz committed himself to important administrative duties, for example, as *Rektor* (chancellor) of his university in 1874, and editor of the celebrated *Journal für die reine und angewandte Mathematik*, duties that Hermite sought to avoid at all costs for his entire life.

Hermite had come to Berlin in the early 1850s, in order to meet some of the most famous representatives of the German mathematical intelligentsia: Peter Gustav Lejeune Dirichlet, Gotthold Eisenstein and Ernst Eduard Kummer in particular,[1] but did not seem to have been acquainted with Lipschitz at this time. Dirichlet, however, was a key figure for both men: Lipschitz's *Doktorarbeit* was written under his supervision, and according to his necrologist, Hermann Kortum: "Lipschitz's mathematical thought was defined through Dirichlet, whose pupil he considered himself," (Kortum 1906, p. 57). As for Hermite, he also described himself on several

[1]See Goldstein (2007, p. 379, n. 8). Mathematics in Berlin at the time is described in Biermann (1973/1988) and Begehr et al. (1998). On this particular circle, see also Pieper (2007).

occasions as one of Dirichlet's disciples.[2] This is not the only point in common between the two men: both were offered a position in Göttingen (which they ultimately both declined), both had a large spectrum of mathematical interests, from number theory to forms to mechanics, as well as a deep commitment to analysis. Their national and international recognition is well attested to, by the number of mathematical journals to which they were both asked to contribute, as well as by their election to prestigious Academies of Sciences; both, for instance, were correspondents of the Accademia dei Lincei in Rome and of the Berlin Akademie der Wissenschaften.

1.2 Correspondence

Another point in common is that both mathematicians were centers of vast correspondence networks, with large areas of overlap. We know that they both exchanged letters with Eugenio Beltrami, Georg Cantor, Jules Hoüel, Leopold Kronecker, Leo Königsberger, Gösta Mittag-Leffler, Henri Poincaré and James Joseph Sylvester, for instance. For both of them, a number of these letters were published as technical articles during their lifetime. However, important differences exist between their correspondence, then and now.

Hermite wrote exclusively in French, and most of his correspondents followed suit. The bulk of his passive correspondence, transmitted after his death to one of his sons-in-law, the mathematician Émile Picard, is said to have been destroyed in a fire. His correspondence with Thomas Stieltjes, edited in 2 volumes as early as 1905 by Benjamin Baillaud and Henry Bourget, is one of the rare cases to offer letters in both directions.[3] On the other hand, Hermite's letters to a variety of mathematicians, deposited in a matching variety of archives and libraries, have been published since the beginning of the twentieth century: for example, those to Paul Du Bois-Reymond, by Emil Lampe in 1916; to Andrei Markoff, by Helen Ogigova in 1967; to Gösta Mittag-Leffler, by Pierre Dugac in 1984–1989; to Ernesto Cesàro (and partially to Eugène Catalan), by Paul Butzer, Luciano Carbone, François Jongmans and Franco Palladino in 2000; to Angelo Genocchi, by Giacomo Michelacci in 2003; and to Georg Cantor, by Anne-Marie Decaillot in 2008. To this must be added the publication of many selected letters, e.g., with Italian mathematicians, by Umberto Bottazzini, or with Sylvester, by Karen Parshall.[4]

---

[2]Just before his death, he wrote for instance to Eugen Jahnke: "I have always been and will be until the end the disciple of your great mathematicians, Gauss, Jacobi, Dirichlet." In an 1853 letter to Dirichlet himself after his trip to Berlin, Hermite even evoked "a law of my destiny not to do anything in arithmetic other than unearth some of the discoveries you made a long time ago", (Goldstein 2007, pp. 399–400).

[3]This edition (Hermite and Stieltjes 1905) is unfortunately bowdlerized… and was only partially completed by Dugac (1983).

[4]See, respectively, Hermite (1916, 1967, 1984/1985/1989, 2003), Butzer et al. (2000), Décaillot (2008), Bottazzini (1977), and Parshall (1998).

Lipschitz's correspondence, on the other hand, includes letters in German, French, English and Italian; Lipschitz also kept drafts of his own letters, which, in some cases, allow for a more complete view of the exchange. Lipschitz's *Nachlass*, at the Universitäts- und Landesbibliothek Bonn (Abteilung Handschriften und Rara), hosts most of the surviving letters, about 600 of them, including 455 letters or cards to Lipschitz from 61 correspondents (Scharlau 2006). Winfried Scharlau has published a selection of them (Lipschitz 1986), consisting of 140 letters or extracts of letters, from a few lines to ten pages long, among which 26 are from Lipschitz. Scharlau's edition includes 13 letters from Hermite and 2 to him. A few isolated letters from or to Lipschitz have also been published in translation.[5]

1.3 The Correspondence Between Hermite and Lipschitz

Most of the surviving correspondence between Hermite and Lipschitz is kept in Lipschitz's *Nachlass*. It contains 148 letters and 9 postcards from Hermite to Lipschitz, the first dated August 19, 1877, and the last July 14, 1900, 6 months before Hermite's death. As expected, they are all in French. We also find in the *Nachlass* 70 drafts of letters from Lipschitz to Hermite, written mostly between 1877 and 1886 (one sole letter is dated 1892), again, all in French, with four exceptions (still) in German. Moreover, Hermite's file in the Archives of the French Academy of Sciences contains two letters from Lipschitz, one of them corresponding to a draft kept in Lipschitz's *Nachlass*. Although it is clear from allusions in the surviving correspondence that several letters are missing, this is by far the most extensive correspondence known from or to Lipschitz (the second most numerous in his *Nachlass* is composed of 57 letters written by Carl Borchardt). Their physical appearance, however, is not very appealing. As noted by W. Scharlau, Lipschitz's drafts are very badly written, with many corrections and deletions, often difficult to decipher (Lipschitz 1986, p. xvi). On some of Hermite's letters, the ink has almost disappeared. To this can be added the fact that the microfilm scans are of extremely poor quality (Fig. 1).

The correspondence starts in 1877, when Lipschitz sends to Hermite the first volume of his *Lehrbuch der Analysis* (textbook on analysis) (Lipschitz 1877–1880). Hermite writes back: "You offer me an opportunity that I eagerly seize to remind you of me, while thanking you for the first volume of your treatise on analysis that you have bestowed on me the honor of sending to me."[6] The acquaintance between the two mathematicians dated back only a few months earlier, when they had both

[5]For instance, some letters exchanged with Richard Dedekind are translated in (Dugac 1976).

[6]Except where otherwise indicated, all the quotes come from Lipschitz's *Nachlass*. In the *Nachlass*, the letters are organized by sender and, for each sender, are numbered independently. Here, we use these numbers, but, for the sake of clarity, add a star to those of letters sent by Lipschitz. Letter 1: *Vous m'offrez une occasion que je saisis avec empressement, de me rappeler à votre bon souvenir en venant vous remercier du premier volume de votre traité d'analyse que vous m'avez fait l'honneur de m'envoyer.*

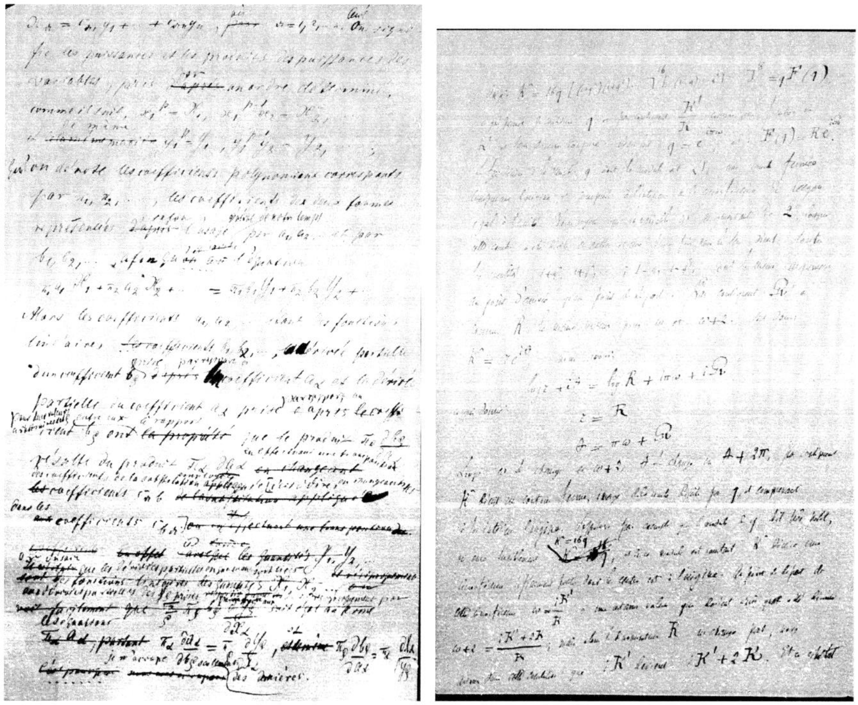

Fig. 1 Two extracts of the correspondence: left, a draft by Lipschitz (6*, 1878); right, a letter from Hermite (109, 1884). Repr. with the kind authorization of the Archives of Universitäts- und Landesbibliothek Bonn

attended the March 30 Göttingen ceremony for the centenary of Carl Friedrich Gauss' birth. This episode is recalled by Hermite on the occasion of New Year's 1878, and again twenty years later, on December 29, 1897: "You call to mind, Sir, our first encounter in Göttingen, during the centennial ceremony in honor of Gauss, which left me with unforgettable memories."[7] The exchange then accelerates rather quickly, culminating in the mid-eighties (see Fig. 2), with one or more letter per month on each side. Several times, letters cross each other, a circumstance duly noted and a cause for another exchange; more than once, Hermite sends a postcard in the immediate aftermath of his regular letter, to signal a formula that needs correction or a reprint he has forgotten to request.

Besides dates, the heading of the letters keeps a trace of Hermite's wanderings. Every year, he spends holidays in his native Lorraine (which had in part been assigned to Germany after the 1870 war), or with one of his married daughters in the western part of France; he also visits these places for family events. After an

[7]Letter 150: *Vous me rappelez, Monsieur, notre première rencontre à Gottingue, lors des fêtes du centenaire de Gauss qui m'a laissé d'inoubliables souvenirs.*

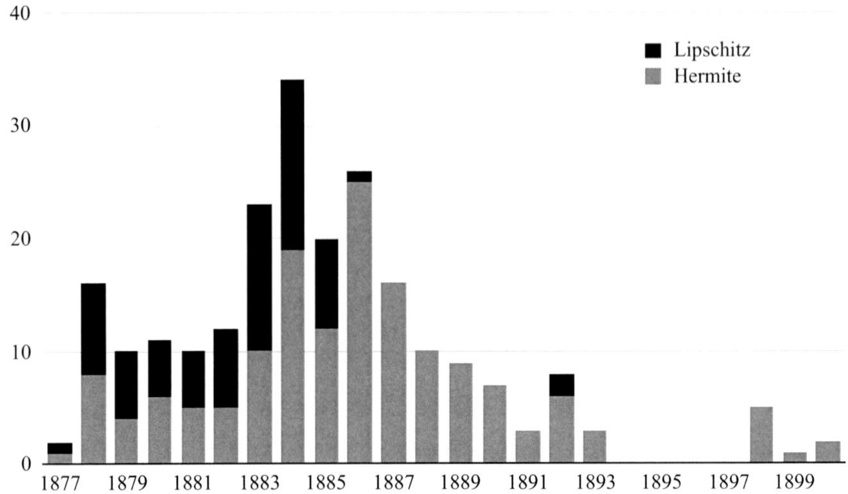

Fig. 2 Distribution by years of the surviving letters between Hermite and Lipschitz: in grey, those authored by Hermite, in black, by Lipschitz

illness, he is also obliged to go to thermal spas. Last, but not least, in November 1886, a short letter announces his upcoming arrival in Bonn for a visit to Lipschitz. These trips do not deter him from writing, nor, as the letters themselves reveal, from working on mathematics, although Hermite often complains of his own laziness or fatigue. On the other hand, all the drafts we have from Lipschitz, except one, are written from Bonn itself (although we know, for instance, that he spends several weeks in Switzerland in 1881 after a serious health problem).

Until 1888, Hermite's letters are simply addressed to "Monsieur Lipschitz, Professeur à l'université, Bonn (Prusse rhénane)" (after 1888, Koenigstrasse 34, Bonn (Prusse rhénane)), a good testimony to the efficiency of mail deliveries at the time and of the status of university professors. The salutations between both men never vary: "Monsieur" (Sir) is used all along in both directions. This contrasts with the "Mon cher Monsieur Schwartz" (my dear M. Schwarz) Hermite uses for Hermann Amandus Schwarz or with "Mon cher ami" (My dear friend), which appears in his letters to Carl Borchardt, Thomas Stieltjes or Gösta Mittag-Leffler. Stieltjes and Mittag-Leffler are much younger than Hermite, but the case of Borchardt, born in 1817, five years before Hermite, shows that it is not only a generational issue; rather, Hermite and Borchardt had met in Paris as young men as early as 1847. Borchardt himself uses "Verehrter Freund" (Esteemed friend) in his numerous letters to Lipschitz, while Helmholtz addresses him as "Bester Freund" or "Lieber Freund" (dear friend) (Lipschitz 1986, pp. 13–24, 120–128, resp.). The appellation Hermite and Lipschitz use for one another thus appears rather formal. Still, it does not hamper the raising of personal issues, nor some emphasis on the emotional importance of the exchange for both men. Lipschitz thanks Hermite for "[his] communications that are so dear to me,"[8] while Hermite ends his letters with "his feelings of the most sincere friendship" (January 30, 1883), expresses regrets that they are not geographically closer and colleagues at the Sorbonne (January 5, 1884) and confides, in 1892[9]:

> I consider as one of the greatest joys of my scientific career that for nearly 20 years I have always had your counsel near at hand.

1.4 A Closer Look: A Year and a Letter

To grasp more concretely the nature of the correspondence, let us consider as an example the year 1881: at the very end of the preceding December, Hermite had announced the marriage of his daughter Marie to his young colleague and protégé Émile Picard. A day before the wedding, on January 3, Lipschitz thanks Hermite for the various bits of news and expresses his best wishes for the young couple. Hermite answers on January 31 with some news of the family, a summary of the way one of his students, Jules Tannery, had constructed, following Karl Weierstrass, a new example of discontinuities of functions expressed as series, and his own comments on the issue; the end of the letter alludes to the upcoming election at the Academy of Sciences, in which Gaston Darboux and Camille Jordan are competing. Because of a long illness, he explains, Lipschitz delays his answer until April 1st, in which he comments informally on a mathematical remark by Hermite on the theory of transformations of quadratic forms; a few weeks later, on April 25, he completes his views on both topics, series and forms, to which Hermite quickly replies, on May 3, urging Lipschitz to publish his new theory. The French mathematician adds some explanations on his current course on analysis at the Sorbonne, concluding with surprise and regret about the silence surrounding Borchardt's death (on June 27, 1880), in particular, in the *Journal für die reine und angewandte Mathematik* of which Borchardt had been chief editor since 1856. The following letter, addressed from Hermite's vacation resort, explains in a self-mockingly desperate tone his failed attempts to extract some results on the Γ function from a formula given by Cauchy,

$$\int_0^{\frac{\pi}{2}} (2\cos x)^{a+b} \cos(a-b)x dx = \frac{\pi}{2} \frac{\Gamma(a+b+1)}{\Gamma(a+1)\Gamma(b+1)} : \qquad (1)$$

"But how, I ask you, how to disentangle what then becomes of the definite integral? The only thing I have seen clearly is that analysis was created for the chastisement

[8]Draft 23, April 25, 1881: *vos communications qui me sont si chères.*

[9]Letter 148, December 30, 1892: *je regarde comme un des plus grands bonheurs de ma carrière scientifique que depuis près de 20 ans j'ai toujours eu près de moi vos conseils.*

of pride and that it inflicts frequent and salutary humiliation."[10] He also proposes to ask Darboux, the editor of the *Bulletin des sciences mathématiques*, to publish a translation of Lipschitz's self presentation of his textbook. In Switzerland at the time for his health, Lipschitz writes as soon as he returns home, on October 30, and gives a direct proof of Cauchy's formula (1); he also attaches his photograph to the letter—a current practice of the time. Due to several deaths in his family, Hermite only reciprocates on December 13th, with his own photograph, as well as a follow-up on the issue of the *Bulletin*; he also transmits some laudatory commentaries on Lipschitz's treatise from various French mathematicians. He compares, in particular, the rigid programs of French courses with the flexible ones he thinks are possible in Germany, which allow professors to introduce innovations more easily. Hermite also regretfully evokes Eduard Heine, who had died on October 21. The last two letters of the year, one from Hermite on December 27 and the grateful answer from Lipschitz on the 30th, are devoted to the ongoing publication of Cauchy's complete works on behalf of the French Academy of Sciences: Hermite, indeed, has obtained one of the coveted books for Lipschitz and another for his Bonn colleague, Rudolf Clausius.

This mixture is quite typical of the whole correspondence. Discussing the editorial problems attached to it, Winfried Scharlau comments[11]:

> Then, the letters most prominently discuss a variety of mathematical questions and it would be necessary for a proper commentary to reconstruct these questions jointly from both the letters and the original papers.

The second half of this comment raises a fundamental issue to which I shall return in Sect. 3 below. But before this, I would like to look more closely at the first assertion through the observation of a concrete example, Hermite's letter of December 5th, 1883; it is composed of seven and a half pages, with about 16 to 19 lines of writing on each page. It begins with the announcement of the death of Hermite's sister-in-law (c. 5 lines), before turning to mathematics, more specifically, to a comparison between the class numbers of properly and improperly primitive binary quadratic forms of determinant $-D$, for $D \equiv +3 \bmod 8$.[12] Hermite recalls

[10]Letter 22, August 4, 1881: *Comment je vous le demande, comment débrouiller ce que devient alors l'intégrale définie ? La seule chose que j'ai vue clairement c'est que l'analyse a été créée pour le châtiment de l'orgueil et qu'elle inflige de fréquentes et salutaires humiliations.*

[11](Lipschitz 1986, p. xvi): *zweitens geht es in den Briefen ganz überwiegend um mathematische Fragen verschiedenster Art und für einen sachgerechten Kommentar wäre es erforderlich, diese Fragen aus den Briefen und Originalarbeiten zusammenhängend zu rekonstruieren.*

[12]In Gauss's normalization, a (binary quadratic) form is an expression of the type $Ax^2 + 2Bxy + Cy^2$, here with integral coefficients A, B, C; the determinant D is $B^2 - AC$. The form is properly (resp. improperly) primitive when $\gcd(A, B, C) = \gcd(A, 2B, C) = 1$, resp. when $\gcd(A, B, C) = 1, \gcd(A, 2B, C) = 2$. Two forms are equivalent when they can be transformed into each other by an invertible linear change of variables of determinant ± 1; forms equivalent to a properly (resp. improperly) primitive form are properly (resp. improperly) primitive form. For a given D, the number of classes of equivalent forms with integral coefficients is finite, thus so

the relation stated by Gauss, as well as a relevant article published by himself in 1862 (13 lines). He then proves the required relation via the series expansion of θ-functions (98 lines). This, Hermite says, will be a part of a future article for the *Bulletin of the Saint Petersburg Academy* that he sketches and which will involve the use of a formula communicated to him by Lipschitz (9 lines).[13] Finally, a post-scriptum (9 lines 1/2) adds[14]:

> Mr. Bischoffsheim, the member of Parliament who spoke to the House about the courses at the Sorbonne in the manner you know [i.e., very critically] is a candidate for a seat as a free member of the Academy of Sciences; you may well imagine that I shall not give him my vote. But his generosity in favor of astronomy and even of the Sorbonne, to which he offered a magnificent portrait of the great physiologist Claude Bernard, will give him many votes and he may well succeed.

Mathematics per se, results and proofs, represent here about three quarters of the letter, in line with W. Scharlau's statement. However, the remaining quarter draws a rich tapestry of nineteenth century scientific life, from family to political intelligence. In order to understand the uses of this correspondence, including the communication of mathematical results, a more systematic overview of the matters at hand in the letters is, in fact, worthwhile. The classification is mine and is mostly proposed for the sake of clarity—the quotes will make clear how intertwined the topics are.

2 A Variety of Topics

2.1 Personal Life

Family deaths, births and weddings are evoked very frequently, either explicitly, or, for the former, through the black frame of official mourning letters. Consolations and congratulations are exchanged on such occasions, as we have already seen. "We have lost Madame Duhamel after a long illness that had deprived her almost

is the number of classes of properly primitive (or improperly primitive) forms. The computation of these numbers was one of the difficult problems nineteenth century number theorists inherited from Gauss.

[13]This article was published in 1884, in volume 29 of the *Bulletin*, and reproduced the same year in the 5th volume of *Acta Mathematica*.

[14]Letter 40 (in the Archives, the numbers 39 and 40, in fact, represent two parts of the same letter): *Mr Bischoffsheim, le député qui a parlé à la Chambre des cours de la Sorbonne comme vous savez, se présente à une place de membre libre de l'Académie des sciences; vous pensez que je ne lui donnerai point ma voix. Mais ses générosités en faveur de l'Astronomie et même de la Sorbonne, à qui il a fait don d'un magnifique portrait du grand physiologiste Claude Bernard lui vaudront beaucoup de suffrages, et peut-être va-t-il réussir.* The banker Raphaël Bischoffsheim (1823–1906) had been elected to Parliament in 1881 and, unlike Hermite, supported republicanism. He had launched attacks against the Sorbonne professors as being old-fashioned and ignorant of the most recent innovations. He was ultimately elected to the Academy of Sciences, but only in 1890…

completely of the use of reason, and we had ceased a long time ago to have any hope of saving her,[15] explains Hermite on June 16, 1878. On Lipschitz's side, in 1882: "On March 11, I received the telegraphic announcement that my mother, who had been attacked by a serious illness since Christmas, who had shared all my interests from my childhood until her final days, and in whom clarity of mind and warmth of feeling could vanish only with life itself, had died on the preceding day."[16] Besides glimpses of family relations, personal services are sometimes required, for instance, when Hermite asks Lipschitz's help in favor of his nephew Georges Bertrand (the son of Alexandre Bertrand), who wants to spend five months in Bonn in order to learn German (letter 70). The intricacy of family and professional links among the members of the nineteenth century European intelligentsia is well-known (Zerner 1991), and is confirmed here by hints at professional discussions over family dinners. We also learn about travel plans and details of the personal organization of work. But some letters also provide information concerning personal relations *between third parties*. For example, on January 20, 1882, Hermite tells Lipschitz that "M. Chasles was M. de Jonquières's friend for a long time, but this bond that nothing should have broken was destroyed over a priority dispute."[17]

2.2 Circulation of Mathematics

The correspondence also documents the usually tacit character of a variety of mathematical collaborations. On the occasion of Eduard Heine's death, in October 1881, Hermite reveals, for instance: "I cannot tell you how afflicted I am by the loss of M. Heine, who was an excellent man, as well as a first-rate mathematician, whose works will remain forever a part of science. I wrote to him frequently and he did important favors for me by teaching me things from Riemann that, though very well-known in Germany, were not so to me."[18] Or on March 12, 1878: "May I venture

[15]Letter 6: *Nous avons perdu Madame Duhamel aprés une longue maladie qui lui avait enlevé à peu prés complètement l'usage de la raison, et depuis longtemps nous ne pouvions plus avoir l'espérance de la conserver.* Virginie Duhamel, the wife of the mathematician Jean-Marie Duhamel, to whose positions Hermite succeeded both at the Sorbonne and at the Polytechnique, was the sister of Joseph and Alexandre Bertrand's father, and thus the aunt of Hermite's wife.

[16]Letter 26*: *Le 11 mars, j'ai reçu l'avertissement télégraphique que ma mère qui était attaquée d'un mal grave depuis la fête de Noël, qui a partagé tous mes intérêts dès mon enfance jusqu'à ses derniers jours, chez laquelle la clarté de l'esprit et la chaleur des sentiments ne se sont évanouis qu'avec la vie même a été décédée le jour précédent.* Lipschitz's French, although good, is not always correct. I leave it untouched in the original, but give a grammatically correct English translation.

[17]Letter 26: *Mr Chasles avait été longtemps l'ami de Mr de Jonquières, mais cette liaison que rien n'aurait dû rompre, a été détruite par une question de priorité.*

[18]Letter 24: *Je ne puis vous dire combien j'ai été affecté de la perte de M. Heine, qui était un excellent homme en même temps qu'un géomètre de premier ordre dont les travaux resteront à jamais dans la science. Je lui écrivais fréquemment et il m'a rendu les plus signalés services en m'apprenant des choses de Riemann, extrêmement connues en Allemagne, et que j'ignorais.*

to ask you to tell me if you know of a paper 'On Rotation' in the *Mathematisches Wörterbuch* of MM. Hoffmann and Natani, written, so M. Borchardt tells me, on the basis of lectures that M. Weierstrass gave at the Berlin University and which present a close analogy to what I myself have just done."[19]

Thanks to the proximity of their mathematical interests, more specialized issues are also tackled. For instance, on February 20, 1878, Hermite asks: "Allow me to call your attention to a question concerning elliptic functions that worries me and on which I would like to have your opinion. You know that M. Rosenhain represents the four fundamental Jacobi functions $\Theta(x)$, $H(x)$, $H_1(x)$ and $\Theta_1(x)$ by $\theta_0(x)$, $\theta_1(x)$, $\theta_2(x)$, $\theta_3(x)$; this notation seems to me of real importance, as it allows us to encompass within a single equation a group of four relations, save for, as you will see, some difficulties that I cannot succeed in overcoming."[20]

2.3 Reflections on Science

Technical mathematics is thus sometimes an incentive to display more general points of view, be it, as here, concerning notation or, more generally, concerning the development of mathematics. When Hermite explains to Lipschitz his ideas on cuts—Hermite's analytical version of Riemann's more geometrical ideas on complex functions—he adds: "How greatly have ideas in analysis been modified because of all these facts, since the time when infinity first seemed to be the only possible discontinuity and the more recent period when the study of Fourier series revealed sudden jumps from one continuous series to another, completely different one."[21] Hermite often expresses his deeply-felt epistemological convictions, for instance, on April 12, 1882[22]:

> I believe that in science, and especially in mathematics, we are less masters than servants of our work. I do not deny free will, but I think that it coexists with the action of a force

[19]Letter 4: *Oserais-je aussi vous prier de me faire savoir si vous avez connaissance d'un article Sur la rotation du Mathematisches Wörterbuch de MM Hoffmann et Natani rédigé m'a dit Mr Borchardt d'après les leçons de M. Weierstrass données à l'Université de Berlin et qui offriraient une grande analogie avec ce que je viens de faire moi-même.*

[20]Letter 3: *Permettez aussi d'appeler votre attention sur une question relative aux fonctions elliptiques qui me préoccupe et sur laquelle j'aimerais avoir votre avis. Vous savez que M. Rosenhain représente les quatre fonctions fondamentales de Jacobi,* $\Theta(x)$, $H(x)$, $H_1(x)$ *et* $\Theta_1(x)$ *par* $\theta_0(x)$, $\theta_1(x)$, $\theta_2(x)$, $\theta_3(x)$ *; c'est cette notation qui me paraît avoir une importance réelle, en permettant de comprendre dans une seule équation, un groupe de quatre relations, sauf toutefois les difficultés que vous allez voir et que je ne puis réussir à lever.*

[21]Letter 19: *Combien les idées en Analyse se sont modifiées en présence de tous ces faits, depuis le temps où l'infini avait paru d'abord la seule discontinuité, et l'époque plus récente où l'étude de la série de Fourier a révélé des sauts brusques d'une série continue, à une autre toute différente.*

[22]Letter 27: *Je crois que nous sommes dans les sciences et tout particuliérement dans les mathématiques, moins les maîtres que les serviteurs de notre œuvre. Je ne nie point le libre arbitre, mais que je pense qu'il coexiste avec l'action d'une force qui naissant par notre fait, agit en dehors de nous, et à notre insu, nous dirige là même où nous ne voudrions pas aller ... je reconnais à la fois que la conception des espaces de Riemann m'effraye, et qu'elle a sa raison d'être.*

> which, while it arises from us, acts outside us, and unwittingly directs us where we would not like to go ... I recognize that the conception of Riemann spaces frightens me, but that it has its utility.

2.4 Publication

On a more material note, we also learn about the concrete functioning of publications at the time. On a 1880 postcard, Hermite summarizes: "I have just finished correcting the proofs of the first part of your article that was presented to the session of the Academy, and I informed M. Gauthier-Villars that you wished to have reprints, but it would be necessary to tell him how many copies you want. Please write him a note to inform him without delay."[23] Or, concerning one of his own texts[24]:

> My lessons at the Sorbonne are being published this year in lithography. They have been written down by a student of the Ecole normale, and with the authorization of the Faculty. But the publisher did not think it relevant to give me a single copy, so that the one I possess I had to buy with my own money. One of my friends told me on this occasion that, for this publisher, I was a goose to be plucked, and I had a good laugh. But the main point is that the students can, with the short draft of the lessons, easily follow a course from which, I have been told, very few were able to profit in preceding years.

French scientific journals published only in French, and translations had to be negotiated. "I had the occasion of discussing you with M. Darboux, who expressed the desire to have a reprint of the note you gave to the Göttingen *Nachrichten*, on your treatise on analysis, in order to give it to one of his coworkers, who will do a translation to appear in the *Bulletin*."[25]

Despite the continuity of the title, the change of editors of the *Journal de mathématiques pures et appliquées* gives rise, on July 23, 1884, to Hermite's comment: "I shall make it the subject of an article that M. Camille Jordan asked

[23]Card 18: *Je viens de corriger les épreuves de la première partie de votre article qui a été présenté à la séance de l'Académie, et j'ai prévenu Mr Gauthier-Villars que vous désiriez en avoir un tirage à part, mais il serait nécessaire de lui faire connaître combien vous voulez d'exemplaires. Permettez-moi de vous prier de lui écrire un mot pour l'en informer sans retard.*

[24]Letter 27: *Les leçons à la Sorbonne se publient cette année sous forme de feuilles lithographiées, qui ont été rédigées par un élève de l'Ecole Normale, et avec l'autorisation de la Faculté. Mais l'éditeur n'a point jugé à propos de m'en donner un seul exemplaire, de sorte que celui que je possède je l'ai acheté de mes deniers. Un de mes amis m'a dit à cette occasion, que j'avais été pour cet éditeur, une poule à plumer, et j'en ai bien ri. Mais l'essentiel c'est que les élèves puissent avec la rédaction sommaire des leçons suivrent (sic) facilement un cours dont très peu profitaient m'a-t-on dit les années précédentes.*

[25]Letter 25: *J'ai eu l'occasion de m'entretenir de vous avec Mr Darboux qui m'a exprimé le désir d'avoir un exemplaire séparé de la notice que vous avez donnée dans les Nachrichten de Gottingue, sur votre traité d'Analyse, afin de le donner à celui de ses collaborateurs qui en fera la traduction destinée à paraître dans le Bulletin.* On the translations into French of foreign scientific articles during the nineteenth century, see Bret and Verdier (2012).

me to publish in the first issue of a *Journal de mathématiques*, of which he will become the main editor. This future *Journal* is that of M. Resal, which continued with mediocre success that of M. Liouville, and which the publisher M. Gauthier-Villars wants to revive and transform."[26]

2.5 Teaching

The publication of textbooks is not the only mention of teaching in this correspondence. The various reforms of the curricula, the difference between the situations in France and Germany, and even specific pedagogical issues are discussed in detail. For instance, Lipschitz explains, on November 13, 1877: "I have come to believe that the understanding of the fundamental theorem of algebraic equations requires from the beginner a particular effort and that a longer way, which leads to the proof while teaching how to find a root of an equation by computation, is to be preferred to a shorter, but less illuminating, way."[27]

Or Hermite, on December 5, 1882: "None of our legislators could imagine that M. Bouquet and myself agreed, after the war, to combine our efforts to raise the level of the teaching of analysis at the faculty, and that, with this objective, I dropped my course on advanced algebra in order to be the assistant in the basic course, on differential and integral calculus."[28]

2.6 Scientific Policy

As shown by the letter of December 5th, 1883, summarized earlier, battles for a position or recruitment to a scientific society occupy a rather prominent place in the letters. On December 28, 1880, Hermite explains that "[t]he geometry section [of the Academy] will have some difficulty in deciding between the two main candidates, M. Camille Jordan and M. Darboux, who both have good credentials, but have unequal chances of success. The first is more favored by more of our colleagues, but I must confess that the second seems to me to have done more and better work,

[26]Letter 52: *J'en ferais le sujet d'un article que Mr Camille Jordan m'a demandé pour paraître dans le premier n^o d'un Journal de Mathématiques dont il sera le rédacteur en chef. Ce futur Journal est celui de Mr Résal, qui a succédé avec un succès médiocre à celui de Mr Liouville, et que l'éditeur Mr Gauthier-Villars veut relever et transformer.*

[27]Letter 1*: *Je suis parvenu à croire que l'entendement du théorème fondamental des équations algébriques exige des commençants un effort tout particulier et qu'un chemin plus long qui mène à la démonstration en apprenant comme on puisse trouver une racine d'une équation par le calcul soit préférable à un chemin plus court, mais moins lumineux.*

[28]Letter 29: *Aucun de nos législateurs ne s'est douté que Mr Bouquet et moi nous sommes convenus, après la guerre, de réunir nos efforts pour relever l'enseignement de l'analyse à la faculté, et que dans ce but j'ai renoncé à mon cours d'algèbre supérieure afin de me faire l'auxiliaire du cours fondamental, de calcul différentiel et de calcul intégral.*

entirely free of the obscurity for which one can only too easily reproach M. Jordan's work on the theory of equations."[29] Or, on February 24, 1885: "A great pitched battle was just fought at the Faculty around the choice of a substitute for the chair of analysis; it was a fight between the former students of the Polytechnique and those of the École normale."[30]

Administrative duties are also commented on in general, at least from Hermite's side, who regularly complains or jokes about them. On October 5, 1889, he describes, for instance, the ceremony of the inauguration of the new Sorbonne that had taken place in August, "in pomp and circumstance, in front of the President of the Republic, several ministers, representatives of the main bodies of the State, eminent characters such as M. Pasteur, M. Duruy, M. Jules Simon, etc., and a thousand students from every part of the world, with their national costumes and the banners of their countries."[31]

2.7 Politics

There is thus just one step from scientific administration to general politics. Hermite, who despised the Third Republic, loses no occasion to express his proximity with Germany and the values he believes to be incarnated by the Prussian state. In December 14, 1882, he declares: "My intimate feeling, which is more an impression than a deduction, is that radicals like M. Paul Bert, M. Laisant, etc., are leading us to imminent and horrible catastrophes. I would go further, I believe that M. Wirchow and his party, who in your country want ministerial responsibility, are thus moving to the revolutionary side. There is no need, it seems to me, of this responsibility in order to resist M. von Bismarck as fully as necessary."[32] Or again, on December

[29]Letter 19: *la section de géométrie aura fort à faire pour se prononcer entre deux candidats principaux Mr Camille Jordan et Mr Darboux qui tout deux ont bien des titres, mais avec des chances inégales de succès. C'est le premier qui est le plus en faveur auprès du plus grand nombre de nos confrères, mais je vous avoue que le second me semble avoir fait plus de travaux et des travaux meilleurs, entièrement exempts de l'obscurité qu'on n'a que trop à reprocher à ceux de Mr Jordan, sur la théorie des équations.* Jordan will nonetheless be elected in 1881, to replace Michel Chasles, who had died on December 18, 1880.

[30]Letter 63: *Une grande bataille rangée vient de se livrer à la Faculté pour le choix d'un suppléant à la chaire d'analyse; c'était la lutte entre les Polytechniciens et les Normaliens.*

[31]Letter 23: *C'est le 8 Aout qu'a eu lieu la cérémonie de l'inauguration de la Sorbonne, en grand apparat, par devant le Président de la République, plusieurs ministres, des représentants des grands corps de l'Etat, d'éminents personnages comme Mr Pasteur, Mr Duruy, Mr Jules Simon, etc. etc., et un millier d'étudiants de toutes les parties du monde avec leurs costumes nationaux, et les bannières de leurs pays.*

[32]Letter 30: *Mon sentiment intime qui est plutôt une impression qu'une déduction, c'est que les radicaux tels que Mr Paul Bert, Mr Laisant, etc. nous conduisent à de prochaines et d'affreuses catastrophes. J'irai plus loin, je crois que Mr Wirchow et son parti qui veulent chez vous la responsabilité ministérielle vont ainsi du côté de la révolution. Point n'est besoin, ce me semble, de cette responsabilité, pour résister autant qu'il est nécessaire à Mr de Bismarck.* The physiologist Paul Bert was Minister of Education in 1881 and 1882, and an advocate of a secular and free school

28, 1899, a year before his death: "At least M. Doumer reassured us with respect to Germany, with whom our relations are better, it is said, so that we should expect an alliance between the two nations, against England, rather than a new war. What a marvellous thing it would be to march into battle with our previous adversaries!"[33] On this topic, the correspondence is not symmetric: as far as we can see from the drafts, Lipschitz remains very circumspect on political issues.

3 Echoes: The Correspondence as a Non-closed Corpus

3.1 The Intricacies of Privacy

Letters are both social and textual links. Scientific letters have sometimes been described as defining open networks of scientific communication, as opposed to those of closed institutions, academies or journals, which require membership or referee processes or entrance examinations (Lux and Cook 1998). Sophie Germain, for instance, could not enter the Polytechnique, nor be a member of the Academy of Sciences, but she could write to Gauss. As we have seen, the exchanges between Hermite and Lipschitz were born in, and supported by, a professional setting. But they were not strictly professional; they were initiated and reinforced by personal encounters, in Göttingen or Bonn, and as such, their correspondence is far from being truly open[34]: it relies on strong ties between the two participants, which at first sight appear to be both personal and scientific, or at least to develop in such a way.

The traditional classificatory dichotomies, private vs. public and personal vs. professional, are particularly called into question.[35] The exchanges, as we have seen, are situated at the margins of the professional world, crossing the frontier regularly, but partially. The letters display a large spectrum of subjects, from strictly confidential matters, both professional and personal (for instance, when comparative

system (he was also in favor of colonization and of a republican racism). The Polytechnician Charles-Ange Laisant was a mathematician who supported *Boulangisme* in the 1880s; he was later a cofounder of the journal *L'Enseignement mathématique* (see Auvinet 2013). Both men sat on the extreme left in Parliament. The pathologist Rudolf Virchow cofounded the radical *Deutsche Fortschrittspartei*; he defended the idea that ministers should be held responsible for state expenditures engaged without authorization of Parliament.

[33]Letter 157: *Au moins Monsieur Doumer nous a rassurés à l'égard de l'Allemagne, avec qui nos rapports sont meilleurs, dit-on, de sorte qu'on devrait plutôt croire à une alliance entre les deux nations, contre l'Angleterre, qu'à une nouvelle guerre. Quelle chose merveilleuse ce serait de marcher au combat avec nos anciens adversaires!* Paul Doumer, a future president of the French Republic, was, at the time, Governor-General of French Indochina.

[34]The difference between a one-to-one exchange and a correspondence network involving several persons, such as that relative to the editing committee of a journal, is decisive. On this point, for another period, see Goldstein (2010). This is true even if one focuses on the exchanges between two persons inside a more collective setting; a good example here is the exchange between Lipschitz and Darboux for the *Bulletin des sciences mathématiques* (Lipschitz 1986, pp. 44–46).

[35]This classification has been extensively discussed for the nineteenth century, see Chartier (1991), Secord (1994), Diaz (1995) and Dauphin (2003).

opinions on candidates for a position are requested or details on a family member are provided), to public matters in the most obvious sense, such as those letters published verbatim in mathematical journals. It can become domestic, or even intimate, as when Hermite vents his regrets or frustration on his own work. But this intimacy is clearly delimited: it does not involve sharing thoughts on their marriages, nor even comments on novels one of them may have read, or concerts they may have attended.[36] From the social point of view, then, this correspondence appears to be both closed and restricted in its content.

From a textual point of view, on the other hand, it is not a closed corpus. This may appear to be a trivial remark: many parts of the letters are, of course, not understandable to the modern reader without a recourse to external information (as illustrated in several footnotes of the preceding sections), information which would have been obvious to any nineteenth century cultured person (that pertaining to politics), or to any nineteenth century mathematician (the current abbreviations for the titles of journals). Allusions to people, political events, recent mathematical results, are of this kind. They delineate tacit knowledge, operating at different scales, from what is shared by all contemporaries to what is shared by the two correspondents; and, in this sense, the corpus is not closed textually, all the more so *because* it is closed socially.

But what I mean to say is different: contextualization through external sources is decisive for a proper understanding of the place of the correspondence in the work and lives of Hermite and Lipschitz themselves, and of its role in the more general scientific communication network. To take an example, the election at the Academy of Sciences to replace Michel Chasles in 1881–1882 diffuses through a variety of writings: the opinion required from Lipschitz—on Darboux's and Jordan's respective merits—appears to be, in fact, a simple sidetrack of the main issue, which was to avoid, at all costs, the election of the engineer Amédée Mannheim (Hermite 1984/1985/1989, I, pp. 99–100, 117–118). Hermite is pushed by some of his colleagues to put Jordan (who, for a variety of reasons, was more likely than Darboux to be elected) alone in the first line, in order to secure the votes against Mannheim. Although they do not mention Mannheim, the letters to Lipschitz express Hermite's resistance to this strategy and his attempts to circumvent it through international support: this point has required external documents to be understood.

3.2 Publication Echoes

The role of their correspondence in the work and life of each mathematician is indeed impossible to evaluate from within the correspondence itself. We have already seen how both expressed their appreciation in the letters. But external

[36]In 1898, however, Lipschitz sends folk and military songs to Hermite after an exchange on the memories of past wars, and Hermite evokes music listened to at Lipschitz's home during his visit, letter 155.

information—a comparison with their other correspondence—has to be used to state that this appreciation did not imply the same familiarity that the two men may have had with others. To decide further if the warm description of their relation inside their correspondence has meaning beyond the basic politeness of the time also requires confirmation from outside. In 1881, in an attempt to have Lipschitz elected to the Berlin Academy, Hermite writes to Kronecker that "among so many distinguished mathematicians [...], I value and love above all M. Lipschitz."[37] To Mittag-Leffler, he mentions how he "greatly treasures" Lipschitz's opinion (Hermite 1984/1985/1989, I, p. 229), while Lipschitz, after his meeting with Hermite in Göttingen, confides to Richard Dedekind how "Hermite's personality especially inspires trust," and later, that he has "developed an affection towards him" (Lipschitz 1986, p. 87, p. 90). In the 1890s, Mittag-Leffler refers to Lipschitz as one of the two German geometers (along with Fuchs) who had the closest relations with Hermite (1984/1985/1989, I, p. 193).

To appreciate the scientific role of the correspondence, we again need external help. Hermite had published in German journals since the beginning of his career, but did not mention Lipschitz before their meeting in Göttingen; among the 92 articles Hermite published after 1877—many of them letters to a variety of people—only 9 articles refer to Lipschitz (with 14 mentions of his name[38]), all published between 1877 and 1887: we find, in particular, two articles cosigned with him in *Acta* and a comment to a note by Lipschitz in the *Comptes rendus*. Hermite often discusses his own results in letters to Lipschitz, but personal communication with Lipschitz (supported by the evidence of the correspondence) is explicitly hinted at in only 3 published papers, while another 3 refer to specific results of Lipschitz.

The situation is not symmetrical. Let us look at the list of Lipschitz's publications provided in Lipschitz (1986, pp. 235–244). Fifty items are listed before 1877, the year the first volume of Lipschitz's treatise appears and the correspondence between Hermite and Lipschitz is launched. Among them, only four appear in French journals—one is a note in the *Comptes rendus* of the French Academy,[39] the three others appear in Darboux's *Bulletin des sciences mathématiques*. The *Bulletin* had published short reviews of Lipschitz's articles, as they did for all papers published in Crelle's *Journal* and in 1872, Darboux explicitly requests from Lipschitz a longer analysis of a series of his papers (Lipschitz 1986, pp. 44–45). Besides this, Lipschitz addressed a letter to the journal to complete and correct bibliographical references.

---

[37]Letter from September 30, 1881: *parmi tant de géomètres éminents [...], j'estime et j'aime surtout M. Lipschitz*. A copy of this letter is kept in Hermite's file in the Archives of the French Academy of Sciences.

[38]Kronecker, on the other hand, is cited in 24 papers, Jacobi in more than 50. Lipschitz is one of 20 authors born between 1830 and 1850 and cited by Hermite after 1880. These data are established and discussed in Goldstein (2012).

[39]There is no indication of the member of the Academy who communicated it, which suggests that Lipschitz sent it directly to the Academy and its *secrétaire perpétuel* Bertrand. This was the course indicated to Lipschitz as being the "most natural" by Borchardt in December 1875 (Lipschitz 1986, p. 21).

The third paper in the *Bulletin* and the note to the Academy only summarize longer contributions published elsewhere.

After 1877, Lipschitz's mode of publication obviously changed. Among the 47 articles published after this date, 26 were published in French journals. Among them, 17 articles are notes in the *Comptes rendus*, including 14 extracts from letters to Hermite! Although not explicitly indicated as letters, two others are communicated by Hermite (and discussed in the correspondence). Only one note on probability, written quite late, in 1898, is communicated by Bertrand. Moreover, among the nine papers in other French journals, four are explicit extracts from letters to Hermite, two can be traced to them, and two are only summaries or translations of work published in German elsewhere. To this can be added two letters to Hermite published outside France (one in *Acta mathematica*, one in Crelle's *Journal*) and the two articles in *Acta mathematica* cosigned by Hermite. To summarize, after his direct acquaintance with Hermite, Lipschitz published 27 articles (out of 47) that are directly connected with the correspondence and his recognition on the French scene significantly improved.

3.3 Views on Mathematical Creation in Proper Perspective

This need to contextualize items of the correspondence through outside sources is not restricted to mathematics per se. It also touches upon epistemic issues. We have quoted Hermite's letter from April 12, 1882:

> I believe that in science, and especially in mathematics, we are less masters than servants of our work. I do not deny free will, but I think that it coexists with the action of a force which, while it arises from us, acts outside us, and unwittingly directs us where we would not like to go.

The local context, in the letter itself, is that of an opposition between German ways of doing mathematics, presented as abstract, and French ones, presented as more concrete; specifically, it concerns Lipschitz's research on the movement of a body in a Riemannian space[40]:

> If I were your colleague and neighbor, I would hold you as hard as I could by the hem of your garment, so that you would not start down such a prodigiously abstract path, while so

[40]Letter 27: *Si j'étais votre collègue et votre voisin, je vous retiendrais autant que je pourrais, par un pan de votre habit, pour ne pas vous engager dans une telle voie si prodigieusement abstraite, lorsque tant de questions s'offrent qui sont d'un intérêt immédiat et plus tangible, si je puis dire. Entre nous ce serait le grand combat des germains et des latins; cependant la lutte impliquerait de ma part une réserve que je vais vous dire. Je crois que nous sommes dans les sciences et tout particulièrement dans les mathématiques, moins les maîtres que les serviteurs de notre œuvre. Je ne nie point le libre arbitre, mais je pense qu'il coexiste avec l'action d'une force qui naissant par notre fait, agit en dehors de nous, et à notre insu, nous dirige là même où nous ne voudrions pas aller ... Quelque chose vous pousse peut-être qui est au dessus de vous et de moi; je reconnais à la fois que la conception des espaces de Riemann m'effraye, et qu'elle a sa raison d'être.*

> many questions present themselves, which are of immediate interest and, if I may say so, more concrete. Between us, it would be the great combat of Germans and Latins; however, the fight would imply on my side a restriction that I will reveal to you. I believe that in science, and especially in mathematics, we are less masters than servants of our work. I do not deny free will, but I think that it coexists with the action of a force, which, while it arises from us, acts outside us, and unwittingly directs us where we would not like to go. [...] Something pushes you forward perhaps, something above you and me; I recognize that the conception of Riemann spaces frightens me, but that it has its utility.

The main issue here seems to be the status of Riemannian geometry. Although Hermite was a main actor in the importation of Riemannian ideas into France (organizing, in particular, the publication in French of Riemann's complete works), he was also a defender of a down-to-earth analysis, against any ad-hoc ontologization, in particular a geometrical one, and thus expressed several times doubts with respect to certain interpretations or uses of Riemann's results (Goldstein 2011). In the letter mentioned above, his reluctance was framed within a national setting. The opposition between Germans and Latins, thus the main theme, is only tempered by the possible intervention of an external force, which may lead mathematics, and almost unwillingly mathematicians, down this new, apparently abstract, path. The occurrence of the "masters vs servant" theme, on the other hand, remains isolated in Hermite's correspondence with Lipschitz.

However, taking into account other known correspondence involving Hermite offers a different picture. In March 1876, Hermite writes to Leo Königsberger: "I reject as totally wrong the idea that mathematicians are the creators of their science" (Goldstein 2011, pp. 156–157). On February 19, 1880, to Genocchi, this time, he adds: " I reject as totally wrong the idea that mathematicians are the creators of their science. Mathematicians seem to me as much servants as masters of their science" (Hermite 2003, p. 25). The expression surfaces once more in a well-known letter to Du Bois-Reymond, on March 24, 1882, a few weeks before that to Lipschitz: "In mathematics, which seems the fruit of the most complete intellectual freedom, we are nonetheless more servants than masters" (Hermite 1916). Or to Mittag-Leffler in 1885: "in the development of mathematics, we are servants, much more than masters" (Hermite 1984/1985/1989, II, p. 100). Each time, the direct, mathematical context of the respective letters is different: a discussion on the role of computations in mathematics, some thoughts about good topics for academic prizes, Cantor's set theory, uniform functions arising from the study of second-order differential equations. The repetition of the theme and its variants, however, and of the words themselves, shows that it is much more than a passing remark; it points to a central conviction in Hermite's view of mathematical creation and development.[41] Echoes, from outside the Hermite-Lipschitz correspondence, are here the warrant of the meaningfulness—and ultimately of the meaning—of the sequence.

[41]On this viewpoint and further contextualization with respect to his contemporaries' positions, see Goldstein (2011).

3.4 Mathematical Links

The correspondence between Hermite and Lipschitz was the main locus of mathematical collaboration between the two mathematicians, and, as such, offers glimpses of the genesis of several articles by each author. To take a simple example, the second volume of *Acta mathematica*, published in 1883, contains two direct extracts of the correspondence (see letter 35 of May 12, 1883, and draft 37* of June 6, 1883) combined as a single contribution, "Sur quelques points dans la théorie des nombres, par Ch. Hermite et R. Lipschitz." Their point of departure is Dirichlet's memoir on mean values of arithmetical functions (Lejeune-Dirichlet 1849), which, as Hermite says, they "both know and admire." In this memoir, Dirichlet evaluates an asymptotic approximation of $F(n) = \sum_{i=1}^{n} \phi(i)$, where $\phi(i)$ designates the number of divisors of the integer i. In order to prove the approximation, Hermite uses a new expression for $F(n)$,

$$F(n) = 2\sum_{i=1}^{E}(\sqrt{n})E(\frac{n}{i}) - [E(\sqrt{n})]^2,$$

$E(x)$ being here the integral part of x (the largest integer less than or equal to x). Lipschitz's answer generalizes Hermite's formula to the arithmetical functions $F_s(n) = \sum_{i=1}^{n} \phi_s(i)$, where $\phi_s(i)$ is the number of divisors of the integer i that are also sth powers (thus $s = 1$ is Hermite's case).

The exchange stimulates research, and the link between the correspondence and the outside world, here the publications, is obvious.

But another kind of external link is illustrated in the letter of January 31, 1881 (letter 20) already mentioned. "It may interest you," writes Hermite, " to know that one of my students, Mr Tannery, has discovered a series that is much simpler than that expressed by Mr Weierstrass as:

$$\begin{aligned}\chi(x) &= \frac{2x}{\pi}\Psi(1, 1, xi) + \frac{2}{\pi x}\Psi(1, 1, \frac{i}{x}) \\ &= \frac{2}{\pi}(x + x^{-1}) + \frac{2}{\pi}\sum[\frac{x}{(1 - 2\nu - 2\nu' ix)(2\nu + 2\nu' x)^2} \\ &\qquad + \frac{x^{-1}}{(1 - 2\nu - 2\nu' ix^{-1})(2\nu + 2\nu' x^{-1})^2}]\end{aligned}$$

and has the same type of discontinuity. This is the following:

$$\frac{1 + x^2}{1 - x^2} + \frac{2x^2}{x^4 - 1} + \frac{2x^4}{x^8 - 1} + \frac{2x^8}{x^{16} - 1} + \dots$$

It has the value +1 or -1 depending on whether the modulus of the variable is smaller or greater than unity."

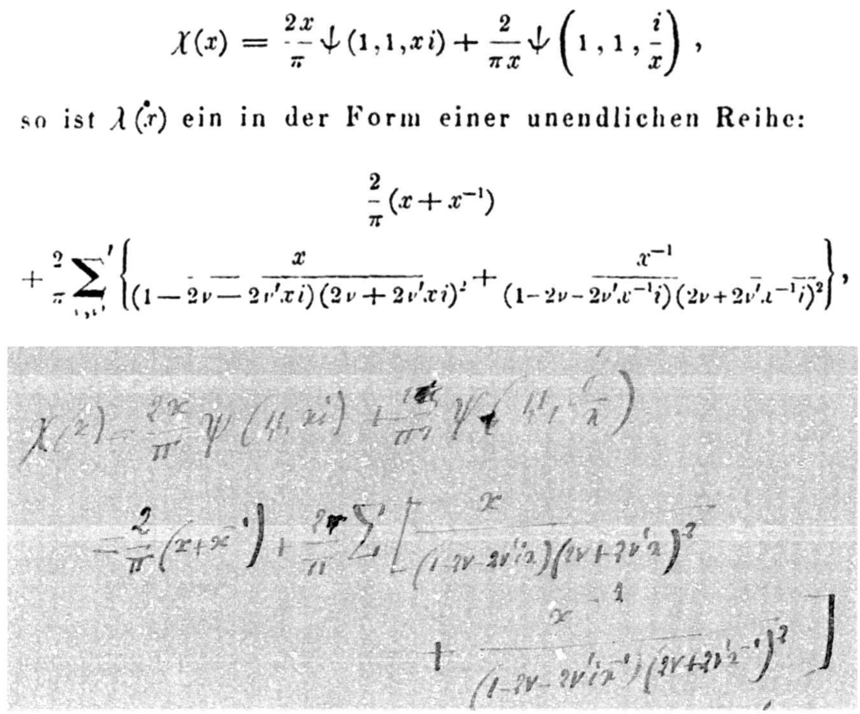

$$\chi(x) = \frac{2x}{\pi}\psi(1,1,xi) + \frac{2}{\pi x}\psi\left(1,1,\frac{i}{x}\right),$$

so ist $\lambda(x)$ ein in der Form einer unendlichen Reihe:

$$\frac{2}{\pi}(x+x^{-1})$$
$$+\frac{2}{\pi}\sum_{\nu,\nu'}{}'\left\{\frac{x}{(1-2\nu-2\nu'xi)(2\nu+2\nu'xi)^2}+\frac{x^{-1}}{(1-2\nu-2\nu'x^{-1}i)(2\nu+2\nu'x^{-1}i)^2}\right\},$$

Fig. 3 Weierstrass's example in (Weierstrass 1880/1881) and in Hermite's letter 20

The issue here is well-known (Kolmogorov and Yushkevich 1996, pp. 265–266): Weierstrass had constructed, by means of elliptic functions, a series that converges to different analytic functions on different domains. If we only take into account the publications, we see that the series appears in a communication by Weierstrass to the Berlin Academy of Sciences, on August 12, 1880 (Weierstrass 1880/1881, p. 735). Its form and its notation are exactly those given by Hermite (see Fig. 3). On February 21, 1881, a second communication reproduces a letter from Jules Tannery to Weierstrass, announcing the possibility of simplifying Weierstrass's example. The two communications, one following the other, were translated by Tannery himself into French for the April 1881 issue of the *Bulletin des sciences mathématiques* (pp. 157–181, and 181–183).

But instructive information on the sequence of events and a closer dating are offered by the correspondence: on November 27, 1880, Darboux asks Weierstrass for permission to publish a French translation of his 1880 communications to the Berlin Academy (Confalonieri 2013, II, p. 51); Tannery is in charge of it (Hermite 1984/1985/1989, I, p. 100). On December 24, in a sequence of letters which involves Weierstrass's, Mittag-Leffler's and Hermite's intertwined results on analytic functions, and their various publications and translations, Hermite comments to Mittag-Leffler: "What marvelous things are these discontinuous series of M. Weierstrass, which represent absolutely different functions in separate domains" (Hermite 1984/1985/1989, I, p. 87). As explained earlier, Hermite's youngest daughter, Marie, had married Picard at the very beginning of January 1881, and on January 22, Hermite explains to Genocchi that he has not been working much lately because of that event, but is now studying several articles by Weierstrass (and among them that of August 1880), in order to include them in his Sorbonne lectures (Hermite 2003, p. 27). There is still no mention of Tannery's simpler example. A letter[42] from Hermite to Tannery, while congratulating him on his result, suggests that he inform Weierstrass directly about it, and indeed, on February 8, 1881, Tannery thanks Weierstrass for the latter's interest in his "little remark," and his permission to translate the relevant articles (Confalonieri 2013, II, p. 54). As we have seen, Hermite writes to Lipschitz about Tannery's example as early as December 31. But it is not until February 13 that he communicates it to Mittag-Leffler, a result which, he "cannot doubt, will please [him] too" (Hermite 1984/1985/1989, I, p. 102). Schwarz's *Nachlass* also contains a letter from Weierstrass to Schwarz, dated March 6 (thus, after the communication of Tannery's letter to the Berlin Academy), stating that "the editor of the Darboux Bulletin, J. Tannery, has communicated to me recently a very interesting remark" that his series can be replaced with a much more elementary one (Confalonieri 2013, III, p. 76).

This episode demonstrates the intricate role of correspondence in the communication of mathematics. If the main figure, Weierstrass, receives Tannery's result first, it is noticeable that Hermite immediately dispatches the news to his own favorite correspondents: Lipschitz receives it even before Mittag-Leffler, though the latter was directly involved in the matter, both mathematically and as one of the intermediates in the translation process. Even before publication, the business of French translations favors the transfer of knowledge through exchanges that organize them, their correction and their development. Mittag-Leffler makes no mistake; on February 19, 1881, he writes to Hermite just after receiving the announcement of Tannery's example: "M. Tannery's series interested me a lot. It is admirably simple and one sees the proof immediately. The translation of M. Weierstrass's memoirs is apparently not unfruitful, it seems, for the French mathematicians."[43] We come full circle with Tannery's own presentation of his result, in his *Notice sur travaux*.[44]

[42]Hermite's file, Archives of the French Academy of Science, Paris. The letter is not dated, but the chronology we have reconstructed here suggests that it was probably written in late January 1881.

[43]Letters from Mittag-Leffler to Hermite, File 53J, Archives of the French Academy of Sciences, Paris: *La série de M. Tannery m'a extrêmement intéressé. Elle est admirablement simple et on voit tout de suite la démonstration La traduction des mémoires de M. Weierstrass n'est pas sans fruit, il parait pour les géomètres français.*

[44](Tannery 1901, p. 19): *En traduisant pour le* Bulletin *les communications "Zur Functionenlehre" de Weierstrass à l'Académie des sciences de Berlin (août 1880), [...] je remarquai qu'on pouvait remplacer, par une série plus simple, une série construite par Weierstrass [...]. L'origine de la série que je crus devoir communiquer à Weierstrass se trouve dans un problème [...] qui m'a été probablement suggéré en étudiant la démonstration du théorème fondamental de l'algèbre que M. Lipschitz a donnée dans son* Lehrbuch der Analysis.

> While translating for the *Bulletin* Weierstrass's communications to the Berlin Academy of Sciences "Zur Functionenlehre" (August 1880) [...] I noticed that one could replace by a simpler series the series Weierstrass had constructed [...]. The origin of the series I allowed myself to communicate to Weierstrass is to be found in a problem [...] which was probably suggested to me while I was studying the proof of the fundamental theorem of algebra that M. Lipschitz had given in his *Lehrbuch der Analysis*.

Published articles are thus only the tip of the iceberg, integrating parts of letters or being integrated into them. Reciprocally, reconstructing this web of texts highlights the value of the correspondence as a place of rapid diffusion between French and German mathematics.

4 Editorial Issues

The correspondence between Lipschitz and Hermite is an important platform for the observation of mathematical milieux in the second half of the nineteenth century, and of the renewal of the French-German relations after the 1870 Franco-Prussian war.[45] As explained above, it plays a decisive role in Lipschitz's manner of publishing. It is thus only natural to raise the issue of a complete edition, and more specifically, to discuss how to properly take into account echoes, some instances of which have been explored above.

Letters have traditionally been edited as texts, organized chronologically (or by sender and addressee). Footnotes or endnotes are then generally used to explain the various allusions made in the letters.[46] This access to the complete text of the letters, without too many disturbing interventions of the editor, is, of course, essential.

However, as shown above, each letter can and should be envisioned as being composed of a number of units of meaning: in order to take into account the richness of the correspondence, both as an informational device and as a linking device, we should be able to connect each of these units to others, internal or external. Examples include parts of letters within parts of articles in certain journals, as well as identical formulas appearing in different correspondences. Recently, electronic editions have presented correspondence as a communication network, in which each letter is a link between the sender and the addressee. Appropriate search functions would thus allow one to easily locate places where the letter was written or received, dates, sometimes persons or matters discussed in each letter; graphical representations could visualize such information.[47]

[45]On this issue, see Thomas (2002).

[46]The difficulties already involved in this simple presentation are well-known for ancient texts and literary manuscripts, but are no less important in contemporary scientific texts. For instance, TEI-encoding offers interesting features for Lipschitz's drafts, as it permits a display of erasures, loose additions between lines and even certain types of links, but its complete compatibility with LaTeX is still a delicate issue (see http://www.tei-c.org/index.xml).

[47]Illuminating examples are the Van Gogh online correspondence, http://vangoghletters.org/vg/, the edition of D'Alembert's letters, http://dalembert.academie-sciences.fr/Correspondance/ and

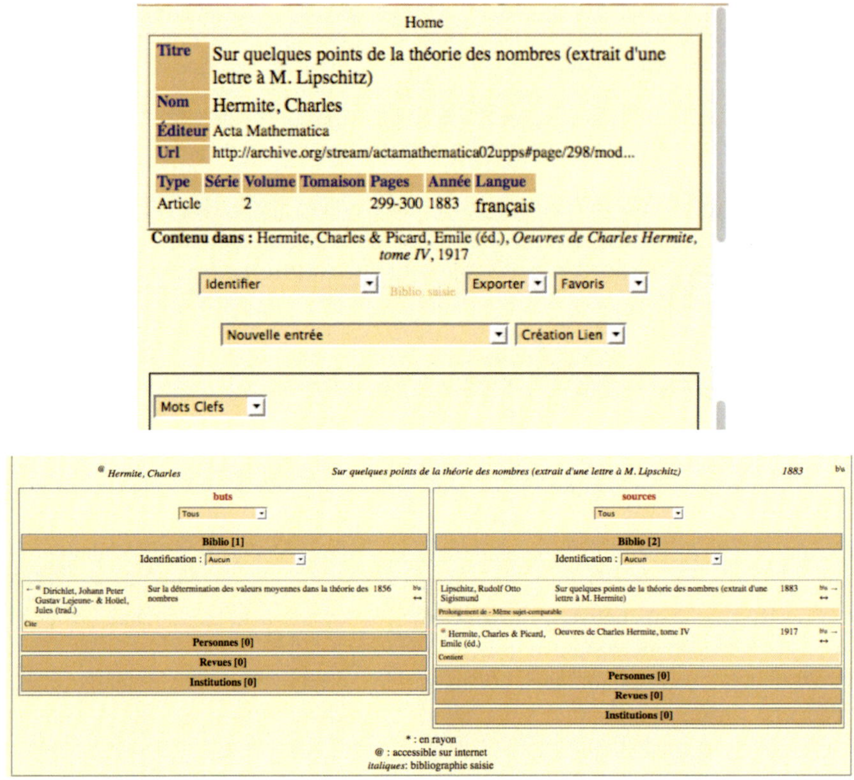

Fig. 4 Thamous database: card for an article by Hermite (extract), and card for links from and to this article

But the relevant links are not only those connecting a letter to its writer or to the institution to which it is sent. Even when they are materialized by chains of characters associated with such entities, the association can be variegated: a proper noun can appear as an addressee, a mathematical author quoted in the letter, a candidate for some position, an enemy. Or let us consider, for instance, the chain of characters "Académie" (for the French Academy of Sciences). It is first a multiple link between Hermite and Lipschitz inside the correspondence: Hermite reports on his efforts to have Lipschitz elected as a corresponding member; he also presents Lipschitz's "Notes" (often extracts of letters) to be inserted into the proceedings of the Academy. But it also constitutes a link between Hermite and Bischoffsheim (Hermite trying to prevent the election of Bischoffsheim), with Lipschitz as a depository of and witness to this link. And of course, internal references are not

Early Modern Letters On Line, emlo.bodleian.ox.ac.uk/. See also http://www.newtonproject.sussex.ac.uk/view/texts/normalized/NATP00225:theproblemofmathematics.

sufficient: as shown in the case of Tannery's example, the echoes of important issues should be traced in the correspondence of others, in the publications of the Academy or in the minutes of its meetings. Mathematical journals and political events, but also mathematical formulas, for instance (see Fig. 3), constitute other possible links to be taken into account. Such links operate at a variety of levels, and thus should themselves be capable of indexation, commentary and labeling, in order to capture the multiplicity of the operational aspects they encapsulate. An inspiring example of such commented and labeled links (but without editing functionality) is provided by the collective database Thamous, constructed by Alain Herreman (see Fig. 4).

To summarize, we need a platform which (i) is *open*, in order to be able to add new documents in real time, for instance newly discovered letters, or new links; (ii) provides *selective display*, to be able to read only the letters themselves or to have access exactly to those linked to a specific concept or reference; (iii) is *homoiconic*, in the sense that, as is partially the case in Thamous, links should be treated as data, as well as the texts of the letters, capable of receiving links and commentaries themselves.

The identification of mathematical activities, through a variety of documents, and the efficient sharing of their multiple echoes so as to understand effective transfer procedures and concrete knowledge dynamics is perhaps the next challenge for the history of mathematics. A key step will be the development of appropriate tools that would allow us to concretize current reflections on sources and their uses.

Acknowledgements I would like to thank Jim Ritter and the anonymous copyeditor for their revision of my English, Alexandre Guilbaud for useful discussions on text encoding, and the personnel of the Universitäts- und Landesbibliothek Bonn for their help during and after my stay at the Archives. My trip to Bonn was funded by the program EMANMATH of the MSH Lorraine, directed by Scott Walter.

References

Archibald, Thomas. 2002. Charles Hermite and German Mathematics in France. In *Mathematics Unbound: The Evolution of an International Mathematical Research Community, 1800–1945*, ed. K. Hunger Parshall, and A.C. Rice, 123–138. Providence: AMS.

Auvinet, Jérôme. 2013. *Charles-Ange Laisant : Itinéraires et engagements d'un mathématicien de la Troisième République*. Paris: Hermann.

Begehr, Heinrich, et al., eds. 1998. *Mathematics in Berlin*. Basel: Birkhäuser.

Biermann, Kurt-R. 1973/1988. *Die Mathematik und ihre Dozenten an der Berliner Universität 1810–1933*. Berlin: Akademie Verlag; 2nd ed.

Bottazzini, Umberto. 1977. Le funzioni a periodi multipli nella corrispondenza tra Hermite e Casorati. *Archive for History of Exact Sciences* 18 (1): 39–88.

Bret, Patrice, and Norbert Verdier. 2012. Sciences et techniques. In *Histoire des traductions en langue française*. XIXe *siècle*, ed. Y. Chevrel, L. D'Hulst, and C. Lombez, 927–1007. Lagrasse: Verdier.

Butzer, Paul Leo, Luciano Carbone, François Jongmans, and Franco Palladino. 2000. Les relations épistolaires entre Charles Hermite et Ernesto Cesàro. *Bulletin de l'Académie royale de Belgique. Classe des sciences* (6) 11: 7–12, 377–417.

Chartier, Roger, ed. 1991. *La correspondance. Les usages de la lettre au* XIXe *siècle*. Paris: Fayard.

Confalonieri, Edoardo. 2013. *Beiträge zur Geschichte der mathematischen Werke von Karl Weierstrass*, Teil II and Teil III, self ed., online.

Dauphin, Cécile. 2003. Écriture de l'intime dans une correspondance familiale du XIXe siècle, *Le Divan familial* 11: 63–73.

Décaillot, Anne-Marie. 2008. *Cantor et la France. Correspondance du mathématicien allemand avec les Français à la fin du* XIXe *siècle*. Paris: Kimé.

Diaz, José-Luis. 1995. Le XIXe siècle devant les correspondances. *Romantisme* 90: 7–26.

Dirichlet, Peter Gustav Lejeune-. 1849. Bestimmung der mittleren Werthe in der Zahlentheorie. *Abhandlungen der Königlich Preussischen Akademie der Wissenschaften zu Berlin*: 69–83.

Dugac, Pierre. 1976. *Richard Dedekind et les fondements des mathématiques*. Paris: Vrin.

Dugac, Pierre. 1983. Partie inédite de la correspondance d'Hermite avec Stieltjes, *Cahiers du séminaire d'histoire des mathématiques* 4: 75–87.

Goldstein, Catherine. 2007. The Hermitian Form of Reading the D. A. In *The Shaping of Arithmetic after C. G. Gauss's* Disquisitiones Arithmeticae, ed. C. Goldstein, N. Schappacher, and J. Schwermer, 375–410. Berlin: Springer.

Goldstein, Catherine. 2010. Liberté, égalité, amitié dans le cercle mathématique de Marin Mersenne. In *L'amitié et les sciences de Descartes à Lévi-Strauss*, ed. J.-C. Darmon, and F. Waquet, 41–58. Paris: Hermann.

Goldstein, Catherine. 2011. Un arithméticien contre l'arithmétisation : les principes de Charles Hermite. In *Justifier en mathématiques*, ed. D. Flament, and P. Nabonnand, 129–165. Paris: MSH.

Goldstein, Catherine. 2012. Les autres de l'un : deux enquêtes prosopographiques sur Charles Hermite. In *Les uns et les autres*, ed. P. Nabonnand, and L. Rollet, 509–540. Nancy: Presses universitaires de Nancy.

Hermite, Charles, and Thomas Stieltjes. 1905. *Correspondance d'Hermite et de Stieltjes*, ed. B. Baillaud and H. Bourget, 2 vols. Paris: Gauthier-Villars.

Hermite, Charles. 1916. Briefe zu P. Du Bois-Reymond aus den Jahren 1875–1888, ed. E. Lampe. *Archiv der Mathematik und Physik* 24 (3): 193–220, 289–310.

Hermite, Charles. 1967. Lettres à A. Markov, 1885–1899, ed. H. Ogigova. *Revue d'histoire des sciences et de leurs applications* 20 (1): 1–32.

Hermite, Charles. 1984/1985/1989. Lettres à Gösta Mittag-Leffler, ed. P. Dugac, *Cahiers du séminaire d'histoire des mathématiques* (I. 1874–1883) 5: 49–285; (II. 1884–1891) 6: 69–217; (III. 1891–1900) 10: 1–82.

Hermite, Charles. 2003. *Le Lettere di Charles Hermite a Angelo Genocchi (1868–1887)*, ed. G. Michelacci, Quaderni matematici II s., 546. Trieste: Dipartimento di scienze matematiche. Repr. *Bollettino di storia delle scienze matematiche* 25–1/2 (2005), 1–196.

Kolmogorov, Andrei, and Adolf-Andrei Yushkevich, eds. 1996. *Mathematics of the Nineteenth Century: Geometry, Analytic Function Theory*. Basel: Birkhäuser.

Kortum, Hermann. 1906. Rudolf Lipschitz. *Jahresbericht der Deutschen Mathematiker-Vereinigung* 15: 56–59.

Lipschitz, Rudolf. 1877–1880. *Lehrbuch der Analysis*, 2 vols. Bonn: F. Cohen.

Lipschitz, Rudolf. 1986. *Briefwechsel mit Cantor, Dedekind, Helmholtz, Kronecker, Weierstrass und anderen*, ed. W. Scharlau, Dokumente zur Geschichte der Mathematik 2. Braunschweig, Wiesbaden: Vieweg and DMV.

Lux David, and Harold Cook. 1998. Closed Circles or Open Networks? Communicating at a Distance During the Scientific Revolution. *History of Science* 36: 179–211.

Parshall, Karen Hunger. 1998. *James Joseph Sylvester: Life and Work in Letters*. Oxford: Oxford University Press.

Pieper, Herbert. 2007. A Network of Scientific Philanthropy: Humboldt's Relations with Number Theorists. In *The Shaping of Arithmetic after C. G. Gauss's* Disquisitiones Arithmeticae, ed. C. Goldstein, N. Schappacher, J. Schwermer, 201–233. Berlin: Springer.

Scharlau, Winfried. 2006. The Mathematical Correspondence of Rudolf Lipschitz. *Historia Mathematica* 13: 165–167.

Secord, Anne. 1994. Corresponding interests: artisans and gentlemen in nineteenth-century natural history. *The British Journal for the History of Science* 27: 383–408.

Tannery, Jules. 1901. *Notice sur les travaux scientifiques*. Paris: Gauthier-Villars.

Weierstrass, Karl. 1880/1881. Zur Funktionlehre. *Monatsberichte der Königlich Preussischen Akademie der Wissenschaften zu Berlin* (1880): 719–743; (1881): 228–230. Repr. in *Mathematische Werke*, Berlin: Mayer & Müller, 1895, vol. 2, 201–234.

Zerner, Martin. 1991. Le règne de Joseph Bertrand (1874–1900). In *La France mathématique*, ed. H. Gispert, 298–322. Paris: SFHST & SMF.

The Correspondences of Luigi Cremona and Placido Tardy in the Libraries of Genoa

Cinzia Cerroni

Abstract
We describe the historical framework and the main issues (biographical, scientific, political, etc.) of the correspondences of Placido Tardy and Luigi Cremona in the libraries of Genoa, which constitute an important contribution to the reconstruction of the History of Mathematics in the Italian "Risorgimento". In particular, we mainly deal with the Cremona-Tardy, Betti-Tardy and Cremona-Guccia correspondences. Tardy's letters are preserved at the Genoa University Library and Cremona's letters at the Mazzini Institute of Genoa.

1 Introduction

The aim of this paper is to analyse the correspondences of Placido Tardy[1] and Luigi Cremona[2] that are included in the archives of the libraries of Genoa. Tardy's

[1]Placido Tardy (Messina 1816–Florence 1914) left Sicily in 1848 for political reasons. From 1851, he was professor of Analytic Geometry and calculus at the Navy School in Genoa, and from 1859, he was professor of Calculus at the University of Genoa. He was Rector of the University of Genoa from 1865 to 1868 and from 1878 to 1888. Even though he was not a mathematician of the first magnitude, he played a key role in the first stages of formation of the Italian School of Mathematics around 1860, as evidenced by the correspondence he held with leading mathematicians of the time.

[2]Luigi Cremona (Pavia 1816–Rome 1903) was one of the leading mathematicians of the Risorgimento. He graduated in engineering in 1853 at Pavia, and in 1860, he became professor of Higher Geometry at the University of Bologna. In 1867, he went to teach at the Polytechnic

C. Cerroni (✉)
Dipartimento di Matematica e Informatica, University of Palermo, Palermo, Italy
e-mail: cinzia.cerroni@unipa.it

M. T. Borgato et al. (eds.), *Mathematical Correspondences and Critical Editions*, Trends in the History of Science, https://doi.org/10.1007/978-3-319-73577-1_11

correspondences are in the Genoa University Library, while the correspondences of Cremona considered here are in the Mazzini Institute in Genoa. We remark, for the sake of completeness, that the main part of Cremona's scientific correspondences, above all, the letters between Cremona and European mathematicians, are preserved in the archive of the Mathematics Department of the "La Sapienza" University of Rome.[3] In the following, we will describe and study these archives. In particular, from an examination of the correspondence, we have identified some issues that are more significant than others for the History of Risorgimento Mathematics. We will analyse these subjects through Cremona's letters and Tardy's letters with their contemporary mathematicians. The two protagonists of the correspondences have been placed among the main figures in the history of the Italian Risorgimento, and especially the founders of a "unitary" school of mathematics.

2 The Libraries of Genoa

We focus our attention on an analysis of some of the correspondences between the mathematicians Luigi Cremona and Placido Tardy, which are housed in two libraries in Genoa: the *University Library* and the *Mazzini Institute*.

In the *University Library* of Genoa, one can find the *Cassetta Loria*.[4] The *Cassetta Loria* contains all of the correspondence donated in 1925 by Professor Gino Loria, from whom it takes its name. It is characterized by the correspondence [784 units] of prestigious Italian and foreign mathematicians with Placido Tardy. This correspondence can help us to understand the connections between the development of Italian mathematics in the second half of the nineteenth century and the main political issues of Italian history.

Tardy's correspondence was first described by Loria, who stressed its importance and remarked that: "*A rapid examination of it induced me to notice some [letters] with such considerable historical and scientific value that I felt it would be opportune to publish them, without ignoring the fact that a more detailed examination of it [the correspondence] could lead to the discovery of others [letters] of no less importance*"[5] (Loria 1915, 516).

of Milan. In 1873, Minister Scialoja called him to Rome to head the School for Engineers, where he taught Graphic Statics until 1877, when he was given the chair of Higher Mathematics at the University. In 1879, he was named a Senator of the Kingdom.

[3]Cf. Millán Gasca (1992), Menghini (1994), Menghini (1996), Nurzia (1999).

[4]A cataloging of the documents contained in it was undertaken by Oriana Cartaregia, Ariella Pennacchi and Maria Teresa Sanguineti (2000–2001) and it can be found at the link http://www.bibliotecauniversitaria.ge.it/it/cataloghi/f_a_s/loria.htm.

[5]Un rapido esame da me fattone m'indusse a notarne alcune dotate di tanto considerevole valore storico e scientifico che reputo opportuno il renderle di pubblica ragione, senza escludere che un più minuto esame di esse possa portare alla scoperta di altre di non minore importanza.

The letters to Tardy mentioned by Loria, published as an appendix to a commemoration of Tardy,[6] are the following: a letter from Angelo Genocchi of December 25, 1866, in which a new pair of *Amicable numbers is described*, two letters from Enrico Betti, dated October 6 and 16, 1863, respectively, in which conversations with Bernhard Riemann are reported, two letters from Ludwig Schläfli of August 17, 1864, and October 4, 1865, respectively, in which appreciations are expressed for papers that Tardy had written.

Another study of this correspondence was made by Umberto Bottazzini[7] in 1980. In particular, he remarked that: "*Such an investigation has revealed the existence of very interesting materials, especially for the history of Italian mathematics in the second half of the* 19th *century and for the scientific biographies of such men as Betti, Genocchi, Brioschi, Cremona, Bellavitis, Tortolini, and Beltrami*" (Bottazzini 1980, 84).

Bottazzini, in his study, identified some relevant topics covered in the correspondence,[8] many of which will be analyzed in Sect. 3.

At the *Mazzini Institute*, surprisingly, one can find *Legato Itala Cremona Cozzolino*, which contains a portion of Cremona's correspondence. The reason why part of Cremona's correspondence is housed at the *Mazzini Institute* is that the testamentary executor of Maria Mazzini, Giuseppe Mazzini's mother, was Napoleone Ferrari, the uncle of Cremona's wife, Elisa Ferrari, who was one of Mazzini's aides. After the death of Napoleone, the documents passed to Cremona's family.[9] The *Legato Itala Cremona Cozzolino* was given to the library of the *Mazzini Institute* by Cremona's daughter, Itala, probably in 1939. This legacy, which contains over 6000 documents, mainly consisting of Cremona's correspondence with scientific and institutional Italian interlocutors, can help us to understand the main political initiatives involved in the development of scientific culture in Italy and the considerable advances that were made in the national organization of science in this period, thanks to the intervention of Cremona.

A description of the Archive of the *Mazzini Institute* was made by Aldo Brigaglia and Simonetta Di Sieno, who were the first to identify this archive: "*We have found further documents relating to Cremona in the library of the Mazzini Institute in Genoa which make us certain that the majority of letters to Cremona are now at the disposal of historians of mathematics*" (Brigaglia and Di Sieno 2011, 98).

In particular, in the description of the research project concerning the *Mazzini Institute*, they observed that: "*A great part of the correspondence of the most important mathematicians of the first* 30 *years after the unification of Italy has been saved. These letters form an impressive corpus that reveals how a small group of young mathematicians were led to create, almost from nothing, a first-class mathematical community in just* 20 *years (about 1858–1878), such as to make*

[6]Cf. Loria (1915, 516).

[7]Cf. Bottazzini (1980).

[8]Cf. Bottazzini (1980).

[9]For more on the history of the legacy, see Brigaglia and Di Sieno (2011).

Darboux state in 1870: 'I think that if things continue to go on in this way, Italians will surpass us in a short time.' Through these letters, we can follow day by day the human, scientific and political happenings of this community. The line of their researches, their contacts with the European mathematicians, their hopes and goals can all be studied in detail" (Brigaglia and Di Sieno 2011, 104).

In conclusion, the description of the archives of the *University Library* of Genoa and the *Mazzini Institute* shows the importance of studying the correspondences contained therein for the history of Italian mathematics.

2.1 The Correspondences Analyzed

Some correspondences conserved in these archives have already been edited in their entirety. The correspondences of Placido Tardy that have already been published, housed at the *University Library* of Genoa,[10] are the following:

- The Beltrami-Tardy[11] correspondence;
- The Bellavitis-Tardy[12] correspondence;
- The Betti-Tardy[13] correspondence;
- The Cremona-Tardy[14] correspondence;

additionally, the Brioschi-Tardy correspondence[15] is currently being printed.

The correspondences of Luigi Cremona that have been published, in addition to the one between Cremona and Tardy, housed at the *Mazzini Institute*,[16] are the following:

- The Battaglini-Cremona[17] correspondence;
- The Brioschi-Cremona[18] correspondence;
- The Chelini-Cremona[19] correspondence;
- The Cremona-Guccia[20] correspondence;

[10]This means the portions of correspondence that concern letters from mathematicians to Tardy.

[11]Cf. Giacardi and Tazzioli (2012).

[12]Cf. Canepa and Fenaroli (2009).

[13]Cf. Cerroni and Martini (2009).

[14]Cf. Cerroni and Fenaroli (2007).

[15]Maria Teresa Borgato analyzed this correspondence.

[16]This means the portions of correspondence that concern letters from mathematicians to Cremona.

[17]Cf. Palladino and Mercurio (2011).

[18]Cf. Palladino et al. (2009).

[19]Cf. Enea and Gatto (2009).

[20]Cf. Cerroni (2013).

additionally, the Masonic[21] correspondences between Cremona and Giosuè Carducci and Francesco Magni are currently being printed.

We will focus our analysis mainly on the following correspondences:

The Betti-Tardy correspondence: the letters from Betti to Tardy are in *Cassetta Loria* at the *University Library* of Genoa. There are 79 letters, covering the period 1850–1891. The letters from Tardy to Betti are in the Library of the *Scuola Normale Superiore* of Pisa. There are 49 letters, covering the period 1850–1889.

The Cremona-Tardy correspondence: the letters from Cremona to Tardy are in *Cassetta Loria* at the *University Library* of Genoa. There are 74 letters, covering the period 1860-post-1888. The letters from Tardy to Cremona are at the *Mazzini Institute*. There are 50 letters, covering the period 1860–1884.

The Cremona-Guccia correspondence: the letters from Guccia to Cremona are at the *Mazzini Institute*. There are 44 letters, covering the period 1878–1900. The letters from Cremona to Guccia are in the Archive of the *Circolo Matematico* of Palermo. There are 14 letters, covering the period 1878–1900.

We will also be dealing with the Beltrami-Tardy, Brioschi-Tardy[22] and Genocchi-Tardy[23] correspondences.

3 Principal Subjects of the Correspondences

From an examination of the correspondence, we have identified certain issues that are more significant than others. They are the following:

- The planning and foundation of the *Annali di Matematica pura e applicata*;
- The foundation and development of the *Circolo Matematico di Palermo*;
- The discussion among the Italian mathematicians over Non-Euclidean geometry;
- The references to the researches of Bernhard Riemann;
- The references to Giuseppe Garibaldi and the wars of independence;
- The references to the politics and organization of the university.

The issue concerning the planning and foundation of the *Annali di Matematica pura e applicata*[24] is mainly present in the letters of Enrico Betti, Luigi Cremona, Francesco Brioschi and Angelo Genocchi.

The previous journal, *Annali di Scienze Matematiche e Fisiche*, also known as *Annali di Tortolini*,[25] was published in Rome starting in 1850. Betti, Brioschi and

[21]Brigaglia and Di Sieno studied these correspondences.

[22]Cf. Lacaita and Silvestri (2000).

[23]Cf. Garibaldi (1991).

[24]For insights into the history of the *Annali di Matematica pura e applicata*, see, also Bottazzini (1994).

[25]They were called this because they were drafted by Barnaba Tortolini (1808–1874).

Genocchi joined the editorial staff in 1858 and the journal was refounded as the *Annali di Matematica pura e applicata*. The project was to create a prestigious journal in which Italian and other European mathematicians could publish, to contribute to the rebirth of Italian mathematical studies.

> [...] I feel pleased that the project of the new journal is to your liking, and that you will also contribute to its success. The purpose of this journal is twofold, both to acquaint people abroad with what we do in Italy and introduce (by means of writings and bibliographic articles, as well as translation) the main memoirs published in foreign journals, in the proceedings of Academies and even in new books, to those lovers of mathematics who do not live in scientific centres [...].[26] [Brioschi to Tardy, December 23 1857]

> [...] We have almost arranged with Brioschi to be in Genoa over the Easter holidays, where Genocchi will perhaps also come. We also want to speak with you about our Journal; and to establish all that is necessary for it to follow the best course. [...][27]
> [Betti to Tardy, February 26 1858][28]

It was at the meeting at Tardy's home that the trip to the Universities of France and Germany was to be planned, the purpose of the trip being to increase their knowledge of European research and broaden their relationships with European mathematicians. The journey was subsequently undertaken by Betti, Brioschi and Felice Casorati on September 20, 1858.[29]

The publication of the *Annali di Matematica pura e applicata* ceased in 1865. Brioschi and Cremona, in 1867, decided to resume publication and transfer its headquarters to Milan. They sought the cooperation of their colleagues in accomplishing this.

> [...]Here, they are thinking about stopping publication of the Annali di Tortolini, and to found a Journal, analogous to the Crelle, in instalments, to be paid for separately with no time limits. I have written to Betti and Genocchi about it on behalf of Brioschi. Betti has already replied affirmatively, and as soon as I have the answer from Genocchi

[26][...] Sento con piacere che il progetto del nuovo giornale è di vostro gradimento, e che contribuirete anche voi pel meglio di esso. Lo scopo di questo giornale è duplice, e di far conoscere al di fuori quanto si sa fare in Italia e di render note (col mezzo di scritti e di articoli bibliografici ed anche di traduzioni) a quei cultori delle matematiche i quali non abitano in centri scientifici le principali memorie pubblicate sui giornali stranieri e negli atti delle Accademie ed anche i nuovi libri [...].

[27][...] Abbiamo quasi fissato con Brioschi di trovarsi a Genova nelle vacanze di Pasqua, dove forse verrà anche Genocchi. Vogliamo parlare un poco anche insieme con Te del nostro Giornale; e stabilire bene tutto ciò che è necessario per il migliore andamento dello stesso [...].

[28]The letter is in Cerroni and Martini (2009).

[29]See Bottazzini (1994).

as well, I will write to Tortolini. But we need the support of friends, have I secured yours?[...][30] [Cremona to Tardy, 10 January 1867][31]

[...] I very warmly recommend to you the Annali just transferred to Milan, both for the collaboration and for the association. I take the liberty of sending you some copies of this circular, asking you to send it to the indicated names, which I have largely found in the note of the Associates given me by Tortolini; but I don't know their addresses[...][32]
[Cremona to Tardy, 10 February 1867][33]

[...] The new Annali di Matematica, to be printed in Milan, seem to me to be seriously delayed. What is our Cremona doing?[...][34] [Genocchi to Tardy, May 28 1867]

The issue concerning the foundation and development of the *Circolo Matematico di Palermo* is present in the correspondence between Luigi Cremona and Giovanni Battista Guccia.

In 1884, Guccia, through personal contributions of resources and labour, founded the *Circolo Matematico di Palermo*,[35] whose journal *Rendiconti del Circolo Matematico di Palermo* became, a few decades later, one of the foremost international journals of mathematics. Guccia was a student of Cremona, and turned to him for advice and support from the Ministry of Education.

[...] The day before yesterday at the institute, Mr. Darboux congratulated me on the foundation of the Circolo of Palermo;[...] And speaking of the Circolo, I believe I have worked well. Everyone, without exception, has taken an interest in the Library of the mathematical society in Palermo and gifts rain down from all parts. Now we will see how you do in Rome!![36] [Guccia to Cremona, October 22 1884][37]

[...] It was then that I had the idea of a Circolo Matematico that would bring together, in a comfortable room equipped with reading and study areas (no fewer than 16 maths periodicals), all people, young and old, from nearby and far away, who had dealings with

[30][...] Qui si pensa di far cessare gli Annali del Tortolini, e fondare qui un giornale, analogo al Crelle, che sia per fascicoli da pagarsi separatamente e senza vincolo di tempo. Ne ho scritto a Betti ed a Genocchi per incarico di Brioschi. Betti ha già risposto assentendo, appena avrò risposta anche da Genocchi scriverò a Tortolini. Ma ci abbisogna l'appoggio degli amici, il vostro ce l'ho assicurato?[...]

[31]The letter is in Cerroni and Fenaroli (2007).

[32][...] Vi raccomando caldissimamente gli Annali trasferiti a Milano, sia per la collaborazione sia per l'associazione. Mi prendo la libertà di mandarvi sotto copia alcune copie di sta circolare, colla preghiera di farla ricapitare ai nomi segnativi, i quali in massima parte ho trovato nella nota degli Associati datami da Tortolini; ma non ne conosco gli indirizzi precisi [...].

[33]The letter is in Cerroni and Fenaroli (2007).

[34][...] I nuovi Annali di Matematica, da stamparsi a Milano, mi sembrano molto in ritardo. Che fa il nostro Cremona? [...]

[35]For more on the history of *Circolo Matematico di Palermo*, see Brigaglia and Masotto (1982).

[36][...] L'altro ieri all'Istituto il Sig. Darboux si mosse al mio incontro per congratularsi a proposito della fondazione del Circolo di Palermo; [...] Ed a proposito del Circolo credo di aver lavorato bene. Tutti, nessuno eccetto, si sono interessati alla Biblioteca della società matematica di Palermo ed i doni piovono da tutte le parti. Vediamo ora cosa sa' fare lei a Roma!!

[37]The letter is in Cerroni (2013).

math.[...] Therefore, the formation of a good library is one of the main objectives of the institution that I am creating. [...] This is the point upon which I take the liberty of appealing to you most urgently.[38] [Guccia to Cremona, April 13 1884][39]

The *Circolo Matematico di Palermo*, starting from a few years after its birth, was enriched by members who were not residents of Palermo; in 1905, it was elevated to the rank of an international society.

[...] The noble initiative of the young mathematicians of Palermo has given Italy an institution that has already earned the recognition of foreign scholars, and I heartily hope that it will prosper and be to the benefit of science in Italy.[...][40]
[Cremona to Guccia, December 31 1887][41]

[...] I had with me for several days Mr. Mittag-Leffler and family. You can imagine the great advantage I derived for the Circolo !!! [...] The work is increasing every hour, the enterprise is colossal,[...] but success is greater and greater! If the Executive Council [...] lives up to the importance of its mandate and is able, when necessary and without disrespect to anyone, to stand firm; if this is possible in Italy, then, but only then, can we be proud of having created a beautiful institution in our country that will honour us abroad[...][42]
[Guccia to Cremona, May 6 1888][43]

[...] At the Congress [of Heidelberg], I met everybody. [...] and all, without distinction, received me very well, which I attribute to the new Italian institution founded by myself with your support, the Circolo Matematico, which, to my great surprise, was already known and appreciated in Germany, more perhaps than it is in France. You can imagine what a great pleasure this was for me! Spontaneously, some German mathematicians came to ask me if they could join our Society. Many told me about the subjects of study that they would

[38][...] Fu allora che mi venne l'idea di un Circolo Matematico che riunisse in un locale comodo e fornito di un buon gabinetto di lettura e di studio (non meno di 16 pubblicazioni periodiche di matematica) tutte le persone, vecchi e giovani, che da vicino e da lontano avessero avuto commercio colle matematiche.[...] È dunque la formazione di una buona biblioteca, uno dei principali obiettivi della istituzione da me creata.[...] Ecco il punto su cui mi permetto di rivolgerle il più caldo appello.

[39]The letter is in Cerroni (2013).

[40][...] La nobile iniziativa de' giovani matematici di Palermo ha dotato l'Italia di una istituzione che già ha meritato il plauso di dotti stranieri, ed alla quale io di cuore auguro che prosperi e che riesca a vantaggio delle scienze in Italia.[...]

[41]The letter is in Cerroni (2013).

[42][...] Ho avuto con me parecchi giorni il Sig. Mittag-Leffler e famiglia. Può immaginare quanto profitto abbia ricavato pel Circolo!!! [...] Il lavoro è ognora crescente, l'intrapresa è colossale, [...] Il successo si afferma sempre più! Se il Consiglio Direttivo [...] s'inspirerà all'importanza del suo mandato e saprà tener fermo, ove accorra, senza riguardi di persone, pur usando delle forme di squisita cortesia; se ciò sarà possibile in Italia, allora, ma soltanto allora, potremo vantarci di aver creata nel nostro paese una bella istituzione che ci farà onore all'estero.[...]

[43]The letter is in Cerroni (2013).

> desire to publish in the Rendiconti, among which was George Cantor (Set theory).[...][44]
> [Cremona to Guccia, October 2 1889][45]

The issue concerning the Italian mathematicians' discussion on non-Euclidean geometry[46] is mainly present in the letters of Eugenio Beltrami, Giusto Bellavitis, Cremona and Genocchi.

In 1866, 10 years after the death of Nikolaj Ivanovich Lobachevsky, Guillaume Jules Hoüel published a French translation of Lobachevsky's geometry[47] together with some of the correspondence on non-Euclidean geometry by Johann Carl Friedrich Gauss. In 1868, Beltrami presented a concrete model of Lobachevsky's geometry.[48]

> [...] Have you read a booklet by Lobachevsky about parallel lines? It was translated into French by Hoüel and printed in the Mémoirs de la Société des Sciences Physiques et Naturalles de Bordeaux (1866), along with some letters from Schumacher to Gauss. It is nothing less than a deconstruction of the famous postulate by Euclid through proof that all geometry can be based on the assumption that that famous postulate is false, and it comes down to entrusting to experience (through astronomical observations) the choice between the Euclidean doctrine and its antithesis. [...] I confess that I do not really accept these new ideas: I do not understand how experience should be a judge of a geometric theory [...] Meanwhile, the work of Lobachevsky and the solemn approval of Gauss seem more appropriate to an increase rather than a diminishing of the importance of attempts at direct demonstration of that postulate. If you have the chance to deal with it, please do and tell me your opinion. [...] So the new doctrine does not find any hitch in geometric applications, but could it not find any, for example, in rational Mechanics, a science that is also based on simple notions of common sense?[49] [Genocchi to Tardy, May 28 1867]

[44][...] Al Congresso [di Heidelberg] conobbi tutti. [...] e tutti, senza distinzione, mi usarono grande accoglienza, che io debbo attribuire alla nuova istituzione italiana da me fondata col suo appoggio, il Circolo Matematico, il quale con mia grande sorpresa, era già conosciuto ed apprezzato in Germania, più di quanto, forse non lo è in Francia, che è quanto dire. Può immaginare se ciò mi ha fatto piacere! Alcuni matematici tedeschi spontaneamente son venuti a chiedermi di far parte della nostra Società. Molti mi hanno sviluppato gli argomenti di lavori che desidererebbero veder pubblicati dai nostri Rendiconti, fra cui George Cantor (teoria degli insiemi)[...]

[45]The letter is in Cerroni (2013).

[46]For further information, see Giacardi (1991).

[47]Cf. Lobachevsky (1866).

[48]Cf. Beltrami (1868).

[49][...] Avete letto un opuscolo di Lobaschevsky intorno alle parallele? fu tradotto in francese da Hoüel e stampato nelle Memorie della Società di scienze fisiche e naturali di Bordeaux (1866) e seguito da alcune lettere di Schumacher a Gauss. Si tratta niente meno che di abbattere il famoso postulato di Euclide provando che tutta la geometria può benissimo stabilirsi sull'ipotesi che quel postulato sia falso, e si finisce col rimettere alla esperienza (alle osservazioni astronomiche) la scelta fra la dottrina euclidea e la contraria. [...] Io confesso che difficilmente accetterei queste nuove idee: non capisco che l'esperienza debba esser giudice d'una teorica geometrica [...] Intanto il lavoro del Lobachevsky e l'approvazione solenne di Gauss mi sembrano atti piuttosto ad accrescere che a togliere l'importanza dei tentativi diretti alla dimostrazione di quel postulato. Se avete agio di occuparvene fatelo e ditemi il vostro avviso. [...] Così nelle applicazioni geometriche

At that time, Beltrami was presenting his model of non-Euclidean geometry to which Genocchi was openly opposed.

> [...] I do not know if you have given any attention to the system of ideas that is now spreading under the name of non-Euclidean geometry, and what judgment you have made. I know that Professor Chelini is decidedly adverse, and that Bellavitis calls it loony geometry: while Cremona believes it questionable and Battaglini approves it without hesitation. I analyzed it a little and I sent to Cremona a confidential exposé of my views, but although these actually lead to an exact interpretation of non-Euclidean theorems, I yet have serious doubt, which comes from the fact that this interpretation seems to have no link with the system of ideas toward which Gauss leaned in giving his consent to the new geometry: or at least none of the laconic phrases contained in his letters to Schumacher leaves legitimate reason to support it. Now, I'm waiting to know what Cremona thinks. [...] I believe that the rigorous and acute Genocchi is unfavorable to non-Euclidean geometry.[50]
>
> [Beltrami to Tardy, November 14 1867][51]

In 1870, Cremona spoke definitively in favor of non-Euclidean geometry:

> [...] The 4th issue of the 3rd volume will soon be published; it will contain, among other things, the important memoir by Riemann on the fundamental hypotheses of geometry, translated by Hoüel. So, in France, they have published the translation of Beltrami's memoirs on the same subject. Analogous ideas have been mentioned by Neumann in a recent discourse on the principles of Galileo and Newton. So, soon only Bellavitis and the foolish old men of the Academy of France will have the triple privilege of opposing the correct ideas of Gauss, of Lobachevsky, of Riemann, etc.[...][52]
>
> [Cremona to Tardy, April 26 1870][53]

la nuova dottrina non trova alcun intoppo: ma non potrebbe trovarne per esempio nella Meccanica razionale, scienza fondata anche essa sopra semplici nozioni di senso comune?

[50][...] Non so se ella abbia accordato alcuna attenzione a quel sistema di idee che ora si va divulgando col nome di geometria non euclidea, e quale giudizio ne faccia. So che il prof. Chelini gli è decisamente avverso, e che il Bellavitis lo chiama geometria da manicomio: mentre il Cremona lo crede discutibile ed il Battaglini lo abbraccia senza reticenze. Io me ne sono un po' occupato ed ho indirizzato al Cremona una esposizione confidenziale delle mie vedute: ma benché queste conducano effettivamente ad una esatta interpretazione dei teoremi non-euclidei, pure mi rimane un grave dubbio il quale proviene da ciò che questa interpretazione non sembra avere alcun nesso col sistema d'idee al quale si appoggiava Gauss nel dare il suo assenso alla nuova geometria: o per lo meno nessuna delle laconiche frasi contenute nelle sue lettere a Schumacher lascia legittima ragione di supporto. Ora sto aspettando di sapere che ne dice il Cremona. [...] Credo che anche il rigoroso ed acuto Genocchi sia poco favorevole alla geometria non-euclidea.

[51]The letter is in Giacardi and Tazzioli (2012).

[52][...] Presto uscirà il 4° fascicolo del tomo 3°, esso conterrà, fra l'altre cose, l'importantissima memoria di Riemann sulle ipotesi fondamentali della geometria, tradotta da Hoüel. Così in Francia hanno pubblicato tradotte le memorie di Beltrami su argomennti analoghi. Idee pure analoghe sono accennate da Neumann in un suo recente discorso sui principi di Galileo e di Newton. Così fra poco, resterà al solo Bellavitis ed ai vecchi rimbambiti dell'Accademia di Francia, il triplo privilegio di combattere le sane idee di Gauss, di Lobachevsky, di Riemann, ecc. [...]

[53]The letter is in Cerroni and Fenaroli (2007).

The references to the researches of Bernhard Riemann are mainly present in the letters of Betti, Cremona, Genocchi and Tardy.

In the spring of 1858, as we have remarked, Betti, Brioschi and Casorati made their famous trip to Europe.[54] The meeting with Riemann was decisive for Betti's research; his memoirs became his new object of study, and when Betti took the Chair of Higher Analysis, at the end of 1859, the theory of elliptic functions became the topic of his lectures.

> [...] I have translated the memoir by Riemann, but I did not make comments. Now, I am engaged in studying the theory of elliptic functions, for lectures. [...][55]
>
> [Betti to Tardy, September 30 1859][56]

> [...] In the next issue, 10 sections of the translation of Riemann's memoir will be published, and in the issue after that, the subsequent ones, and then a Monograph, which I am expounding upon in my lessons, above the elliptic functions, that will also be a comment on Riemann's memoir.[...][57] [Betti to Tardy, January 3 1860][58]

> [...] I had started a paper on Riemann's researches. It is a beautiful subject. I have a slightly different starting point that makes the exposition easier and clearer.[...]"[59]
>
> [Betti to Tardy, February 19 1862][60]

In October of 1863, Riemann began his stay in Pisa, for health reasons. During that time, Riemann interacted with Italian mathematicians, as evidenced by the famous letters[61] by Betti describing the conversations with "the eminent mathematician":

> [...] I spoke again with Riemann about the Connection of Spaces, and I got an exact idea. [...] Riemann also spoke with me about his ideas in mathematical physics, but it is necessary that we talk about it. I'm very sorry that you are not yet here and I remember with great pleasure the very few days when, after talking with Riemann, we spent time talking together about what we had understood to help each other to get an exact idea.[...] And we

[54]For further information about the contacts with the European mathematicians, see Neuenschwander (1978/1979), Neuenschwander (1983) and Neuenschwander (1998).

[55][...] Ho tradotta la memoria di Riemann, ma non ho fatto commenti. Ora Io sono impegnato a studiare nella teoria delle funzioni ellittiche, per le lezioni.[...]

[56]The letter is in Cerroni and Martini (2009).

[57][...] Nel prossimo numero saranno pubblicati 10 paragrafi della traduzione della Memoria di Riemann, nell'altro numero i seguenti, e poi una Monografia, che sto esponendo nelle mie lezioni, sopra le funzioni ellittiche, che servirà anche in gran parte di commento alla Memoria di Riemann. [...]

[58]The letter is in Cerroni and Martini (2009).

[59][...] Aveva cominciato un lavoro sopra le cose di Riemann. È un bel soggetto. Prendo un punto di partenza un poco differente che rende più facile e più chiara l'esposizione[...]

[60]The letter is in Cerroni and Martini (2009).

[61]See Sect. 2.

were so lucky that he came to see us. We must not miss the chance and take advantage of this opportunity.[...][62] [Betti to Tardy, October 6 1863][63]

[...] much more than I hope that you will write to me here again and that you will communicate to me something of your very interesting conversations with the incomparable Riemann. My stay in Florence left me with sweet memories, both for the hours spent with you and for the precious knowledge that I got from Riemann, and I envy you greatly, as you are lucky enough to have him all to yourself for the whole winter in Pisa. If only I could come too and enjoy the presence of such a great geometer in Italy! [...][64]
[Tardy to Betti, October 14 1863][65]

[...] Riemann demonstrated with much ease that one can reduce any space to being simply connected, by means of straight sections and surface sections, simply connected. [...][66]
[Betti to Tardy, October 16 1863][67]

Cremona and Tardy were both involved in the study and geometric interpretation of Riemann's theory, through the reading of the works published at that time on the subject, namely *Vorlesungen über Riemann's Theorie der Abel'schen Integrale*[68] (1865) by Carl Neumann and especially *Theorie der Abelschen Functionen*[69] (1866) by Alfred Clebsch and Paul Gordan.

[...] I too received the work of Neumann, and I am enchanted by it. I didn't have time to get too far into it, but the exposition in the part that I did study seems to me admirable for its clarity and accuracy. Oh, if only everyone wrote like that! You told me to wait for a work by Clebsch on the same subject. Will it be a separate book, or some memoir that will be

[62][...] Ho parlato nuovamente con Riemann della connessione degli spazi, e me ne sono fatta una idea esatta. [...] Riemann mi ha parlato anche delle sue idee in fisica matematica, ma bisogna che ne parliamo ancora. Mi dispiace molto che tu non sia ancora qui e rammento con desiderio i pochissimi giorni nei quali dopo aver parlato con Riemann andavamo insieme parlando di quello che avevamo inteso aiutandoci a farcene un'idea esatta.[...] e noi abbiamo avuta la fortuna che è venuto a trovarci egli stesso. Bisogna approfittarne e non lasciar fuggire l'occasione.[...]

[63]The letter is in Cerroni and Martini (2009) and in Loria (1915).

[64][...] molto più che spero che tu mi scriva un'altra volta quassù e mi comunichi qualche cosa delle tue interessantissime conversazioni con l'impareggiabile Riemann. Il mio soggiorno a Firenze mi ha lasciato dolcissima ricordanza, e per le ore passate con te e per la preziosa conoscenza che ho fatta di Riemann, e t'invidio grandemente la fortuna di averlo tuo per tutto l'inverno a Pisa. Potessi venire anch'io e godere del soggiorno di un si' grande geometra in Italia!![...]

[65]The letter is in Cerroni and Martini (2009).

[66][...] Riemann dimostra con molta facilità che si può ridurre uno spazio qualunque ad essere semplicemente connesso, mediante sezioni lineari e sezioni superficiali semplicemente connesse. [...]

[67]The letter is in Cerroni and Martini (2009) and in Loria (1915).

[68]Cf. Neumann (1865).

[69]Cf. Clebsch and Gordan (1866).

published in the Crelle journal? After reading the Treatise by Neumann, I will try to read the writings of Riemann himself. [...][70] [Tardy to Cremona, January 3 1866][71]

[...] I am reading the work of Clebsch, but it is not as clear and easy as that of Neumann. When will Casorati publish his lectures that you told me about?[...][72]
[Tardy to Cremona, January 9 1867][73]

[...] I have been in Pavia for days. I often see Casorati, because he makes frequent trips to Milan. He is printing his work on the general theory of functions: it will be a large book and, I think, very well done and very important. The printing will not be complete for another six months. I hope you will not want to compare the book by Clebsch with that of Neumann. The latter has a most basic formulation, but this is the work of a much higher genius: it is a wonderful book, at least it produces this effect on me. [...][74]
[Cremona to Tardy, January 10 1867][75]

[...] It also seems difficult to me to read Clebsch's book, and indeed, having found some difficulties with the first pages, I stopped, putting off to quieter days the continued study of this work.[...][76] [Genocchi to Tardy, June 4 1867]

From these letters emerges the esteem that Cremona had for Clebsch, with whom he was in correspondence at that time,[77] as well as the difficulty, shared by Tardy and Genocchi, in reading the book by Clebsch and Gordan. In addition, they were all waiting for the publication of the book "*Teorica delle funzioni di variabili complesse*,"[78] which came out in 1868. In it, Casorati reported on the lectures held in Pavia during an extraordinary course of Higher Analysis of Riemann's theory.

[70] [...] ò ricevuto anch'io l'opera di Neumann, e ne sono incantato. Non ò avuto il tempo di andare molto innanzi, ma nella parte che ò fin qui studiata l'esposizione mi sembra mirabile per chiarezza e precisione. Oh se tutti scrivessero così! Voi mi dicevate aspettare un lavoro di Clebsch sullo stesso argomento. Sarà un libro a parte, o qualche memoria che uscirà nel giornale di Crelle? Dopo la lettura del trattato di Neumann mi proverò a leggere gli scritti di Riemann stesso.[...]

[71] The letter is in Cerroni and Fenaroli (2007).

[72] [...] Sto leggendo l'opera di Clebsch, ma non è così chiara e facile come quella di Neumann. Casorati quando pubblicherà quelle sue lezioni di cui mi parlaste?[...]

[73] The letter is in Cerroni and Fenaroli (2007).

[74] [...] Sono stato da giorni a Pavia. Vedo spesso Casorati perché egli fa corse frequenti a Milano. Egli sta stampando il suo lavoro sulla teoria generale delle funzioni: sarà un grosso volume, e, ritengo, fatto bene e molto importante. La stampa non sarà compiuta prima di sei mesi ancora. Spero che non vorrete paragonare il libro di Clebsch con quello di Neumann. Questo ha una formula più elementare, ma quello è il lavoro di un ingegno decisamente superiore: è un libro meraviglioso, almeno a me fa questo effetto. [...]

[75] The letter is in Cerroni and Fenaroli (2007).

[76] [...] Anche a me pare difficile a leggere il libro di Clebsch, anzi avendo trovato qualche intoppo dalle prime pagine mi arrestai rimettendo a giorni di maggior agio a quiete lo studio continuato di quest'opera.[...]

[77] Cf. Menghini (1996).

[78] Casorati (1868).

In 1869, Brioschi, Casorati and Cremona held their famous lectures on the theory of elliptic and Abelian functions, for the purpose of understanding the "*Riemannian mysteries*":

> [...] The three of them will hold a Complementary Course of Mathematics, and they will expound upon the researches of Clebsch on the Abelian functions. Cremona will do the geometric part, Casorati the part that refers to the theory of the functions of the complex variable, and the remaining part will be done by Brioschi. Brioschi is enthusiastic about Clebsch's book: I too admire the power of analysis of this beautiful talent, but it doesn't arouse in me the enthusiasm that I get from the study of the researches of the kind of deep thinker that Riemann was, and I hope this year that I will have young students able to follow me in the exposition that I intend to give them of his theory of the Abelian functions. [...][79]
>
> [Betti to Tardy, October 31 1868][80]

> [...] The book that most engrosses me is Clebsch's, in which, if I am not mistaken, I was able to simplify and complete some important points: and I will perhaps have an opportunity to publish a few copies on the subject. Also, Casorati will soon print a paper on periodicity, having succeeded in simplifying much of the 4th Section of Clebsch. I attend Casorati's lectures, and so I am beginning to be able to see a little inside the Riemannian mysteries. It is really true that there is strength in numbers: alone, I would never have succeeded in making these studies, of which I now also see their true importance for geometry. [...][81]
>
> [Cremona to Tardy, February 20 1869][82]

References to the wars of independence and Garibaldi are present mainly in the letters of Cremona and Tardy. As is well known, Betti, Brioschi, Cremona and Tardy were involved in the Italian Risorgimento, to varying degrees, starting in 1848.

Betti took part in the famous battle of Curtatone with the Tuscan battalion commanded by Fabrizio Mossotti; Brioschi played an active part in the five days of Milan; Tardy, after the first riots in Reggio Calabria in 1847, had to leave Sicily to repair to Tuscany, where he met and became friends with Betti; Cremona participated for over a year in the heroic defence of Venice. These patriotic feelings never left them.

---

[79][...] Faranno un Corso Complementare di Matematica in tre, ed esporranno i lavori di Clebsch sulle funzioni abeliane. Cremona farà la parte geometrica. Casorati quella parte che si riferisce alla teoria delle funzioni di una variabile complessa e il resto Brioschi. Brioschi è entusiasmato del libro di Clebsch: anch'io ammiro la potenza dell'analisi di questo bell'ingegno, ma non mi desta l'entusiasmo che mi da lo studio dei lavori di quel pensatore profondo che era Riemann e io questo anno spero che avrò giovani capaci di seguirmi nella esposizione che ho intenzione di far loro della sua teoria delle funzioni abeliane. [...]

[80]The letter is in Cerroni and Martini (2009).

[81][...] Il libro che più mi occupa è quello di Clebsch dove m'è riuscito, se non m'inganno, di semplificare e completare alcuni punti importanti: e forse avrò occasione di pubblicare qualche copia in proposito. Anche Casorati stamperà presto un articolo sulla periodicità, essendogli venuto fatto di semplificare moltissimo il 4° Abschnitt del Clebsch. Assisto alle lezioni di Casorati, e così comincio a vedere un po' entro ai misteri riemaniani. È proprio vero che l'unione fa la forza: da me solo non sarei mai riuscito a fare questi studi, dei quali ora vedo tutta l'importanza anche per la geometria.[...]

[82]The letter is in Cerroni and Fenaroli (2007).

Cremona wrote to Tardy, during the first meeting of the Italian Parliament on February 18, 1861, as follows:

> [...] You will have the fortune to attend, in a few days, the opening of the first Italian Parliament, where Vittorio Emanuele will be proclaimed the first king of united Italy! What a Solemn moment! [...]"[83] [Cremona to Tardy, February 18 1861][84]

The correspondence shows that Cremona and Tardy lived through the tragic and exciting events of the third war of independence and the subsequent actions by Garibaldi with great trepidation.[85] In particular, there emerges a pattern of alternating hopes and painful disillusionment:

> [...] How is it possible to study with the mind so worried about the political and financial conditions of our country? I already feel old, because at other times, I would have been full of enthusiasm, and today, I look at the future with dismay, and I don't have trust in the intelligence of our men of state. God help us, and save Italy! [...][86]
> [Tardy to Cremona, May 1 1866][87]

> [...] The war is ready to break out: all Italians are in agreement and are pleased the government is pushing ahead armaments with great energy.[...][88]
> [Cremona to Tardy, May 23 1866][89]

> [...] Dear Cremona, write me and give me courage. I do not despair yet, but this first defeat makes my heart bleed.[...][90] [Tardy to Cremona, June 26 1866][91]

All of which led to a conclusive painful outburst:

[83][...] Voi avrete la fortuna d'assistere fra pochi giorni all'apertura del primo Parlamento italiano, ove si proclamerà Vittorio Emanuele primo re dell'Italia una! Qual momento Solenne! [...]

[84]The letter is in Cerroni and Fenaroli (2007).

[85]During the Third War of Independence, Italy suffered the tough losses of Lissa and Custoza, but obtained the liberation of Venice and of the Veneto. The following year, the Roman expedition of Garibaldi suffered a disastrous defeat by French troops at Mentana.

[86][...] Ma come si fa a studiare con l'animo così agitato per le condizioni politiche e finanziare del nostro paese? Io mi sento già vecchio, perché in altri tempi sarei stato pieno d'entusiasmo, cd oggi guardo l'avvenire con sgomento, e non ò fiducia nell'intelligenza de' nostri uomini de' stato. Iddio ci aiuti, e salvi l'Italia! [...]

[87]The letter is in Cerroni and Fenaroli (2007).

[88][...] La guerra sta per iscoppiare: tutti gli italiani sono concordi e lieti: il governo spinge gli armamenti con somma energia [...]

[89]The letter is in Cerroni and Fenaroli (2007).

[90][...] Scrivetemi e fatemi coraggio, caro Cremona. Non dispero ancora, ma questa prima sconfitta mi fa sanguinare il cuore.[...]

[91]The letter is in Cerroni and Fenaroli (2007).

> [...] I will not speak about politics: the madness of Garibaldi and the dishonesty of Rattazzi have brought us to the brink of the abyss. In order not to get dizzy, you should not think about it, since it is not in our hands to remedy it.[...][92]
>
> [Cremona to Tardy, November 9 1867][93]

The references to politics and to the organization of the university are present mainly in the letters of Betti, Brioschi, Cremona and Tardy.

Already before the unification of Italy, Betti, Brioschi, and Tardy, as well as Cremona, who was not even a university professor, played a collective leading role in education policy,[94] but their role in Italian politics, especially, of course, in the field of Education, grew after the proclamation in 1861 of the Kingdom of Italy.[95]

From the correspondence, it emerges, for example, that Cremona, after transferring to the University of Rome and becoming the director of the School of Engineering in 1873, was so completely absorbed by the institutional activities[96] that he tried to escape from these tasks so as to return to research and that he suffered tremendous pressure to stay:

> [...] Obviously, if you went away, the consequences would be very serious from every point of view. I do not know what would remain of the School of Application. I do not know what would remain of the Academy. It would be such a terrible mess that if you think about the consequences for even a moment, any desire to leave should disappear altogether. If scientists do not want to be in Rome, if they do not bear any inconvenience that may occur in regard to the needs of the country, they declare, by their conduct, that they are of

[92][...] Di politica non vi parlo: la follia di Garibaldi e la malafede di Rattazzi ci hanno tratti sull'orlo dell'abisso. Per non avere le vertigini, non bisogna pensarci, dacché non è in nostra mano metterci rimedio.[...]

[93]The letter is in Cerroni and Fenaroli (2007).

[94]In 1859, Brioschi was a member of the Committee for the preparation of the "*Casati*" law (the "Legislative Royal Degree, November 13, 1859, n. 3725" of the Kingdom of Sardinia, which came into force in 1860 and was later extended, with the unification, to all of Italy. The law, which took its name from the Minister of Education, Gabrio Casati, organically reformed the entire educational system.); in 1860, Cremona was deeply involved in the review of school mathematics programmes. Thus, in fact, Genocchi, who was involved in this operation, wrote to him: "[the ministerial programmes] *are yours, neither more nor less.*"

[95]Brioschi, among other things, was a Deputy (from 1861), the rector of the Polytechnic of Milan from its foundation in 1863, and a Senator (from 1865). He was then the vice president of the Higher Council of Education. Betti was a Deputy (from 1862), the Director of the "Scuola Normale Superiore" of Pisa in 1865, the Secretary-General of Education (i.e. the undersecretary) from 1874 to 1876, a Senator (from 1884), and the vice president of the Higher Council of Education. Cremona was director of the School of Engineering of Rome in 1873, a Senator from 1879 and the vice president of the Senate in 1897–1898, as well as Minister of Education for 1 month in 1898.

[96]To learn more about the political activities of Luigi Cremona, see Brigaglia and Di Sieno (2009) and Brigaglia and Di Sieno (2010).

> the idea... horribile dictu ... that we have to give Rome back to the Pope. So, I make the warmest appeal to your patriotism, and even your love of science.[...][97]
>
> [Sella to Cremona, October 4 1877]

The correspondence is particularly useful in allowing for the reconstruction of the ephemeral but significant life of the commission established by the Minister Terenzio Mamiani to renew and supplement the Casati law. The commission was established before the proclamation of the Kingdom of Italy (July 18, 1860), in order to "write up a unitary school code to apply to the new kingdom."

Of course, after the events of May to September 1860, in view of the proclamation of the Kingdom of Italy, the Minister, on January 12, 1861, ordered an extension "so as to represent the whole of Italy."[98] Tardy was consulted by the Minister regarding this extension and, on January 16, he gave notice of it to Betti:

> [...] I am writing you two lines to tell you that I have proposed to Mamiani that he appoint you to a commission of which I will also be part for the compilation of a law on public education. I don't know if you have had official notice of it yet, neither do I know when the commission will meet.[...][99] [Tardy to Betti, January 16 1861][100]

Cremona wrote to Tardy, with high expectations, a few days after the inauguration of the Committee, on February 8, as follows:

> [...] I am pleased that the new Commission for constituting the law on Public education includes Betti and you. Nobody prays more than I do that this future law will be established with haste, the conditions of this university are so miserable![...][101]
>
> [Tardy to Cremona, February 8 1861][102]

---

[97][...] Evidentemente se voi ve ne andate, le conseguenze sarebbero gravissime sotto ogni punto di vista. Non so ciò che rimarrebbe della Scuola di applicazione. Non so ciò che rimarrebbe dell'Accademia. Sarebbe uno scompiglio così grave che se pensate anche solo un momento alle conseguenze ogni voglia di andarvene debba scomparire del tutto. Se gli uomini di scienza non vogliono stare a Roma, se non sopportano qualche inconveniente, che vi possa essere in vista delle necessità della patria, essi dichiarano con la loro condotta, che sono d'avviso ... horribile dictu... che si deve ridare Roma al papa. Io faccio quindi il più caldo appello al vostro patriottismo, ed anche al vostro amore per la scienza. [...]

[98]Cf. Ciampi (1983) and also Polenghi (1993).

[99][...] Ti scrivo due righe per dirti che ò proposto a Mamiani di nominarti in una commissione della quale faccio parte anch'io per la compilazione di una legge sulla pubblica istruzione. Non so se tu ne abbia ancora avuto avviso ufficiale, né so quando la commissione si riunirà.[...]

[100]The letter is in Cerroni and Martini (2009).

[101][...] Sono lieto che la nuova Commissione per formare la legge della pubblica Istruzione comprenda voi e Betti. Nessuno più di me affretta coi voti questa futura legge, tanto sono miserabili le condizioni di questa università![...]

[102]The letter is in Cerroni and Fenaroli (2007).

The Commission met regularly starting on February 5 and began collecting opinions from universities throughout Italy, on the basis of directives from the Minister. The members of the committee were, besides Betti and Tardy, Quintino Sella and Francesco De Sanctis (who, a few months later, would replace Mamiani).[103] The outcome of the committee's work was almost nil. As reported by Tardy:

> [. . .] Several times, I made a resolution to write to you, but in Turin, I was busy, and then I waited for our famous commission to achieve something to acquaint you with. [. . .] The comments you made then on the teaching of mathematics were absolutely right. [. . .] But all these arguments remained without any result, and I do not think that the new minister wants to make a general law on public education, because then its discussion in Parliament would go on indefinitely.[. . .][104] [Tardy to Cremona, June 11 1861][105]

And as Mamiani himself said : "The poor law, certainly not yet well-defined and accomplished, is a dead embryo, apparently."[106]

4 Conclusions

We have seen that through the correspondences analysed here, it is possible to describe the historical framework and the main interests and political positions of Placido Tardy, and also to enhance existing information on the political role of Luigi Cremona. Furthermore, we can study the human, scientific and political events of the mathematical community. In particular, their line of research, their contacts with other European mathematicians, and their hopes and goals can all be analysed in detail. Hence, these correspondences constitute an important contribution to the reconstruction of the History of Risorgimento Mathematics.

References

Beltrami, Eugenio. 1868. Saggio di interpretazione della Geometria Non-Euclidea. *Giornale di Matematiche* 6: 284–322.

Brigaglia, Aldo, Simonetta Di Sieno. 2009. L'opera politica di Luigi Cremona attraverso la sua corrispondenza. Prima parte. *La Matematica nella Società e nella Cultura* 2: 353–388.

Brigaglia, Aldo, Simonetta Di Sieno. 2010. L'opera politica di Luigi Cremona attraverso la sua corrispondenza. Seconda parte. *La Matematica nella Società e nella Cultura* 3: 137–179.

[103] Cf. Romizi (1902).

[104] [. . .] Più volte aveva fatto il proponimento di scrivervi ma in Torino era occupatissimo, e poi aspettava che in quella nostra famosa commissione si fosse concluso qualche cosa per rendervene conto.[. . .] Le osservazioni che allora mi facevate sull'insegnamento matematico erano giustissime. [. . .] Ma tutti questi discorsi sono rimasti senza risultato, e non credo che il nuovo ministro pensi a fare una legge generale per l'istruzione pubblica, perché poi la discussione di essa in Parlamento ci porterebbe alle calende greche.[. . .]

[105] The letter is in Cerroni and Fenaroli (2007).

[106] Romizi (1902, 62).

Brigaglia, Aldo, Simonetta Di Sieno. 2011. The Luigi Cremona Archive of the Mazzini Institute of Genoa. *Historia Mathematica* 38: 96–110.
Brigaglia, Aldo, Guido Masotto. 1982. *Il Circolo Matematico di Palermo* Bari: Edizioni Dedalo.
Bottazzini, Umberto. 1980. Tardy's letters and library in Genoa. *Historia Mathematica* 7: 84–85.
Bottazzini, Umberto. 1994. *Và pensiero. Immagini della matematica nell'Italia dell'ottocento.* Bologna: il Mulino.
Canepa, Giuseppe, Giuseppina Fenaroli. 2009. *Il carteggio Bellavitis-Tardy (1852–1880).* Milano: Mimesis.
Casorati, Felice. 1868. *Teorica delle funzioni di variabili complesse.* Pavia: Fratelli Fusi.
Cerroni, Cinzia, Giuseppina Fenaroli. 2007. *Il carteggio Cremona-Tardy (1860–1886).* Milano: Mimesis.
Cerroni, Cinzia, Laura Martini. 2009. *Il carteggio Betti-Tardy (1850–1891).* Milano: Mimesis.
Cerroni, Cinzia. 2013. *Il carteggio Cremona-Guccia (1878–1900).* Milano: Mimesis.
Ciampi, Gabriella. 1983. *Il governo della scuola nello stato postunitario.* Roma: Edizioni di Comunità.
Clebsch, Alfred, Paul Gordan. 1866. *Theorie der Abelschen Functionen.* Leipzig: B.G. Teubner.
Enea, Maria Rosaria, Romano Gatto. 2009. *Le carte di Domenico Chelini dell'Archivio Generale delle Scuole Pie e la corrispondenza Chelini-Cremona.* Milano: Mimesis.
Garibaldi, Antonio Carlo. 1991. Sui rapporti tra Angelo Genocchi e Placido Tardy. In *Angelo Genocchi e i suoi interlocutori scientifici. Contributi dell'epistolario*, ed. Alberto Conte e Livia Giacardi, 281–292. Torino: Deputazione subalpina di storia patria.
Giacardi, Livia. 1991. La polemica Genocchi-Beltrami sulle superficie pseudosferiche: una tappa nella storia del concetto di superficie. In *Angelo Genocchi e i suoi interlocutori scientifici. Contributi dell'epistolario*, ed. Alberto Conte e Livia Giacardi, 155–209. Torino: Deputazione subalpina di storia patria.
Giacardi, Livia, Rossana Tazzioli. 2012. *Le lettere di Eugenio Beltrami a Betti, Tardy e Gherardi.* Milano: Mimesis.
Lacaita, Carlo G., and Andrea Silvestri, eds. 2000. *Francesco Brioschi e il suo tempo (1824–1897).* Milano: Franco Angeli.
Lobachevsky, Nikolai Ivanovich. 1866. *Études géométriques sur la théorie des parallèles. Traduit de l'allemand par J. Hoüel; suivi d'un extrait de la correspondence de Gauss et de Schumacher.* Paris: Gauthier-Villars.
Loria, Gino. 1915. Commemorazione del Socio Placido Tardy. *Rendiconti della Reale Accademia Nazionale dei Lincei* 24 (5): 505–531.
Menghini, Marta, ed. 1994. La corrispondenza di Luigi Cremona (1830–1903), Vol. II. *Quaderni della Rivista di Storia della Scienza* 3, Roma.
Menghini, Marta, ed. 1996. La corrispondenza di Luigi Cremona (1830–1903), Vol. III. *Quaderni P.RI.ST.EM.*, Università Bocconi, Milano.
Millán Gasca, Ana Maria, eds. 1992. La corrispondenza di Luigi Cremona (1830–1903), Vol. I. *Quaderni della Rivista di Storia della Scienza* 1, Roma.
Neuenschwander, Erwin. 1978/1979. Der Nachlaß von Casorati (1835–1890) in Pavia. *Archive for History of Exact Science* 19 (1) :1–89.
Neuenschwander, Erwin. 1983. Der Aufschwung der italienischen Mathematik zur Zeit der politischen Einigung Italiens und seine Auswirkungen auf Deutschland. In *Symposia mathematica* 27, 213–237. Roma: Istituto Nazionale di Alta Matematica.
Neuenschwander, Erwin. 1998. Documenting Riemann's Impact on the Theory of Complex Functions. *The Mathematical Intelligencer* 20: 19–26.
Neumann, Carl. 1865. *Vorlesungen über Riemann's Theorie der Abel'schen Integrale.* Leipzig: B.G. Teubner.
Nurzia, Laura, ed. 1999. Per l'Archivio della corrispondenza dei Matematici italiani. La corrispondenza di Luigi Cremona (1830–1903), Vol. IV. *Quaderni P.RI.ST.EM.*, Università Bocconi, Milano.
Palladino, Nicla, Anna Maria Mercurio. 2011. La corrispondenza di Giuseppe Battaglini a Luigi Cremona. *Rendiconto dell'Accademia delle Scienze Fisiche e Matematiche* 78(iv):7–67.

Palladino, Nicla, Anna Maria Mercurio and Franco Palladino. 2009. *Le corrispondenze epistolari Brioschi-Cremona e Betti-Genocchi*. Firenze: Olschki.

Polenghi, Simonetta. 1993. *La politica universitaria italiana nell'età della destra storica* (1848–1876). Brescia: Editrice La Scuola.

Romizi, Augusto. 1902. *Storia del Ministero della Pubblica Istruzione, Parte II*. Milano: Alberighi, Segati, & c. Editori.

The Renewal of Mathematical Research in Italy: The Correspondences Between Brioschi-Betti (1857–1890) and Brioschi-Tardy (1853–1893)

Maria Teresa Borgato and Iolanda Nagliati

Abstract

The Italian Risorgimento and the subsequent unification of Italy took place during a period of not only political, but general renewal of the country, and Italian scientists linked themselves to the more advanced fields of European research. This paper focuses on the central figure of Francesco Brioschi and his correspondences with Enrico Betti and Placido Tardy. The scientific themes discussed in their letters are analyzed: the theory of invariants of binary forms, the resolution of fifth degree algebraic equations by elliptic functions, and the theory of fractional integrals. The epistolary relationships of these three mathematicians with their foreign colleagues are also described, which allows for a reconstruction of the frequent journeys undertaken by Italian scholars abroad, as well as those of foreign scholars to Italy. Of particular importance, for the number of letters and their contents, are Brioschi's correspondences with Felix Klein and Charles Hermite. The main collections come from the Polytechnic of Milan, the Scuola Normale of Pisa, the University Library of Genoa, and the Historical Archives of Göttingen University.

1 Introduction

The second half of the nineteenth century saw the return of Italian mathematical studies to the European scene after several decades during which it had been in decline. The renewal was sustained by a progressive internationalization of mathematical studies, which had already begun in the pre-unification period and was now undergoing a new burst of energy in the crucial years of the birth of a new

M. T. Borgato (✉) · I. Nagliati
Department of Mathematics and Computer Sciences, University of Ferrara, Ferrara, Italy
e-mail: bor@unife.it

M. T. Borgato et al. (eds.), *Mathematical Correspondences and Critical Editions*, Trends in the History of Science, https://doi.org/10.1007/978-3-319-73577-1_12

national state. This process was made possible by the creation and consolidation of a network of personal relations among Italian mathematicians and leading European scholars, whose correspondences allow us to make a fairly accurate analysis of the dynamics involved in this process, which, moreover, brings out the mutual influences and genesis of some of their discoveries.

However, the number of letters and the many links among the correspondents are such that they do not allow us to refer to a single person or to make an exhaustive reconstruction in just one edition. For this reason, choices have been made and partial correspondences have been published in various phases, the final goal being to reconstruct a general picture.

Much research has already been carried out on this period by historians of mathematics, and their studies were given more impetus in 2011 on the occasion of the 150th anniversary of the Unification of Italy. Several of these studies were presented at the Conference held in the Scuola Normale, while others have been published in the form of monographs or articles.[1]

We shall mainly be focusing on two correspondences, shortly to be published, namely, the Brioschi-Betti correspondence (1857–1890) and the Brioschi-Tardy correspondence (1853–1893), expanding the study to include letters between these mathematicians and their foreign colleagues. Of particular importance are the letters that were exchanged between Francesco Brioschi and, respectively, Charles Hermite and Felix Klein.

From an examination of unedited documents and previously published material, we may deduce how a progressive opening towards the English and German scientific milieu went hand in hand with the traditional relationship with the French School, and how, after 1860, the German model largely prevailed as far as organisation of studies and research was concerned.

1858 turned out to be a crucial year in this process, as two fundamental events took place: firstly, the journey that Enrico Betti, Francesco Brioschi and Felice Casorati undertook to visit the leading universities in Europe, and secondly, the commencement of the publication of the *Annali di matematica pura ed applicata*, which, under the editorial guidance of Betti, Brioschi, Barnaba Tortolini and Angelo Genocchi, constituted the first Italian journal entirely devoted to mathematical research.

After a presentation of the correspondences, we will focus mainly on the following aspects:

- The international relationships between Italian mathematicians and their foreign colleagues, focusing particularly on the latter's various visits to Italy for reasons of study and research and those made by Italian scholars abroad;

[1]Pepe (2012), Cerroni and Fenaroli (2007), Cerroni and Martini (2009), Carbone et al. (2006), Gatto (1995, 1996, 2004), Nagliati (2000), Palladino et al. (2009). For a general overview of the Italian mathematics of that period, see: Giusti and Pepe (2001), Bottazzini (1994), and Borgato (2009).

- The most significant moments of the research carried out by Betti, Brioschi and Tardy, as discussed in their letters, in particular, the resolution of the equations of fifth degree, the theory of invariants, and fractional calculus;
- The role that the publishing of scientific journals played in this process of renewal.

On the other hand, we will only cite the important issues of national interest that emerge from these correspondences concerning political participation in the ideals of the Risorgimento, commitment to the institutions of a united Italy, and the organization of higher education and mathematical research.

2 Presentation and Substantialness of the Correspondences

The two figures Enrico Betti (1824–1897) and Francesco Brioschi (1824–1897) may be compared not only for their outstanding scientific discoveries but also for the important public offices they held (in Parliament, in the Ministry of Education), as well as their leading positions at two of the most prestigious Italian scientific institutes, the *Scuola Normale* of Pisa and the *Politecnico* of Milan, respectively.[2] We shall not be discussing here the roles played by other important scientists of the post-Unification period in Italy (the period in our history that saw the greatest involvement of scientists in Italian politics), men like Luigi Cremona (1830–1903), whose correspondences constitute the subject of other contributions to this volume.

A third figure in this picture is Placido Tardy (1816–1914), professor of infinitesimal calculus and Rector of the Genoa University, who, despite having a smaller mathematical output with some interesting elements, was at the very centre of a wide network of scientific correspondences, which makes it impossible not to include him in a study of this period.[3]

2.1 The Brioschi-Betti Correspondence

The Brioschi-Betti correspondence has often been cited for the scientific importance of its authors and their role in the scientific community in Italy. It is made up of 60 letters and notes from Brioschi (preserved in the Betti Archive of the *Scuola Normale* in Pisa) and 4 letters from Betti (preserved in the Brioschi Archive of the Polytechnic in Milan), thus Brioschi is more strongly represented (Borgato 2012). The first series of 16 letters, dated between 1857 to the end of 1859, is full of

[2] Brioschi's mathematical works have been published in five volumes (Brioschi 1901–1905); essays on various aspects of his activity, inventories of correspondence and other documents and official speeches have been collected in three volumes edited by Lacaita and Silvestri (2000–2003). The mathematical works of Enrico Betti have been published in two volumes (Betti 1903–1913).

[3] See, for example: Cerroni (2012) and Bottazzini and Nastasi (2013).

references and details concerning Betti and Brioschi's research works, which, at that time, were closely connected, in the field of elliptic functions, binary forms and algebraic equations. Of particular interest are those written in the year 1858, when Brioschi, after Hermite and in concomitance with Kronecker, provided the resolution to the fifth degree equation in terms of elliptic functions. Betti, for his part, was working on the same problem by means of modular equations. Mention is also made of the birth of the first Italian scientific journal devoted exclusively to mathematics: the *Annali di Matematica pura ed applicata*, an evolution of Tortolini's *Annali di scienze matematiche e fisiche*, with the addition of Brioschi, Betti and Genocchi to the editorial staff. The letters written between 1860 and 1890 are less frequent and shorter, and rarely touch on mathematical topics: under the influence of Riemann, Betti had changed the focus of his research. These letters, however, concerned important issues, like the editorship of the *Annali di Matematica*, which had been transferred to Milan from Rome in 1867 under the direction of Brioschi and Cremona, education reform, university policies, academies and scientific publications. Concerning the university, Betti and Brioschi were supporters of the creation of a limited number of centres of excellence, around which specialisations could be built. Brioschi, moreover, taking his inspiration from the German model, supported the establishment of independent polytechnics that would provide both preparatory courses and applied specialization courses, which were in line with the specific social and productive situations of the various regions of the country. Brioschi, as is well known, was the founder and first director of the *Istituto Tecnico Superiore*, later called the *Politecnico* of Milan, and as a member of the State Board of Education (Consiglio Superiore della Pubblica Istruzione) for over 30 years, he was in a good position to influence education reforms at all levels. The creation of a National Academy was also part of the post-Unification renewal plans for the country, and Brioschi, who was president of the *Società dei XL*, did everything he could to bring about the union with the *Accademia dei Lincei*, after the annexation of Rome as the capital of the Kingdom of Italy, in 1870.

2.2 The Brioschi–Tardy Correspondence

The Brioschi–Tardy correspondence is almost entirely made up of letters from Brioschi (63 in all from 1853 to 1893), preserved in the University Library of Genoa, with Brioschi's research works also being a prominent feature in this case, particularly those carried out in the years 1857–1860, in which his scientific studies focused on algebraic themes. The correspondence continued over the years, providing interesting references to political, academic and institutional life. Besides the resolution of algebraic equations of the fifth degree discussed in this correspondence, other matters concerning contemporary studies by Hermite, Cayley and Sylvester are debated: for example, the theory of invariants and the partition of numbers. Tardy was directly involved in the latter theme, because he had studied a particular problem of partition in a short note from 1851 (Tardy 1851a). Another scientific topic that emerged from the correspondence is linked to one of Tardy's studies on differentials with non-integer indices (fractional calculus). Tardy had used an integral operator with a fractional index in one of his works on hydrodynamics, and had written a paper on these operators that Brioschi intended to publish. Although Tardy's main reference is to Liouville, the starting point is different: while Liouville refers to the complex exponential function and, therefore, leads the definition of fractional derivation of a whatever function to that of its series expansion, Tardy's method of defining the fractional differentiation by means of the fractional integration, which makes use of the Gamma function, has similarities with the line initiated by Euler.

2.3 Letters Betti Received from Foreign Scholars

The archive trust related to Enrico Betti containing the correspondences is preserved at the *Scuola Normale Superiore* in Pisa. Some uncatalogued letters from foreign mathematicians were found among the material set aside after the First Exhibition of the History of Science held in Florence in 1929, and others have been identified among those classified as unknown correspondents.

As for the scientific correspondence with foreign scholars, letters addressed to Betti from Bernhard Riemann (n. 9), Rudolf Clausius (n. 1), Carl Borchardt (n. 1) and Hermite (n. 2) have already been published.[4]

Fifty-seven unedited letters from foreign mathematicians and physicists (from 1852 to 1892) have now been transcribed (Nagliati 2014). The number of German correspondents is noticeably high: the highest numbers of unpublished correspondence came from Felix Klein (7), Leopold Kronecker (8), Ernst Julius Schering (6), James Joseph Sylvester (6), Gosta Mittag-Leffler (7), and François Moigno (4). Among the correspondents can be found other French, German, British, Austrian, Swedish, Dutch, Polish and American scholars: Charles Hermite (1), Jules Hoüel (1), Gottlieb Adler (1), Franz Meyer (2), Carl Neumann (1), Friedrich Emil Prym (3), Hermann Amandus Schwarz (2), Arthur Cayley (1), Thomas Craig (1), Huber Anson Newton (1), Hjalmar Holmgren (1), Thomas Joannes Stieltjes (1), Gustav Viedemann (1), Emil Weyr (1) and Wilhelm Dörpfeld (1).

2.4 Letters Brioschi Received from Foreign Scholars

These letters, 66 in all, can be found in the Brioschi Archive of the Polytechnic in Milan, recently transferred together with the complete archive trust of the Polytechnic to the new headquarters of Campus Bovisa-Durando. The already-published catalogue for these letters has now been updated. The most substantial correspondence is that with Charles Hermite (9 letters), Felix Klein (17 letters) and Leopold Kronecker (4 letters).

[4]In Bottazzini (1977), Tazzioli (2000) and Nagliati (2000), respectively.

Hermite's letters (1882–1897), which are conserved here, refer to a late phase of the scientific life of the two mathematicians. Even though they do not contain important mathematical details, they do report interesting observations on the mathematical developments of the day, such as the research works on elliptic functions and the work by Pincherle and Mittag-Leffler, as well as providing comments on the cultural and political situation in France and Europe in general.

The letters sent by Klein, mostly belonging to the period from January 1876 to December 1878, with the exception of one from 1884, deal essentially with research on the resolution of the quintic equation, which Klein achieved through the icosahedral group by expanding results obtained by Hermite, Kronecker and Brioschi himself. They also mention Klein's stay in Italy in 1878, during which he met Betti in Florence and Brioschi in Milan. This correspondence is completed by 10 letters from Brioschi to Klein, from the Göttingen University library, six from the crucial period between October 1878 and June 1879, and the other four from 1880 to 1888. On the theme of hyperelliptic functions and the solution of algebraic equations, we also quote a letter from Heinrich Maschke (December 30, 1888), which refers to the general equation of the sixth degree, whose solution was accomplished by Brioschi with the contribution of Maschke in those years.

To these, we have added a lengthy letter and two notes written at the end of 1895, sent by Brioschi to Max Noether, housed at the Staatsbibliothek in Berlin, in which Brioschi thanks Noether for sending him the work, *Ueber den gemeinsamen Factor zweier binären Formen*, and informs him of a theorem on the same topic. The letter, in fact, deals with the theory of elimination and contains the extension, to the roots common to two or more equations, of a theorem about which Brioschi had already informed the Berlin Academy of Science in the month of October, 1895.

Here follows the list of foreign correspondents with the numbers of letters they wrote: Carl Wilhelm Borchardt (1), Arthur Cayley (1), Elwin Bruno Christoffel (2), Johannes Eggenberger (3), Carl Friedrich Geiser (1), Paul Gordan (1), Johann-Heinrich Graf (2), Edward Jan Habich (2), Charles Hermite (9), Arthur Hirsch (1), Camille Jordan (2), W. S. Kirkland (1), Benno Klein (1), Felix Klein (17), Martin Krause (3), Leopold Kronecker (4), Ernst Eduard Kummer (1), Aimé Laussedat (1), Henry Léauté (1), Carl Ludvig (1), Edmond Maillet (3), Carl Johan Malmsten (1), Victor Mertens (1), Heinrich Maschke (1), Franz Meyer (2), Gösta Mittag-Leffler (1), Max Noether (2), Theodor Mommsen (1), Henri Ruelle (1), Pieter Hendrik Schoute (1), Henry Y. D. Scott (1), Ernst Eduard Wiltheiss (1), and Gustav Zeuner (1).

2.5 Letters Tardy Received from Foreign Scholars

The main part of Tardy's correspondence is preserved at the University Library of Genoa, having been donated by Gino Loria in 1925. It consists of 784 documentary units involving 45 Italian and foreign correspondents during the years 1837–1904.

Thirty still-unpublished letters from foreign mathematicians to Tardy have now been transcribed (Nagliati 2014) (two other letters from Schläfli were included by Loria himself in Tardy's obituary in 1905). Here follows the list of foreign correspondents with Tardy and their respective number of letters: Christian Heinrich Friedrich Peters (7), James Joseph Sylvester (6), Thomas Hirst (5), Ludwig Schläfli (2), Arthur Cayley (2), Carl Pelz (2), Johann August Grunert (2), George Biddell Airy (1), Bernhard Riemann (1 post card), Felix Klein (1), Peter Gustav Lejeune Dirichlet (1), and William Spottiswoode (1).

3 The Trips of Italians Abroad and Those of Foreign Colleagues Visiting Italy

Regarding the theme of trips abroad for the purpose of study, Lagrange may be considered a precursor, with his lengthy travel from Turin to Paris and London before he settled down in Berlin (Borgato and Pepe 1990). Many young men followed his example: we may recall the two Tuscans, Gaetano Giorgini and Alessandro Manetti, who studied at the *Ecole Polytechinique* at the beginning of the nineteenth century and went on to be leading figures in the scientific and technological milieu.

Of interest is what we may call the "forced internationalization" caused by political exile, which involved many scientists in the period of uprisings that took place during the Italian Risorgimento. Insurrectional attempts from 1820–1821 until 1848 caused the emigration of a large group of scientists, such as Ottaviano Fabrizio Mossotti, Guglielmo Libri, Francesco Orioli, Macedonio Melloni, Carlo Matteucci, Faustino Malaguti, and Agostino Codazzi. A good representative example is Ottaviano Fabrizio Mossotti who, from Milan, where he worked at the liberal journal *Il Conciliatore* (1818–1819), repaired first to Geneva (1825) and then to London and Buenos Aires (1827). In Argentina, he created the first astronomical and meteorological observatories. After returning to Italy, he emigrated again, becoming a professor of the Ionia Academy in Corfu (1839) and later arriving, in 1841, at the University of Pisa, where he began to spearhead the scientific activity of Enrico Betti.

Guglielmo Libri, whose parents had enthusiastically adhered to revolutionary principles, after a brilliant debut at the University of Pisa and the uprising of 1831, found asylum in France, where he met Mazzini and came into contact with the liberal circles (François Guizot) (Del Centina and Fiocca 2010; Fiocca and Nagliati 2009).

Welcomed at many foreign universities, Italian scientists established scientific contacts with those in environments more steeped in European culture and were able to draw up, after the unification of Italy, a program of scientific renewal that had great relevance. They also had relationships with other political exiles (Mazzini, Gioberti, Garibaldi etc.) (Pepe 1998).

Tardy left Sicily in 1838, spending a long period of formation in France, where he met Libri, in Belgium and Great Britain, where he was introduced into the scientific milieu by Mossotti, and, lastly, in Ireland.

A substantial group of scholars from Turin also spent periods studying abroad, especially after the second congress of Italian scientists held in Turin in 1840; among these, we can mention Amedeo Avogadro, Giovanni Plana, and Angelo Genocchi. Francesco Faà di Bruno also spent an important part of his scientific formation in Paris on two separate occasions, firstly between 1849 and 1851, at the end of which time he graduated with a degree in mathematical science from the Sorbonne, and secondly, from 1854 to 1856, during which he obtained his doctorate under the guidance of Cauchy. Their contribution, and that of their pupils, was directed at a renewal of the structure and organisation of studies and research, first in Piedmont and then throughout the Kingdom (Roero 2012).

Between September and November 1858, the aforementioned journey took place, during which Betti, Brioschi and his assistant, Felice Casorati, visited the University of Göttingen, where they met Dedekind and Riemann, Berlin, where Weierstrass, Kronecker and Kummer worked, Leipsig, Dresden, and Paris, where they met Hermite and Bertrand. The mathematicians they would have liked to meet, but did not, were Dirichlet in Göttingen and Liouville in Paris. One possible explanation for their not visiting England during the journey may be the fact that neither Cayley nor Sylvester were university teachers at the time; the former was still a lawyer and would not hold a university chair in Cambridge until 1863, while the latter had been a teacher at the Woolich Military Academy since 1854, and did not become part of the Oxford University teaching staff until 1883, after a working experience in the United States.

This stable network of scientific contacts gave young students a regular opportunity to go abroad so as to widen their studies during their formation. A large number of Betti's students were sent abroad: Ulisse Dini (1845–1918) studied for one year in Paris under Hermite before taking up a teaching post in Pisa in 1866; Gregorio Ricci Curbastro (1853–1925) was awarded a grant in 1877 to study at the Institute of Higher Technology in Munich where Klein was working; Alberto Tonelli (1849–1921), a graduate of the University of Pisa in 1871, spent a period in Göttingen, and later went on to teach in Palermo and Rome; Luigi Bianchi (1856–1928) spent many years studying abroad, first in Munich and then in Göttingen under Klein.

Brioschi, too, integrated his students into this international network: Felice Casorati, after the journey of 1858, went back to Berlin in 1864, where he studied with Kronecker and Weierstrass; Brioschi himself, Cremona and Beltrami planned another journey to London to meet Boole and Sylvester, to be undertaken in 1870, although it did not ultimately happen. However, many journeys to Zurich, organised by Brioschi, did take place for the betterment of students of the Polytechnic in Milan.

Throughout the nineteenth century, many European mathematicians, in their turn, came to Italy for their studies. The Congress of Italian Scientists may be seen as a significant event; it commenced in 1839 with the First Meeting in Pisa and continued in various cities of the peninsula up to 1847, until they were interrupted by Risorgimento uprisings. It was one of the first events testifying to the existence of a national scientific community at a time when there was no political unity. Babbage, who had already visited Italy several times, took part in the second congress with

Hamilton in Turin in 1840, presenting his calculating machine; Jacobi participated in the Lucca congress in 1843, accompanied by Dirichlet, Steiner and Kummer, and then undertook an extensive journey through Italy, traveling to Sicily, passing through Palermo and Messina, where he was welcomed by Tardy; Borchardt was also present at the congress.

Other important journeys for study were undertaken in Italy, as is clear from documents that refer to contacts with a substantial group of foreign scholars. In 1843–1844, Carl Borchardt spent a year in Italy with Carl Jacobi, mostly in Naples and Rome, were he met Lejeune Dirichlet and Jacob Steiner. The astronomer Christian Peters was in Sicily in 1844, where he remained for some time, taking part in uprisings that prevented him from being given the post of Head of the Naples Observatory. In 1852, Holmgren visited Pisa, during his search for scientific articles for Malmsten. James Joseph Sylvester had his first stay in Italy in 1854, and while in Florence, he had the opportunity to meet Betti. He was back in Italy for the Christmas holidays in 1856, and during the first months of 1857, he visited various cities, including Pisa, Genoa, where he met Tardy, and Naples, where he established contact with Nicola Trudi. He was yet again in Italy during the first few months of 1862, when he once more met with both Betti (he arrived in Pisa on February 9) and Brioschi.

Thomas Hirst first visited Italy in August 1858, staying for several months, during which time he worked with Tortolini in Rome. He returned in the summers of 1863 and 1864, staying for some time in Bologna, where he worked with Cremona, who had a profound effect on his studies.

Bernhard Riemann, whom Betti and Brioschi had met during the 1858 journey, later spent long periods in Italy, which have been closely studied from the point of view of their influence on the research of the day, particularly that of Betti. He came to Italy for reasons of health, going first to Pisa from 1862 to 1863, and then visiting again from 1864 until his death two years later while he was at Lake Maggiore; Tardy had also met him during his first stay in Florence.

In 1865, Prym (in Italy after Riemann) visited Tuscany and Pavia, where he met Casorati. In 1872, Kronecker first visited Pisa, returning there in 1884. Klein went to Pisa, Naples and Florence in 1872, returning to Italy again in 1878.

The American, Hubert Anson Newton, was in Florence in 1880, and in the same period, Mittag-Leffler undertook a lengthy journey across Italy, visiting Pisa, Florence, Venice, Milan (where he met Brioschi), Pavia (with Casorati and Beltrami), and Bellagio.

4 Mathematical Topics Under Discussion

Among the various themes of mathematical research debated in the correspondences are the theory of binary forms, the resolution of algebraic equations, and fractional calculus. We provide a brief description of these important contents.

4.1 The Invariant Theory of Binary Forms

The first topic of a mathematical nature to appear in the letters Brioschi wrote to both Betti and Tardy concerns invariants and covariants of binary forms. This was a leading theme in international research, in its various applications, for which Brioschi and Betti were the main Italian exponents in the second half of the nineteenth century (in fact, many of the almost 300 printed works by Brioschi are devoted to the theory of invariants, including his last work, published 2 months before his death).

At the beginning of the nineteenth century, the work by Gauss on binary quadratic forms (*Disquisitiones Arithmeticae*, Lipsia 1801, Chap. 5) contained the first observations on invariant algebraic properties, but the main developments within this theory were to come in the 1840s, from George Boole in England, Otto Hesse in Germany, and later, from their students: Cayley, Sylvester and Salmon to start with, followed by Aronhold. The application of these theories involved various fields, but it was the study of linear systems of curves in projective geometry that brought about its success.

A binary form is a homogeneous polynomial in two variables (*binary quantic*) :

$$f(x, y) = a_0x^m + a_1x^{m-1}y + a_2x^{m-2}y^2 + \cdots + a_my^m.$$

Under a linear change of variables, $f(x, y)$ is transformed into another binary form in the new variables X and Y defined by[5]:

$$F(X, Y) = A_0X^m + A_1X^{m-1}Y + A_2X^{m-2}Y^2 + \cdots + A_mY^m.$$

The main problem of the invariant theory was to determine all homogeneous polynomials in the coefficients of f (*invariants*) and, more generally, all homogeneous polynomials in the coefficients and variables of f (*covariants*) that remain unchanged under such linear transformations:

$$I(A_0, A_1, A_2, \ldots, A_m) = KI(a_0, a_1, a_2, \ldots, a_m),$$

$$C(A_0, A_1, A_2, \ldots, A_m; X, Y) = KC(a_0, a_1, a_2, \ldots, a_m; x, y),$$

where the factor K is linked to the determinant of the linear transformation.

Between 1846 and 1854, Cayley and Sylvester (separately, even though they were in contact) had laid down the bases of the theory and, by 1854, had established the concepts and fundamental theorems, above all the *degree*

[5]A *linear change of variables* is a linear transformation of the variables x and y such that the determinant is nonzero.

(degree of homogeneity with respect to the coefficients), the *order* (degree of homogeneity with respect to the variables), and the *weight*.[6]

It was Cayley who expressed the main problem of the theory, or rather the determination of the number of invariants of a given form, and when studying techniques to obtain them, he also discovered that various invariants of the same form can satisfy opportune relationships. In 1853, Sylvester adopted the term "syzygy" to indicate these linear relationships between invariants and covariants of the same degree and order. The problem may then be re-formulated in the study of a minimal set of invariants that would allow for complete construction, through opportune use of the relative syzgies, of the complete system of invariants of a given form.

In the 1850s, binary forms appeared to allow a complete solution to the determination of invariants.

The interest in the theory of invariants went on throughout the second half of the nineteenth century. The studies carried out by Cayley, Sylvester, Hermite and Brioschi belong to the phase that Hilbert called 'ingenuous' (1893), in which partial results were constructed on the basis of known algorithms, without specific formalization, which was later developed with symbolic calculus by Aronhold, Clebsch and Gordan.

Brioschi's letters to both Betti and Tardy of June 1858, make reference to the attempts to find relations between invariants and covariants of fifth degree binary forms (*quintics*):

> I found a way to express the four invariants very simply, but for three, it had already been done. [...] Given the form $(a_1, a_2, \ldots, a_5)(x, y)^5$ and assuming
>
> $$A = 2\left(a_0a_4 - 4a_1a_3 + 3a_2^2\right), \quad B = a_0a_5 - 3a_1a_4 + 2a_2a_3, \quad C = 2\left(a_1a_5 - 4a_2a_4 + 3a_3^2\right),$$
>
> $$\alpha = a_0a_2a_4 + 2a_1a_2a_3 - a_0a_3^2 - a_1^2a_4 - a_2^3,$$
>
> $$3\beta = a_1a_2a_4 + a_0a_2a_5 - a_0a_3a_4 + a_1a_3^2 - a_2^2a_3 - a_1^2a_5,$$
>
> $$3\gamma = a_1a_3a_4 + a_0a_3a_5 - a_1a_2a_5 + a_2^2a_4 - a_2a_3^2 - a_0a_4^2,$$
>
> $$\delta = a_1a_3a_5 + 2a_2a_3a_4 - a_1a_4^2 - a_2^2a_5 - a_3^3,$$

[6]The weight of a term of an algebraic form is a numeric value determined by assigning single numerical values to coefficients and variables: for example, the term $a_3^2a_4^3x^6y^2$ has the weight: $3 \times 2 + 4 \times 3 + 1 \times 6 + 0 \times 2 = 24$. An important property of invariants and covariants is the *isobaric property* ,i.e., all of their terms have the same weight; if θ represents the degree, and μ the order of a covariant C of a binary form of m degree, C has the constant weight: $\frac{1}{2}(m\theta + \mu)$. See: Crilly (1986, 2006), and Hunger (1989, 1998, 2006).

we have (ignoring numerical factors)

$$I_4 = AC - B^2, \quad I_8 = A\left(\beta\delta - \gamma^2\right) + B\left(\beta\gamma - \alpha\delta\right) + C\left(\alpha\gamma - \beta^2\right), \quad I_{12} = \begin{vmatrix} \alpha & 2\beta & \gamma & 0 \\ 0 & \alpha & 2\beta & \gamma \\ \beta & 2\gamma & \delta & 0 \\ 0 & \beta & 2\gamma & \delta \end{vmatrix}.$$

I now put: $L = A\gamma - 2B\beta + C\alpha$, $M = A\delta - 2B\gamma + C\beta$.
We have $I_{12} = AM^2 + CL^2 - 2BLM$, $I_{18} = \alpha M^3 - 3\beta LM^2 + 3\gamma L^2 M - \delta L^3$.
So, I found a great number of relations between covariants and invariants, among which those of Hermite for invariants [...] It means that I_{18} is the result of the elimination of the ratio $x : y$ from the cubic covariant of the third degree $ax^3 + 3\beta x^2 y + 3\gamma xy^2 + \delta y^3$ and the linear covariant of the fifth degree $Lx + My$.[7]

Besides some particular results aimed at extending those provided by Cayley, Hermite and Sylvester (for example, we may cite Hermite's law of reciprocity: *To every covariant of a form of degree m, of degree p in relation to its coefficients, there corresponds a covariant of degree m in relation to the coefficients, of a form of degree p*),[8] Brioschi's main work during those years was the elaboration of the theory of covariants, in his treatise *La teorica dei covarianti e degli invarianti delle forme binarie e le sue principali applicazioni*, which was published in the *Annali di Matematica*, divided into various parts, between 1858 and 1861.[9] The plans for his treatise on covariants of binary forms were frequently discussed in his correspondence with Betti and Tardy up to the end of 1857. It was part of a wider plan to provide a series of university treatises updated with recent research.

While Betti, influenced by Riemann, at a certain point was to change the focus of his research, Brioschi devoted all of his life to his research on the theory of invariants. After his 1858 journey through Europe, particularly the meeting with

[7] *Ho trovato il modo di esprimere i quattro invarianti assai semplicemente, il che però per tre era già fatto.* [...] *Considerata la forma:* $(a_1, a_2, \ldots, a_5)(x, y)^5$ *posto:*
$A = 2\left(a_0a_4 - 4a_1a_3 + 3a_2^2\right)$, $B = a_0a_5 - 3a_1a_4 + 2a_2a_3$, $C = 2\left(a_1a_5 - 4a_2a_4 + 3a_3^2\right)$
$\alpha = a_0a_2a_4 + 2a_1a_2a_3 - a_0a_3^2 - a_1^2a_4 - a_2^3$, $3\beta = a_1a_2a_4 + a_0a_2a_5 - a_0a_3a_4 + a_1a_3^2 - a_2^2a_3 - a_1^2a_5$,
$3\gamma = a_1a_3a_4 + a_0a_3a_5 - a_1a_2a_5 + a_2^2a_4 - a_2a_3^2 - a_0a_4^2$, $\delta = a_1a_3a_5 + 2a_2a_3a_4 - a_1a_4^2 - a_2^2a_5 - a_3^3$.
Si ha (tralasciando fattori numerici):

$$I_4 = AC - B^2, \quad I_8 = A\left(\beta\delta - \gamma^2\right) + B\left(\beta\gamma - \alpha\delta\right) + C\left(\alpha\gamma - \beta^2\right), \quad I_{12} = \begin{vmatrix} \alpha & 2\beta & \gamma & 0 \\ 0 & \alpha & 2\beta & \gamma \\ \beta & 2\gamma & \delta & 0 \\ 0 & \beta & 2\gamma & \delta \end{vmatrix}.$$

Pongo ora: $L = A\gamma - 2B\beta + C\alpha$, $M = A\delta - 2B\gamma + C\beta$.
Si hanno: $I_{12} = AM^2 + CL^2 - 2BLM$, $I_{18} = \alpha M^3 - 3\beta LM^2 + 3\gamma L^2M - \delta L^3$.
Così ho trovato un grandissimo numero di relazioni tra covarianti ed invarianti fra le quali quelle dell'Hermite per gli invarianti [...] *cioè* I_{18} *è il risultato dell'eliminazione del rapporto x:y dal covariante cubico di terzo grado:* $ax^3 + 3\beta x^2y + 3\gamma xy^2 + \delta y^3$ *e dal covariante lineare di quinto grado:* $Lx + My$.

[8] Brioschi (1856a, b).

[9] Brioschi (1858–1861).

Aronhold in Berlin, Brioschi proceeded with his research, using, however, the symbolic representation.

The subject of partition of numbers was also connected to that of the research on invariants of a binary form. Indeed, linked to the partition of numbers is the problem of the number of irreducible covariants of homogeneous polynomials (*quantics*) that Cayley deals with in his "superb" (according to Brioschi) memoirs, *Upon Quantics* (ten in all, three of which are from the period 1854–1856,[10] and other three in 1858). Besides showing how to generate the covariants of a binary quantic, he also demonstrated that the number of irreducible (or algebraically independent) covariants can be traced back to the number of ways in which certain integers may be partitioned, which are linked to the order of the polynomial and the degrees of homogeneity of the covariants, with respect to coefficients and variables.

In *Upon Quantics* (1856b), his second memoir, Cayley established how to count the number of covariants in terms of degree, order and weight: if A is a polynomial of degree θ and weight $\frac{1}{2}(m\theta - \mu)$ in relation to the coefficients of a binary form and the coefficients of A satisfy certain conditions of linear independence, the number of linearly independent covariants of order μ and degree θ is equal to the number of terms [of the binary form] of degree θ and weight $\frac{1}{2}(m\theta - \mu)$, less the number of terms [of the binary from] of degree θ and weight $\frac{1}{2}(m\theta - \mu) - 1$.

Therefore, the determination of linearly independent covariants is linked to the problem of determining the numbers of ways in which the integers $\frac{1}{2}(m\theta - \mu)$ and $\frac{1}{2}(m\theta - \mu) - 1$ can be obtained as an integer partition of θ parts, with repetitions allowed.

It should be noted that Cayley had therefore provided an algorithm for the calculation of the number of independent invariants and covariants of given order and degree for a binary form of fixed degree, but from this, it did not necessarily follow that this number was finite for binary forms of degree five or more. It is common knowledge that Cayley mistakenly believed that this set was infinite. The theorem (*the finiteness theorem*) that invariants and covariants of binary forms have a finite basis would be demonstrated by Paul A. Gordan in 1868.[11]

General partition problems may be expressed as follows: Given the positive integers $a_1, \ a_2, \ \ldots, a_r; \ n$, find the number of positive integer solutions of the equation $a_1x_1 + a_2x_2 + \ldots + a_rx_r = n$.

By indicating the number sought as $S_{r,n}$, the problem may be traced back to the integration of finite difference equations:

$$S_{r,n} = S_{r-1,n} + S_{r,n-a_r}.$$

According to a result already given by Pietro Paoli (1784), who goes back to Euler (*Introductio in Analysin infinitorum*, Chap. XVI), $S_{r,n}$ coincides with the

[10]Cayley (1854, 1856a, b).

[11]See: Crilly (1988), and Hunger (1989).

coefficient of x^n in the power series expansion of

$$\frac{1}{(1-x^{a_1})(1-x^{a_2})\dots(1-x^{a_r})}.$$

In 1855–1857, Sylvester had provided the analytical expression of $S_{r,\,n}$ (*quotity*), so that it could be easily calculated in particular cases (Sylvester 1857a, b, c). Brioschi then provided a justification by means of residue calculus (Brioschi 1857). Sylvester published other works in those years on the general problem of partitions, on the problem of the Virgins and compound partition, as well as providing a general framework for the theory in a series of lectures at King's College (1859), which were not published until almost 40 years later (Sylvester 1897).

4.2 The Resolution of Quintic Equations

The question related to the resolution of fifth degree equations is dealt with in Brioschi's correspondence with Betti and Tardy in a bulky series of letters from 1858, a crucial year, since various mathematicians accomplished the solution in different ways. The explicit determination of the solution was reached separately by Hermite and Kronecker, and Brioschi played a prominent role in the research studies that led to it, as well as in the clarification of certain mathematical aspects, which were to influence the subsequent work of Felix Klein.[12] Although Tardy did not personally provide any important contribution to the specific question, he did play a fundamental role, since he was the one who steered the young Betti, and Brioschi himself, towards the theory of elliptic functions, which was established with the works of Abel and Jacobi. Tardy had published an important memoir on the division of elliptic functions (Tardy 1851b).

The first systematic attempt at a general solution method by radicals for fifth degree (*quintic*) equations was undertaken in 1683 by Ehrenfried Walther von Tschirnhaus and was based on quadratic substitutions that cause coefficients to disappear. Then, independently and at different times, Bring and Jerrard obtained the reduced normal form, using a Tschirnhaus transformation involving only quadratic and cubic roots, so that the general quintic can be reduced in radicals to the equation $t^5 + t - a = 0$, where a is a real number.

After the demonstration that there is no algebraic solution—that is, solution by the algebraic operations and the extraction of radicals—to the general polynomial equations of degree five or higher, given by Ruffini in 1799 and completed by Abel in 1824, it was necessary to find new algorithms that were provided by elliptic functions.

[12]The question has been the focus of many studies, among which are Franci and Toti-Rigatelli (1979), Zappa (1997, 1999), and Borgato (2012).

Évariste Galois developed new techniques for determining whether a given equation could be solved by radicals, which gave rise to the group theory and the Galois theory. If the Galois group of a quintic is unsolvable, the quintic is not solvable by radicals.

Betti was the first to complete some of Galois' demonstrations and to use the Galois theory in connection with algebraic equations. In a group of memoirs from the early years of the 1850s, Betti proved the necessary and sufficient conditions for an algebraic equation whose degree is a power of a prime number to be solved by radicals and determined the form of the solution, extending Abel's and Kronecker's results (Betti 1851a, b, 1852, 1855a, b).

The starting point was the same, later followed by Hermite, that is the use of Jerrard's transformation for obtaining the equation $y^5 + y = \lambda$.

Besides the general characterization of the algebraic equations being solvable by radicals, Betti highlighted the links between this question and the group theory, and conjectured the use of elliptic functions to solve the general quintic. In fact, starting in 1851, he had raised the issue of determining "the irrationals, which, in equations of higher degrees, can replace the radicals)" (letter to Mossotti, May 15, 1851)[13] and had considered the idea of using elliptic functions (Mossotti's reply, May 30. 1851).[14]

In 1853, he applied the Galois theory to the reduction of the degree of modular equations, and in 1854, he obtained an initial result on the 'analytic' solution of quintic equations, but he did not go forward with this research, and was thus preceded by Charles Hermite (Betti 1854).

This abandonment was the result of a number of difficulties: he was a high school teacher, looking for an academic position; additionally, he found technical obstructions in his calculations and needed to clarify questions in group theory (permutations). He hoped to overcome the obstacles using invariant theory, which was showing a growing importance in the application of elliptic functions for the analytic solution of quintic equations in the determination of resolvents. He gradually turned to a different type of problem, until his translation of Riemann's famous work on elliptic functions in 1859.

1858 was a momentous year for the resolution of the general quintic.

In March 1858, in a note on the *Comptes Rendus*, Hermite resolved the fifth degree equation. Starting from the Jerrard normal form, and returning to the elliptic modular equation, he found a resolvent of degree 5 whose solutions were expressed in terms of elliptic modular functions (Hermite 1858a).

In the month of May, in the *Annali di Matematica*, Brioschi derived some properties of the multiplier equation, and in the month of September, he provided

---

[13][...] *gli irrazionali che nelle equazioni dei gradi superiori facciano l'ufficio dei radicali* (Nagliati 2000).

[14]Ottaviano Fabrizio Mossotti (1823–1829) was Betti's teacher at the University of Pisa, and his main figure of reference at the beginning of his scientific research.

the solution through similar means, demonstrating how, from the multiplier equation of degree 6, one can deduce a resolvent of degree 5 (Brioschi 1858a, c).

Brioschi to Tardy (11th May, 1858)[15]:

> I have abandoned the path taken by Hermite i.e., modular equations for those equations that I would call the multiplier equations. Jacobi has stated the main property of the roots of those equations in the 3rd Tome of the Crelle Journal, p. 308; this property, which, to my amazement, seemed so abstruse to Hermite (Jacobi's Works T. 2nd, p. 249), can easily be demonstrated and is a source of many important consequences, one of which may be that of the resolution of fifth degree equations, which is made relatively simple in this way.

In the month of June, in *Comptes Rendus* a letter from Kronecker to Hermite was published in which he suggested a direct method for reducing the solution of the general quintic to a resolvent of degree six, solvable by elliptic functions (Kronecker 1858).

Kronecker's idea, which is based on Galois' group theory and only sketched, was explained in a letter from Betti to Brioschi dated August 13:

> It seems to me that the thread of reasoning that led Kronecker to his fine work on the equations of 5th degree is the following.
>
> The sixth degree equations whose roots are elliptic functions of $\frac{k+rk'}{5}$ all have a group of 60 permutations. The equations of the fifth degree, however, generally have a group of 120. But this is decomposable in many ways into two equal groups G of 60. To turn these into those, we must therefore: 1st Add an irrational quantity of second degree invariant under the substitutions of G. 2nd Decompose the group G into six similar groups H, H_0, H_1, H_2, H_3, H_4. 3rd Build the equation of degree six whose roots are the values of a function invariant under the substitutions of H.
>
> 1st The decomposition of the general group G of the equation of fifth degree into two equal G is the following.
>
> All of the substitutions on five letters, as I found (Ann. Tortolini 1851), are: $\begin{pmatrix} ax+b \\ x \end{pmatrix}\begin{pmatrix} (cx+d)^3+e \\ x \end{pmatrix}$.
>
> The G groups have the only substitutions in which a is residue and c is non-residue of 5.
>
> 2nd The decomposition of G into six similar groups, Kronecker did this by taking the group H to which the substitutions $\begin{pmatrix} ax+b \\ x \end{pmatrix}$ with $\left(\frac{a}{5}\right) = 1$ belong and applying the

[15] *Io ho abbandonato la strada dell'Hermite cioè le equazioni modulari per le equazioni, che denominerei, del moltiplicatore. Jacobi ha enunciato nel Tomo 3° Giornale di Crelle pag. 308, la proprietà principale delle radici di quelle equazioni; questa proprietà, la quale non so come parve tanto astrusa all'Hermite, (Jacobi, Opere T. 2° pag. 249) dimostrasi facilmente ed è fonte di molte importanti conseguenze; una delle quali può essere quella della risoluz. delle equazioni del quinto grado; la quale in questo modo rendesi relativamente molto semplice.*

substitutions to it[16]:

$$\begin{pmatrix} x \\ x \end{pmatrix}, \begin{pmatrix} (2x)^3 \\ x \end{pmatrix}, \begin{pmatrix} (2x)^3+1 \\ x \end{pmatrix}, \begin{pmatrix} (2x)^3+2 \\ x \end{pmatrix}, \begin{pmatrix} (2x)^3+3 \\ x \end{pmatrix}, \begin{pmatrix} (2x)^3+4 \\ x \end{pmatrix}.$$

3rd The function, invariant under the substitutions of the group H, which must include an irrational of the second degree, invariant under the substitutions of G, is

$$f^2 = \left(\{\varphi(m,1) - \varphi(m,4)\}\ sin\frac{2\pi}{5} + \{\varphi(m,2) - \varphi(m,3)\}\ sin\frac{4\pi}{5} \right)^2,$$

with

$$\varphi(m,n) = \sum_0^4 \left(x_m x_{m+n}^2 x_{m+2n}^2 + \nu x_m^3 x_{m+n} x_{m+2n} \right).$$

---

[16]*Mi pare che il filo dei ragionamenti che hanno condotto Kronecker al suo bel lavoro sopra l'equazioni di 5° grado sia il seguente.*

L'equazioni di 6° grado che hanno per radi[ci] funzioni ellittiche di $\frac{k+rk'}{5}$ *hanno tutte un Gruppo di 60 permutazioni. L'equazioni di 5° grado hanno invece in generale un gruppo di 120. Ma questo è decomponibile in più modi in due Gruppi eguali G di 60. Per trasformare queste in quelle bisogna dunque 1° aggiungere una quantità irrazionale di 2° grado invariabile per le sostituzioni di G. 2° Decomporre il gruppo G in 6 Gruppi H,* H_0, H_1, H_2, H_3, H_4, *simili, 3° Costruire l'equazione di 6° grado che ha per radici i valori di una funzione invariabile per le sostituzioni di H.*

1° La decomposizione del Gruppo generale dell'equazione di 5° grado in due G eguali è la seguente.

Tutte le sostituzioni sopra 5 lettere sono, come trovai (Ann. di Tortolini 1851):

$$\begin{pmatrix} ax+b \\ x \end{pmatrix} \begin{pmatrix} (cx+d)^3+e \\ x \end{pmatrix}$$

i Gruppi G hanno le sole sostituzioni, nelle quali $\underline{a}$ *è residuo e* $\underline{c}$ *non residuo di 5.*

2° La decomposizione di G in 6 gruppi simili, Kronecker l'ha fatta prendendo il Gruppo H a cui appartengono le sostituzioni $\begin{pmatrix} ax+b \\ x \end{pmatrix}$ *con* $\begin{pmatrix} a \\ 5 \end{pmatrix} = 1$ *e applicando ad esso le sostituzioni*

$$\begin{pmatrix} x \\ x \end{pmatrix}, \begin{pmatrix} (2x)^3 \\ x \end{pmatrix}, \begin{pmatrix} (2x)^3+1 \\ x \end{pmatrix}, \begin{pmatrix} (2x)^3+2 \\ x \end{pmatrix}, \begin{pmatrix} (2x)^3+3 \\ x \end{pmatrix}, \begin{pmatrix} (2x)^3+4 \\ x \end{pmatrix}$$

3° La funzione invariabile per le sostituzioni del Gruppo H, che deve contenere un irrazionale di 2° grado, invariabile per le sostituzioni di G, è:

$$f^2 = \left(\{\varphi(m,1) - \varphi(m,4)\}\ \sin\frac{2\pi}{5} + \{\varphi(m,2) - \varphi(m,3)\}\ \sin\frac{4\pi}{5} \right)^2$$

dove $\varphi(m,n) = \sum_0^4 \left(x_m x_{m+n}^2 x_{m+2n}^2 + \nu x_m^3 x_{m+n} x_{m+2n} \right)$.

Per assoggettare $\underline{\nu}$ *ad essere invariabile per le sostituzioni di G ha posto* $f^2 + f_0^2 + f_1^2 + f_2^2 + f_3^2 + f_4^2 = 0$ *essendo* f_0, f_1, f_2, f_3, f_4 *i valori di f dopo le sostituzioni* $(2x)^3$, $(2x)^3 + 1$, ...

Il rimanente è affare di calcolo forse assai complicato; io non ho cercato ancora di effettuarlo.

To impose that ν be invariable by the substitutions of G, he assumes

$$f^2 + f_0^2 + f_1^2 + f_2^2 + f_3^2 + f_4^2 = 0,$$

f_0, f_1, f_2, f_3, f_4 being the values of f by the substitutions $(2x)^3$, $(2x)^3 + 1$, ...
The remainder is a question of a perhaps very complicated calculation; I have not even tried to do it.

In a letter to Hermite dated July 31, and published in the *Comptes Rendus,* Brioschi evidenced a series of equations of the sixth degree, analogous to modular equations and solvable by elliptic functions (Brioschi 1858b).[17] At the end of November 1858 (in an article in the *Memorie dell'Istituto Lombardo*), Brioschi completed and amplified Kronecker's rather short note, also showing how the multiplier equation and Kronecker's resolvent were particular cases of a more general class of resolvents, called 'Jacobian' (Brioschi 1858d).

Following the first success in the resolution of the quintic, Brioschi (like Hermite and Kronecker) thought that these techniques could be extended to seventh degree equations,[18] and he wrote about his considerations to Betti (December 21, 1858)[19]:

[17]See also Hermite (1858b, c). On Hermite's work linked to the Galois theory, see Goldstein (2011).

[18]See Hermite (1859) and Kronecker (1859).

[19][...] *ti dirò brevemente dei miei lavori e della lettera dell'Hermite. I miei lavori diretti da prima a dimostrare i risultati di Kronecker si ampliano strada facendo, e posso dire di avere con molta semplicità stabilita una teoria di una classe di risolventi delle equazioni di quinto grado. Questa classe di risolventi, alla quale appartiene come caso particolare quella di Kronecker, è definita dalla proprietà scoperta da Jacobi per l'equaz. del moltiplicatore nella trasformazione di quinto ordine. Non entro in dettagli giacché spero mandarti fra non molto una nota sull'argomento che feci pubblicare negli Atti dell'Istituto Lombardo; ti dirò solo che mi fu di grande giovamento il tuo teorema e le tue espressioni analitiche delle sostituzioni per le funzioni di cinque lettere cioè le:*

$$\begin{pmatrix} r \\ ar+b \end{pmatrix}, \begin{pmatrix} r \\ (ar+b)^3+c \end{pmatrix}$$

Calcolai anche una di queste risolventi mediante gli invarianti delle forme di quinto grado; ma ora vorrei calcolarne un altra in cui entrasse una quantità indeterminata; se non mi fossi ammalato vi sarei giunto da alcuni giorni.

La lettera dell'Hermite verte sull'abbassamento dell'equaz. modulare di ottavo grado; ti trascrivo la forma dell'equaz. di settimo grado alla quale giunge:

$$z^7 - 4^2 7^2 \sqrt{-7} \cdot \alpha k k'^4 z^4 - 4^4 7^4 (\alpha - 3) k^2 k'^8 z + 4^6 7^3 \sqrt{-7} k k'^8 \left(1 - k^2 + k^4\right) = 0$$

essendo $\alpha = \frac{1-\sqrt{-7}}{2}$; *ed una seconda equazione nella* φ *cambiando il segno di* $\sqrt{-7}$.

L'Hermite ha trovato per le funzioni fra sette lettere espressioni analitiche delle sostituzioni analoghe alle tue (egli anzi credeva di averle trovate anche per quelle fra cinque lettere, ma io rispondendo oggi alla sua lettera lo farò avvertito che tu le hai trovate già da tempo). Eccole:
$$\begin{pmatrix} x_r \\ x_{ar+b} \end{pmatrix} \begin{pmatrix} x_r \\ x_{a\theta(r+b)+c} \end{pmatrix}$$
attribuendo alla funzione θ *queste diverse forme:*

[...] I will briefly tell you about my works and Hermite's letter. My works initially directed towards demonstrating Kronecker's results will be extended along the way, and I can say that I have very simply established a theory of a class of resolvents for fifth degree equations. This class of resolvents, to which Kronecker's belongs as a particular case, is defined through the property discovered by Jacobi for the multiplier equation in the transformation of order five [...] I hope to send you soon a note on this subject I published in the Atti dell'Istituto Lombardo; [...] your theorem and analytic expressions of the substitutions for the functions of five letters were very useful to me, i.e.:

$$\begin{pmatrix} r \\ ar+b \end{pmatrix}, \begin{pmatrix} r \\ (ar+b)^3+c \end{pmatrix}.$$

Hermite's letter concerns the reduction of the modular equation of the eighth degree; the form of the seventh degree equation he obtains is as follows:

$$z^7 - 4^2 7^2 \sqrt{-7} \cdot \alpha k k'^4 z^4 - 4^4 7^4 (\alpha - 3) k^2 k'^8 z + 4^6 7^3 \sqrt{-7} k k'^8 \left(1 - k^2 + k^4\right) = 0,$$

where $\alpha = \frac{1-\sqrt{-7}}{2}$, and a second equation for φ by changing the sign of $\sqrt{-7}$.

Hermite found analytical expressions, similar to yours, for the functions of seven letters (on the contrary, he believed he had also found them for those of five letters, but in my answer to his letter today, I will let him know that you had already found them some time ago). Here they are:

$$\begin{pmatrix} x_r \\ x_{ar+b} \end{pmatrix} \quad \begin{pmatrix} x_r \\ x_{a\theta(r+b)+c} \end{pmatrix}$$

by attributing to the function θ these different forms:

$$\theta(r) = -r^6 \pm 2r^2, \quad \theta(r) = 3r \pm r^4, \theta(r) = r^5 + ar^3 + 3a^2 r \qquad (a \text{ whatever}),$$

$$\theta(r) = r^5 + ar^3 \pm r^2 + 3a^2 r \qquad (a \text{ non residual of } 7).$$

He did not seem to attach much importance to these expressions, but I believe the results I obtained for the functions of five letters will reveal just how important they are. Now, I hope to make use of these to tackle the seventh degree equations.

The general solution of equations of degree greater than five cannot be reduced to that of equations related to the division of circular or elliptic functions. About 30 years later, Brioschi, with the contribution of Heinrich Maschke's results, a pupil of Klein's, reached the general solution of the sixth degree equation, by means of hyperelliptic functions, that is, the transcendental functions generated by the inversion of hyperellitpic integrals of the first type.

In relation to the general solution of the quintic, new insights into the Jacobian function theory were developed by Kronecker, Brioschi and Hermite in the 1860s.

$\theta(r) = -r^6 \pm 2r^2, \quad \theta(r) = 3r \pm r^4, \theta(r) = r^5 + ar^3 + 3a^2 r$ (*a qualunque*)
$\theta(r) = r^5 + ar^3 \pm r^2 + 3a^2 r$ (*a non residuo di 7*)

Egli sembra non attaccare molta importanza a queste espressioni, ma io credo che i risultati che ho ottenuto per le funzioni di cinque lettere mostreranno di quanta importanza esse siano. Spero ora di giovarmi di queste per attaccare le equazioni di settimo grado.

In the meantime, other authors (Gordan, Clebsch) developed new methods for solving the general equation of degree five, by taking the invariants of the binary form associated with the algebraic equation of the same degree into consideration. Felix Klein, on the other hand, in 1877, accomplished the solution of the general quintic through the so-called icosahedral irrationality. Klein told Brioschi the results of his research in letters dated November 3, 11 and 15, 1876, and Brioschi immediately reported them to the Accademia dei Lincei (Brioschi 1877).

The correspondence between Klein and Brioschi continued on this theme until December 1878. Brioschi gave a systematic exposition of the results he had obtained up to then in an extensive essay in German (Brioschi 1878).

Within his research, Brioschi arrived at a reduced form for the quintic equation, known as the Brioschi normal form (Brioschi 1888):

$$y^5 - 10Z\,y^3 + 45Z^2y - Z^2 = 0,$$

in which the coefficients are expressed in terms of a single parameter Z. This can be derived from a principal quintic by using a rational Tschirnhaus transformation, which is a good degree simpler than the one used to obtain the Bring–Jerrard form.

The Brioschi normal form is important to Klein's solution of the general quintic in terms of hypergeometric functions. In fact, Klein's icosahedral equation has a Jacobian resolvent of degree six, which, in its turn, has a quintic resolvent reduced to the Brioschi form, hence the general quintic is solvable by radicals and by an icosahedral inverse represented by hypergeometric series.[20]

4.3 Fractional Calculus

Fractional calculus is the extension of differential calculus to non-integer order derivatives and integrals: in other words, it studies the possibility of taking real number powers or complex number powers of the differentiation operator.

This theory was the subject of research studies carried out by many scholars in the nineteenth century. The theme, which was later abandoned for a long period, has, for the past few decades, returned to the foreground with applications in various contexts, like fractal theory, viscoelasticity and the modeling of human tissues, bringing about the publication of monographs and journals devoted to the subject.

It appears among the themes of the Brioschi–Tardy correspondence, in relation to the contribution of the latter, and in constant relation to the developments being carried out abroad. Brioschi intervened in two letters from April 1858 on a work by Tardy that was to be published in the new *Annali*. Brioschi and Betti were

[20]Brioschi's contributions to the resolution of the quintic equation are extensively described by Felix Klein himself, together with Hermite's and Kronecker's techniques, in his *Vorlesungen über das Ikosaeder* (1884, English translation 1888). On Klein's icosahedral equation and Brioschi's normal form, see also: Gray (2000, ch. IV), King (2009), and Shurman (1997).

very interested in this work, which reproduced a presentation given by Tardy at the Congress of Italian Scientists in Milan in 1844. Tardy was seeking to publish a short treatise on the new calculus: "We hope to be able to offer, as soon as circumstances allow us, a short treatise on the principles and applications of this new calculus, which would seem to promise important results,"[21] but it was only after Brioschi's encouragement that Tardy published his contribution with practically no modifications (Tardy 1858).

In 1847, Tardy had taken up the subject once more in a work of hydrodynamics, in which this type of calculus contributed to the determination of arbitrary functions in the integral of partial differential equations (Tardy 1847). In this memoir, Tardy addressed the issue of the motion of a fluid in a cylindrical vessel and, after suitable transformations, reached the integration of the differential equation in two variables:

$$\frac{d^2\Phi}{dr^2}+\frac{1}{r}\frac{d\Phi}{dr}-a\frac{d^2\Phi}{dz^2}=0,$$

which Tardy integrated by means of a fractional operator

$$\Phi=K\int^{\frac{1}{2}}\frac{\varphi\left(z+r\sqrt{a}\right)+\varphi\left(z-r\sqrt{a}\right)}{\sqrt{a}}\,da^{\frac{1}{2}},$$

where

$$\int^{\mu}x^n\,dx^{\mu}=\frac{\Gamma\left(n+1\right)}{\Gamma\left(n+\mu+1\right)}\,x^{n+\mu}.$$

The problem of derivatives of fractional order originated in a letter from Leibniz to John Wallis (May 28, 1697) and the question had already been debated 2 years earlier in the correspondences between Leibniz and both l'Hôpital and Johann Bernoulli.

In 1730, Euler had devoted a memoir to the series whose summation cannot be effected purely algebraically, in which, when trying to interpolate the factorial series, he introduced the fractional derivative. He defined $[x]$ as $\int_0^1\left(\ln\left(\frac{1}{s}\right)\right)^s ds$, and then, substituting $s=e^{-t}$ $(t=-\ln(s))$, he found what is known as the Gamma function:

$$\Gamma\left(z+1\right)=\int_0^{\infty}e^{-t}t^z dt.$$

[21]"Speriamo potere offrire, appena le circostanze ce lo permetteranno, un trattatello de' principi e delle applicazioni di questo nuovo calcolo, che ci sembra essere fecondo di risultati importanti"

In the last paragraph of the memoir, Euler extended the significance of the derivative operator by reducing it to the Gamma function:

$$\frac{d^n x^k}{dx^n} = \frac{k!}{(k-n)!} x^{k-n} = \frac{\int_0^1 (-\ln t)^k dt}{\int_0^1 (-\ln t)^{k-n} dt} x^{k-n} = \frac{\Gamma(k+1)}{\Gamma(k+1-n)} x^{k-n},$$

and concluded with the case of $n = ½$:

$$\frac{d^{1/2}\sqrt{x}}{dx^{1/2}} = \frac{\sqrt{\pi}}{2}.$$

From this point, the studies developed along two main directions with different ways of reducing the fractional derivative to the Gamma function.[22]

Laplace defined a fractional derivative by means of an integral in 1812, Lacroix took up Euler's theory in his treatise on differential and integral calculus (1819) and, in 1822, Fourier proposed an operative definition based on his integral transform, while the first application to physics (tautochrone problem) can be found in the work by Abel in 1823.

Tardy's intervention at the Milan Congress made reference to one of the most significant contributions of Liouville, who had, in 1832, devoted three long memoirs[23] to differential calculus with non-integer indices (the term 'fractional calculus' was not used at that time), with applications to the theory of the potential.

In an attempt to identify the basic principles, Liouville was the first to reunite a series of results from previous works into a single doctrinal corpus. To extend the derivative formula to whatever power μ, Liouville referred to the complex exponential function whose derivative is $\frac{d^\mu e^{mx}}{dx^\mu} = m^\mu e^{mx}$, and then reported the fractional derivative of a function y to its series expansion, $A_1 e^{m_1 x} + A_2 e^{m_2 x} + A_3 e^{m_3 x} + \cdots$, obtaining

$$\frac{d^\mu y}{dx^\mu} = \sum A_m e^{mx} m^\mu.$$

Riemann, too, studied the question, in a work[24] written in 1847, although only published after his death, a work that Tardy evidently was not familiar with.

Another simple way was to define the fractional derivative through fractional integration by means of the Gamma function. Starting from the integral operator

$$J^n y(x) = \int_0^x \int_0^{x_{n-1}} \cdots \int_0^{x_1} y(x_0)\, dx_0 \ldots dx_{n-2}\, dx_{n-1} = \frac{1}{(n-1)!} \int_0^x \frac{1}{(x-t)^{1-n}}\, y(t)\, dt,$$

[22]On the development of the fractional calculus, see Ross (1977). On Tardy's contribution see Bottazzini (1978).

[23]Liouville (1832a, b, c).

[24]Riemann (1876).

where $J^0 y(x) = y(x)$, and substituting the discrete factorial with the continuous Gamma function, which satisfies the condition $\Gamma(n) = (n-1)!$, one can define

$$J^{\alpha} y(x) = \frac{1}{\Gamma(\alpha)} \int_0^x \frac{1}{(x-t)^{1-\alpha}}\, y(t)\, dt,$$

with x and α real and positive. Thus, the fractional derivative can be defined through a composition of integer derivatives and fractional integrals:

$$D^{\alpha} y(x) = D^n J^{n-\alpha} y(x) \ \text{ or : } \ D_*^{\alpha} y(x) = J^{n-\alpha} D^n y(x),$$

(while the former operator is nowadays known as the Riemann-Liouville operator, the latter is called the Caputo differential operator).

In 1855, Liouville took the matter up again for the last time, writing an article after the publication of a memoir by Barnaba Tortolini in which he demonstrates the analogy between Tortolini's formula and his own.

Tardy's memoir of 1858 begins with a reference to Leibniz, and later to Euler, Laplace, Fourier, Brunacci, and Liouville, to whom he submits his own work, ultimately referring to more recent contributions, particularly those by mathematicians from the English school (Peacock, Greatheed, Kelland, Center), who had approached the question after the publication of Liouville's memoirs.

Tardy's starting point was in the formula for a fractional integral operator, which he obtained from an analogous formula for integer powers, by iterated integration by parts. This formula can be derived from Laplace's formula by changing the sign in derivative operators, and thus finding integral operators; in this way, the author returns to Euler's formula of fractional derivatives for power functions.

Integrating the iterated integral by parts, Tardy obtained, through application of the property of the Gamma function, the formula

$$\int^{\mu} \varphi(x)\, dx^{\mu} = \frac{x^{\mu}}{\Gamma(\mu)} \left\{ \frac{1}{\mu}\varphi(x) - \frac{x}{\mu+1}\varphi'(x) + \frac{1}{2}\frac{x^2}{\mu+2}\varphi''(x) - \dots \right\} =$$

$$= \frac{\Gamma(1-\mu)}{2\pi} x^{\mu} \int_{-\pi}^{\pi} \varphi\left(x + xe^{yi}\right) e^{\mu yi}\, dy.$$

The calculus relative to elementary functions then follows.

The weakness in the justification of the formula was underlined by Brioschi, who, while praising Tardy's memoir, suggested justifying it by demonstrating that, in this case as well, the fundamental property

$$\int^{m} \left(\int^{n} \varphi(x) dx^{n} \right) dx^{m} = \int^{m+n} \varphi(x) dx^{m+n}$$

is preserved.

The doubts that Brioschi expressed in his two letters of 1858 induced Tardy to add a *postscriptum* to his published memoir, in which this property was verified in the case of fractional orders. On this question, Brioschi wrote to Tardy:

> I immediately sent your P.S. to Tortolini. I am more and more convinced that it is an extremely useful complement to your work; inasmuch as I agree with you that the function you assigned is one form of the fractional integral, and not the form of the same; yet I have proof that that form is exact, and so I feel assured in using it, but I suspect that a second, third, etc., form may be reducible to that; it would not be the first case in which this has occurred.[25]

In the same year, 1858, Tardy's memoir on hydraulics was commented upon by Angelo Genocchi in a substantial bibliographic report that appeared in the *Annali.* Brioschi, too, in the preface to a posthumous memoir by Gabrio Piola, had provided a description of Tardy's research, inserting it into a wide historical context. Tardy's treatment also caught the attention of the Swiss mathematician Ludwig Schäfli.

In 1868, Baldassarre Boncompagni invited Tardy to publish, in his famous journal of history of science, the *Bullettino,* a final memoir on the topic, in which he reconstructed the history of fractional differentials (Tardy 1868). This memoir by Tardy also had a separate edition and aroused a certain interest abroad: it was presented, in fact, by Borchardt at the Science Academy in Berlin, as well as being translated into French (Tardy 1869).

5 The *Annali di Matematica Pura ed Applicata*

A new journal was included in the first projects for which Brioschi wanted Betti's collaboration [letter dated April 28, 1857]. As Brioschi pointed out, the important Italian scientific journal founded in 1850 and edited by Barnaba Tortolini, namely, the *Annali di Matematica e Fisica,* was no longer effective in disseminating Italian mathematical production abroad or in providing an update of research being carried out outside of Italy. Brioschi's meeting with Angelo Genocchi in Turin at the beginning of April had already produced plans for a new journal, which Brioschi proposed to Betti.

Still to be published in Rome under a new editing team comprising Tortolini, Genocchi, Betti and Brioschi, it was to be entirely devoted to mathematics, replacing Tortolini's Annals with a new title; the *Annali di Matematica Pura ed Applicata* was to come out with an issue every 2 months, divided into two parts: the first for original memoirs specifically concerning mathematics, physics or, more generally, pure and

[25]"Ho mandato subito il tuo P.S. al Tortolini. Io sono sempre più persuaso che esso è un utilissimo complemento al tuo lavoro; giacché sebbene io sia con te d'accordo che la funzione da te assegnata sia una forma dell'integrale ad indice fratto, e non la forma del medesimo; pure ho intanto una prova che quella forma è esatta, e questo mi acquieta nel farne uso, e anche mi fa nascere il sospetto che una seconda, una terza etc. forme siano riducibili a quella; non sarebbe il primo caso in cui questo fatto si verifica." (15th May 1858)

applied mathematics; the second for bibliographic articles and extracts of mainly English and German memoirs, which were not so well known in Italy.

It must be remembered that the first regular periodicals containing mathematical memoirs were those from scientific academies of renown, which continued to publish articles on mathematics and remained the most effective means of dissemination, even after the establishment of mathematical societies and specialized periodicals.

In the 1850s, besides the periodicals of celebrated national scientific academies, like the *Comptes Rendus* of the Science Academy of Paris, the *Philosophical Transactions* of the Royal Society of London, and the *Monatsberichte* of the *Preussische Akademie der Wissenshaften* of Berlin, and the publications of famous universities or institutes of higher education, such as the *Journal de l'Ecole Politechnique*, specialized independent journals also figured among the most important means of dissemination of mathematical research. Publications of mathematical societies were only to appear from 1865 on.

Among the most prestigious journals were the *Nouvelles Annales des Mathématiques Pures et Appliquées*, re-founded in Paris by Liouville in 1836 to carry on the *Annales* published for 20 years by Gergonne (between 1810 and 1831), and known as the *Journal de Liouville*, as well as the journal founded by Leopold Crelle in Berlin in 1826, inspired by Abel: the *Journal für die reine und angewandte Mathematik*, universally known as *Crelle's Journal*. After the death of Crelle, the journal was edited by Borchardt (1856), with an editorial staff that included Kummer and Weierstrass.

The British mathematical periodical, the *Cambridge Mathematical Journal*, later called the *Cambridge and Dublin Mathematical Journal* (1837–1854), had become the *Quarterly Journal of Pure and Applied Mathematics* in 1855. From 1842, Orly Terquem published the *Nouvelles Annales de Mathématiques* in Paris, and, in 1856, Schlömilch commenced publication of the *Zeitschrift für Mathematik und Physik* in Leipzig.

Later, there appeared the *Mathematische Annalen*, founded by Clebsch and Neumann in 1868, through which new mathematical research of the Bernard Riemann school was disseminated, and the *Acta Mathematica* of Mittag-Leffler, published in Stockholm starting in 1882. Of a more elementary nature was the *Giornale di Matematiche*, edited by Battaglini, which was founded in Naples in 1863.

Although the *Cambridge Journal* had published biographical articles of important mathematicians, like Arthur Cayley, and a special bibliographical section had been inserted into Schlömilch's journal, there was still no journal either partially or completely devoted to reviews, like the *Bulletin des sciences mathématiques*, founded by J. G. Darboux and G. J. Hoüel in 1870, which dealt mostly with reviews, translations, and ongoing works, or periodicals containing only abstracts, like the *Jahrbuch über die Fortschritte der Mathematik*, published from 1871, and the *Revue semestrielle des publications mathématiques* in Amsterdam, of later publication starting in 1892.

Brioschi was the promoter of this renewal, gathering inspiration from the leading European scientific journals; in particular, he took the idea of an editing team from Crelle's Journal, whose mathematicians (Borchardt, Kummer, and Weierstrass) did not all come from Berlin, even if they were all working together in Berlin from 1856.

At that time, periodicals containing only abstracts had not yet been published, but, in introducing a special section devoted to reviews, Brioschi took inspiration from the *Cambridge Mathematical Journal* and the *Quarterly Journal*, which had occasionally included reviews, and from Schlömilch's journal.

Tortolini was only contacted later, since Brioschi viewed him with a certain diffidence, which hinted at a strong sense of anticlericalism, as may also be gathered from his correspondence [May 6, 1857]. However, Tortolini was willing to participate (Brioschi to Betti, May 25, 1857) and an announcement was drawn up to introduce the journal to the public.

Brioschi thought it would be wise to sign an agreement in order to avoid future conflict with Tortolini [Brioschi to Betti, June 27, 1857], and the project was finally presented and discussed with Betti and Placido Tardy, who had been corresponding with Betti since 1853.

Some tension with Tortolini did, however, arise towards the end of the year, which seemed to call for a different solution. In the meantime, an alternative project of publication in Pisa had been carried forward by Matteucci and Mossotti, and Brioschi threatened to abandon the project to publish in Rome ("which is not a scientific centre," like Turin, Florence, or Milan [Brioschi to Betti, December 4, 1857]), but the desire to create "a single mathematical journal for the whole of Italy," with an editing team that represented the entire country, won the day.

From 1858 on, the *Annali di Scienze Matematiche e Fisiche* became the *Annali di Matematica Pura ed Applicata*, and the journal acquired a new editorial staff composed of Brioschi, Betti, Tortolini and Genocchi (who all belonged to different Italian states). The programme of the new journal, endorsed by the four editors, was published in the first issue (January–February 1858). Every 2 months, an issue of the journal was devoted, in part, to original works and, in part, to reviews, translations or abstracts of articles or books. The section called *Rivista bibliografica* was inserted into all of the volumes of the first series.

Brioschi's letters to Betti also mirrored the editing work, and contained interesting annotations to the works by Lamé, Jacobi, Kronecker, Raabe, Hermite, Cayley, Weierstrass, Riemann, and Sylvester.

Tortolini remained in charge of the *Annali* until 1865: under his leadership, seven volumes of the *Annali* were published, which constituted the journal's first series. In 1867, the headquarters of the journal were transferred from Rome to Milan when Brioschi took over the editorship, first with Cremona, and then, after the latter's departure, alone. In the second series of the *Annali*, the bibliographic section was removed, leaving only an index of recent publications at the end of each issue.

Over the following 30 years, until Brioschi's death, 26 volumes of the *Annali* were published, testifying to the development of Italian mathematical research, with contributions from leading Italian and European scientists.[26]

6 Conclusions

Analysis of the scientific correspondence reveals a gradual increase in the number of countries involved in scientific relations; among Betti's correspondents were scholars working in France, Germany, England, Holland, Sweden, and even as far away as the United States of America, where Sylvester had contributed to the formation of a mathematical school. Betti's range of contacts was also favoured by the fact that he had become Head of the *Scuola Normale* in 1865, which underwent great reform after Unification, from being a school for the training of teachers to a centre of excellence for research.

As far as mathematical research is concerned, some correspondences (in particular, Brioschi's with Tardy, Hermite and Klein) reveal precious details on the development of contemporary theories, as well as the mutual influence between Italian and foreign protagonists.

Entrance into scientific societies abroad in the second half of the century is further evidence of the recognition of the value of Italian researchers; in 1871, Betti was elected a foreign member of the London Mathematical Society, which numbered very few Italians among its members. In 1880, Betti, Brioschi and Beltrami received nominations to become correspondent members of the Berlin Academy, followed by Casorati and Cremona in 1886. From other material related to these events, it also emerges how well-versed this new generation of mathematicians was with German culture, which was widespread among Italian intellectuals of the period.

The *Annali di Matematica* became one of the leading international mathematical journals, publishing avant-garde works carried out by both Italian and foreign researchers. Brioschi and the Polytechnic, which he founded, aligned Italy's technical education with that of countries like Germany and Switzerland.

The *Annali*, devoted to mathematical research, together with the foundation of the Polytechnic of Milan, constituted the two advanced and specialized aspects of a scientific and cultural reform programme that was carried out after the Unification of Italy by Betti, Brioschi, Cremona, Battaglini, Cattaneo and others regarding all levels of state education (university, academies and technical schools) and the dissemination of a technical-scientific culture that went hand in hand with the industrialization and development of infrastructures in the country.

The programme was vast, and to give the reader an idea of how vast, one only has to recall, for middle school teaching, Giovanni Novi's translation of treatises by

[26]U. Bottazzini, *Brioschi e gli "Annali di Matematica"*, in: Lacaita and Silvestri (2000), *I. Saggi*, pp. 71–84.

Amiot and Bertrand regarding elementary geometry and arithmetic; Betti's edition of Bertrand's treatise on elementary algebra; Brioschi, Betti and Cremona's edition of Euclid's *Elements*; the improvement of journals, like the *Giornale di Battaglini*, devoted to teachers and scholars of mathematics; for technical formation, like the *Politecnico* of Cattaneo; for scientific research, like the generalist academic journals such as the *Annali delle Università Toscane*, the *Rendiconti dell'Istituto Lombardo*, the *Rendiconti dell'Academia dei Lincei*, etc.; and other specialized journals besides the *Annali di Matematica*, like the *Gazzetta Chimica* for chemistry, and the *Nuovo Cimento* for physics. For these aspects mentioned in the correspondences, we refer the reader to specific and more in-depth studies. Historical research is paradoxically complicated by the wealth of documents and correspondences at scholars' disposal and by the vast network of correspondences, the overall picture of which (still to be completely defined) is inevitably limited by the choices made regarding their publication.

References

Betti, E. 1851a. Sopra la risolubilità per radicali delle equazioni algebriche irriduttibili di grado primo. *Annali di Scienze Matematiche e Fisiche* 2: 5–19 (*Opere matematiche di E. Betti*, I, 17–27).

———. 1851b. Un teorema sulle risolventi delle equazioni risolubili per radicali. *Annali di Scienze Matematiche e Fisiche* 2: 102–103 (*Opere matematiche di E. Betti*, I, 28–29).

———. 1852. Sulla risoluzione delle equazioni algebriche. *Annali di Scienze Matematiche e Fisiche* 3: 49–115 (*Opere matematiche di E. Betti*, I, 31–80).

———. 1853. Sopra l'abbassamento delle equazioni modulari delle funzioni ellittiche. *Annali di Scienze Matematiche e Fisiche* 4: 81–100 (*Opere matematiche di E. Betti*, I, 81–95).

———. 1854. Un teorema sulla risoluzione analitica delle equazioni algebriche. *Annali di Scienze Matematiche e fisiche* 5: 10–17 (*Opere matematiche di E. Betti*, I, 96–101).

———. 1855a. Sopra la teorica delle sostituzioni. *Annali di Scienze Matematiche e Fisiche* 6: 5–34 (*Opere matematiche di E. Betti*, I, 102–123).

———. 1855b. Sopra la più generale funzione algebrica che può soddisfare una equazione il grado della quale è potenza di un numero primo. *Annali di Scienze Matematiche e Fisiche* 6: 260–272 (*Opere matematiche di E. Betti*, I, p. 126–135).

———. 1903–1913. *Opere matematiche di Enrico Betti pubblicate per cura della R. Accademia dei Lincei*, vol. 1–2. Milano: Hoepli.

Borgato, M.T. 2009. Continuity and discontinuity in Italian mathematics after the unification: from Brioschi to Peano. *Organon* 41: 219–231.

———. 2012. Ricerca matematica e impegno politico nella corrispondenza Brioschi-Betti. In *Europa matematica e Risorgimento italiano*, ed. L. Pepe, 139–179. Bologna: Clueb.

Borgato, M.T., and L. Pepe. 1990. *Lagrange. Appunti per una biografia scientifica*. Torino: La Rosa.

Bottazzini, U. 1977. The mathematical papers of Enrico Betti in the Scuola Normale Superiore of Pisa. *Historia Mathematica* 4: 207–209.

———. 1978. Ricerche di P. Tardy sui differenziali a indice qualunque 1844–1868. *Historia Mathematica* 5: 411–418.

———. 1994. *Và pensiero. Immagini della matematica nell'Italia dell'ottocento*. Bologna: Il Mulino.

Bottazzini, U., and P. Nastasi. 2013. *La patria ci vuole eroi. Matematici e vita politica nell'Italia del Risorgimento*. Bologna: Zanichelli.

Brioschi, F. 1856a. Sul principio di reciprocità nella teoria delle forme. *Annali di Scienze Matematiche e Fisiche* 7: 303–312 (*Opere Matematiche di F. Brioschi*, I, 233–240).

———. 1856b. Sulla teoria dei covarianti. *Giornale dell'I. R. Istituto Lombardo di Scienze, Lettere ed Arti* 8: 329–333 (*Opere Matematiche di F. Brioschi*, III, 137–141).

———. 1857. Sulla partizione dei numeri. *Annali di Scienze Matematiche e Fisiche* 8: 5–12 (*Opere Matematiche di F. Brioschi*, I, 24 1–246).

———. 1858–1861. La teorica dei covarianti e degli invarianti delle forme binarie e le sue principali applicazioni. *Annali di Matematica Pura ed Applicata* 1(1): 296–309, 349–361; 2: 82–85, 265–277; 3:160–168; 4: 186–194 (*Opere Matematiche di F. Brioschi*, I, 349–414).

———. 1858a. Sulle equazioni del moltiplicatore per la trasformazione delle funzioni ellittiche. *Annali di Matematica Pura ed Applicata* 1(1): 175–177 (*Opere matematiche di F. Brioschi*, I, p. 321–324).

———. 1858b. Sur divers équations analogues aux équations modulaires dans la théorie des fonctions elliptiques (Extrait d'un lettre à M. Hermite). *Comptes Rendus de l'Académie des Sciences de Paris* 47: 337–341 (*Opere matematiche di F. Brioschi*, IV, 327–330).

———. 1858c. Sulla risoluzione dell'equazione del quinto grado. *Annali di Matematica Pura ed Applicata* 1(1): 256–259, 326–328 (*Opere matematiche di F. Brioschi*, I, p. 335–341).

———. 1858d. Sul metodo di Kronecker per la risoluzione delle equazioni di quinto grado. *Atti dell'I. R. Istituto Lombardo di Scienze, Lettere ed Arti* 1: 275–282 (*Opere matematiche di F. Brioschi*, III, 177–188).

———. 1877. Sopra alcuni risultati ottenuti dal sig. Klein nella risoluzione delle equazioni del quinto grado. *Rendiconti della Reale Accademia dei Lincei* 1(3): 31–34 (*Opere Matematiche di F. Brioschi*, III, 357–360).

———. 1878. Ueber die Auflösung der Gleichungen von fünften graden. *Mathematische Annalen* 13: 109–160.

———. 1888. Sopra una trasformazione delle equazioni del quinto grado. *Annali di Matematica* Pura *ed Applicata* 16(2): 181–189 (*Opere Matematiche di F. Brioschi*, III, 21–30).

———. 1901–1905. *Opere matematiche di Francesco Brioschi, pubblicate per cura del Comitato per le onoranze a Francesco Brioschi*, vols 1–5. Milano: Hoepli.

Carbone, L., A. M. Mercurio, F. Palladino, and N. Palladino. 2006. La corrispondenza epistolare Brioschi-Genocchi. *Rendiconto dell'Accademia delle Scienze Fisiche e Matematiche Napoli* 73(4): 263–386.

Cayley, A. 1854. An introductory memoir upon quantics. *The Philosophical Transactions of the Royal Society of London* 144: 245–258.

———. 1856a. A second memoir upon quantics. *The Philosophical Transactions of the Royal Society of London* 146: 101–126.

———. 1856b. A third memoir upon quantics. *The Philosophical Transactions of the Royal Society of London* 146: 627–648.

Cerroni, C. 2012. La figura scientifica ed umana di Placido Tardy. In *Europa matematica e Risorgimento italiano*, ed. L. Pepe, 331–348. Bologna: Clueb.

Cerroni, C., and G. Fenaroli. 2007. *Il Carteggio Cremona-Tardy (1860–1886)*. Milano: Mimesis.

Cerroni, C., and L. Martini. 2009. *Il carteggio Betti-Tardy (1850–1891)*. Milano: Mimesis.

Crilly, T. 1986. The rise of Cayley's invariant theory (1841–1862). *Historia Mathematica* 13: 241–254.

———. 1988. The decline of Cayley's invariant theory (1863–1895). *Historia Mathematica* 15: 332–347.

———. 2006. *Arthur Cayley: Mathematician Laureate of the Victorian Age*. Baltimore: The Johns Hopkins University Press.

Del Centina, A., and A. Fiocca. 2010. *Guglielmo Libri matematico e storico della matematica L'irresistibile ascesa dall'Ateneo pisano all'Institut de France*. Firenze: Olschki.

Fiocca, A., and I. Nagliati. 2009. Le "Cours d'histoire des sciences" de Guglielmo Libri au Collège de France (1833). *Nuncius* 24: 127–171.

Franci, R., and L. Toti Rigatelli. 1979. *Storia della teoria delle equazioni algebriche*. Milano: Mursia.

Gatto, R. 1995. Lettere di Giuseppe Battaglini ad Enrico Betti. Documenti per una storia della facoltà di scienze dell'Università di Napoli. *Nuncius* 10: 196–222.

———. 1996. Lettere di Luigi Cremona a Enrico Betti. La corrispondenza di Luigi Cremona (1830–1903). *Quaderni Pristem per l'archivio della corrispondenza dei matematici italiani* 3: 7–90.

———. 2004. Il carteggio Betti – Cremona presso la Scuola Normale di Pisa. In *La corrispondenza epistolare tra matematici italiani dall'Unità d'Italia al Novecento e la figura e la corrispondenza epistolare di Federico Amodeo*, 61–80. Napoli: Vivarium.

Giusti, E., and L. Pepe. 2001. *La matematica in Italia (1800–1950)*. Firenze: Il Giardino di Archimede.

Goldstein, C. 2011. Charles Hermite's Stroll through the Galois Field. *Revue d'histoire des mathématiques* 17: 211–270.

Gray, J.J. 2000. *Linear Differential Equations and Group Theory from Riemann to Poincaré*. 2nd ed. Boston: Birkäuser.

Hermite, C. 1858a. Sur la résolution de l'équation du cinquième degré. *Comptes Rendus de l'Académie des Sciences de Paris* 46: 508–515.

———. 1858b. Sur la résolution de l'équation de quatrième degré. *Comptes Rendus de l'Académie des Sciences de Paris* 46: 715–722.

———. 1858c. Sur quelques théorèmes d'algèbre et la résolution de l'équation de quatrième degré. *Comptes Rendus de l'Académie des Sciences de Paris* 46: 961–967.

———. 1859. Sur l'abaissement de l'équation modulaire du huitième degré. *Annali di Matematica Pura ed Applicata* 2 (1): 59–61.

Hunger Parshall, K. 1989. Toward a history of nineteenth-century invariant theory. In *The History of Modern Mathematics*, ed. D.E. Rowe and J. McCleary, 1, 157–206. Boston: Academic Press.

———. 1998. *James Joseph Sylvester: Life and Work in Letters*. Oxford: Oxford University Press.

———. 2006. *James Joseph Sylvester: Jewish Mathematician in a Victorian World*. Baltimore: The Johns Hopkins University Press.

King, R.B. 2009. *Beyond the quartic equation*. Boston: Birkäuser (Reprint of the 1996 edition).

Klein, F. 1877. Weitere Untersuchungen über das Ikosaëder. *Mathematische Annalen* 12: 503–560.

———. 1884. *Vorlesungen über das Ikosaeder und die Auflösung der Gleichungen vom fünften Grade*. Leipzig: Teubner.

———. 1888. *Lectures on the Ikosahedron and the Solution of Equations of the Fifth Degree* (trans: Morrice, G.G.). London: Trübner (2nd ed. 1913, reprint 2007 New York: Cosimo).

Kronecker, L. 1858. Sur la résolution de l'équation du cinquième degré. Extrait d'une lettre adressée à M. Hermite. *Comptes Rendus de l'Académie des Sciences de Paris* 46: 1150–1152.

———. 1859. Extrait d'une lettre de M. Kronecker à M. Brioschi. *Annali di Matematica Pura ed Applicata* 2 (1): 131.

Lacaita, C.G., and A. Silvestri, eds. 2000–2003. *Francesco Brioschi e il suo tempo (1824-1897)*, I. *Saggi*; II. *Inventari*; III. *Scritti e discorsi*. Milano: FrancoAngeli.

Liouville, J. 1832a. Mémoire sur quelques questions de géométrie et de mécanique, et sur un nouveau genre de calcul pour résoudre ces questions. *J. Ecole Polytechnique* 13: 1–69.

———. 1832b. Mémoire sur le calcul des différentielles à indices quelconques. *J. Ecole Polytechnique* 13: 71–162.

———. 1832c. Mémoire sur l'intégration de l'équation $(mx^2 + nx + p)\ d^2x/dx^2 + (qx + pr)\ dy/dx + sy = 0$ à l'aide des différentielles d'indices quelconques. *J. Ecole Polytechnique* 13: 163–186.

———. 1855. Note sur une formule pour les différentielles à indices quelconques, à l'occasion d'un Mémoire de M. Tortolini. *Journal de mathématiques pures et appliquées* 20 (1): 115–120.

Nagliati, I. 2000. Le prime ricerche di Enrico Betti nel carteggio con Mossotti. *Bollettino di Storia delle Scienze Matematiche* 20: 3–85.

———. 2014. *L'internazionalizzazione degli studi matematici in Italia a metà dell'Ottocento*, PhD Thesis, Supervisor Prof. M. T. Borgato, University of Ferrara.

Palladino, N., A.M. Mercurio, and F. Palladino, eds. 2009. *Per la costruzione dell'unità d'Italia: le corrispondenze epistolari Brioschi-Cremona e Betti-Genocchi*. Firenze: Olschki.

Pepe, L. 1998. Matematici italiani rifugiati politici nel Risorgimento. *Bollettino dell'Unione Matematica Italiana Sez. A, La Matematica nella Società e nella Cultura* 1 (8): 289–305.

———., ed. 2012. *Europa matematica e Risorgimento italiano*. Bologna: Clueb.

Riemann, B. 1876. *Versuch einer Auffassung der Integration und Differentiation, in Gesammelte Werke*, 331–344. Leipzig: Teubner.

Roero, C.S. 2012. Politica e istruzione scientifica a Torino nell'età del Risorgimento. In *Europa matematica e Risorgimento italiano*, ed. L. Pepe, 219–242. Bologna: Clueb.

Ross, B. 1977. The development of fractional calculus: 1695–1900. *Historia mathematica* 4: 75–89.

Shurman, J. 1997. *Geometry of the quintic*. New York: Wiley.

Sylvester, J.J. 1857a. On a discovery in the partition of numbers. *Quarterly Journal of Mathematics* 1: 81–84.

———. 1857b. On the partition of numbers. *Quarterly Journal of Mathematics* 1: 141–152.

———. 1857c. Sulla partizione dei numeri. *Annali di Scienze Matematiche e Fisiche* 8: 12–21.

———. 1897. Outlines of seven lectures on the partitions of numbers. *Proceedings of the London Mathematical Society* 28: 33–96.

Tardy, P. 1847. *Sopra alcuni punti della teoria del moto dei liquidi*. Firenze: Marzoni.

———. 1851a. Solution générale de la question 78. *Nouvelles annales de mathématiques* 10: 319–324.

———. 1851b. Sulla risoluzione algebrica di alcune equazioni. *Annali di Scienze Matematiche e Fisiche* 2: 197–205.

———. 1858. Sui differenziali a indice qualunque. *Annali di Matematica Pura ed Applicata* 1: 135–148.

———. 1868. Intorno ad una formula del Leibniz. *Bullettino di bibliografia e di storia delle scienze Matematiche e Fisiche* 1: 177–186.

———. 1869. Note sur une formule de Leibnitz. *Nouvelles annales de mathématiques* 8 (2): 69–77.

Tazzioli, R, 2000. Per una storia della fisica matematica italiana (1860–1880). In *Atti del XIX Congresso nazionale di storia della fisica e dell'astronomia (Como 1999)*: 97–108. Milano: Università degli Studi.

Zappa, G. 1997. Storia della risoluzione delle equazioni di quinto e sesto grado, con particolare rilievo sui contributi di Francesco Brioschi. *Rendiconti del seminario matematico e fisico di Milano* 65: 89–107.

———. 1999. *Francesco Brioschi e la risoluzione delle equazioni di quinto grado*. In *Francesco Brioschi (1824–1897). Convegno di Studi matematici*, 95–108. Milano: Istituto Lombardo di Scienze e Lettere.

Democratization of Mathematics Through Cremona's Correspondence with Foreign Colleagues (1860–1901)

Giorgio Israel, Ana Millán Gasca, and Luigi Regoliosi

Abstract

From the mid-nineteenth to the early twentieth century, many common traits were shared by national mathematical communities, which were not only separated geographically (from the Czech lands to Japan), and culturally (from northern to southern Europe), but which also varied from the point of view of the dynamism of original research (from Germany to the United States). Societies and journals were launched in the national languages, thanks to the widening of the social platform of mathematics and the emergence of a national leadership; the development of state school systems increased mathematical knowledge; and furthermore, mathematics played a role in and received encouragement from the processes of social and economic modernization and the evolution of state institutions. Intellectual competition among nations, very much a part of the spirit of the nineteenth century, seems to prevail over early Modern European universalism. A panorama of almost planetary dissemination of Western mathematics resulted from this evolution, leading eventually to a reinforcement of the international circulation of knowledge, which survived two world wars.

The collection of letters written to Luigi Cremona conserved at the Sapienza University of Rome casts light on several aspects of this evolution. The letters offer a "backstage" point of view, in contrast with official proclamations; they show the interplay between national leaders and the mathematical circles in the capitals as well as mathematicians working in isolation; moreover, they show a variety of connected activities—research, institutional commitments, and the fostering of culture, including translations and textbooks. International dialogue grew out of this hive of initiatives, driven by both national passion and philosophical and political convictions, in contrast with the present European trend of entrusting the

G. Israel · A. Millán Gasca (✉) · L. Regoliosi
Department of Education, Università Roma Tre, Roma, Italy
e-mail: anamaria.millangasca@uniroma3.it

M. T. Borgato et al. (eds.), *Mathematical Correspondences and Critical Editions*, Trends in the History of Science, https://doi.org/10.1007/978-3-319-73577-1_13

circulation of ideas—and the production of knowledge—to initiatives governed from the top, standardized (design, funding and assessment) far beyond what is needed. The edition (in the Académie Internationale d'Histoire des Sciences series "De diversis artibus") has been carried out by a European team directed by Giorgio Israel.

A collection of 1122 letters written to Luigi Cremona (1830–1903) from foreign colleagues is conserved at the Sapienza University of Rome Mathematics Department Library, in Italy. 1860–1901 represents 40 years of European scientific exchanges among mathematicians, a period in which a mathematical profession emerged and in which science, and technology, showed their potential in the creation of a modern society.

The collection offers a nice sample of the mathematicians in the second half of the nineteenth century, including their political feelings, mathematical interests (especially in the area of geometry) and cultural aims.

Several points deserve attention: who was Cremona? Who were his correspondents? What are the contents of the letters and the overall meaning of this archival source in regard to nineteenth century mathematics and science? An edition of this collection is forthcoming in the series of studies of the International Academy of History of Science "De diversis artibus," so the contents will be available for further study. In this paper, we will discuss some aspects of the edition (two volumes, almost 2000 pp., see Israel 2017, CLC from hereon). Moreover, we would like to emphasize how these letters, aside from their mathematical contents, offer a point of view on the "backstage" of an evolution that can be considered to be a process of democratization of mathematics as a human enterprise, as a part and in the context of the evolution of nineteenth century liberal democracy.

An Italian PRIN national project in 2011 provided the financial support for the edition; in fact, in 2011, Italy commemorated the 150th anniversary of the unity of the country, and Cremona played an important role in the institutional building of the modern Italian state.[1]

1 Luigi Cremona, the Geometer, the Senator

The first two letters of the collection, in chronological order, were written by Carl Wilhelm Borchardt (1817–1880), editor of Crelle's Journal, and by Cremona's peer, Alfred Clebsh (1833–1871); both date back to 1860, an important year in the life of Cremona and in the history of Italy. Cremona, who had graduated in 1853 from the university of his native town Pavia (in northern Italy, close to Milan) as a civil engineer and architect, after several years giving private lessons and working as

[1]A first overall presentation of this edition was discussed by the first two authors with the title "Luigi Cremona's network of foreign correspondents (1860–1901): a testimony to the evolution of the "Europe of science" in the late nineteenth century" at the *International Conference Mathematical Schools and National Identity (sixteenth to twentieth cent.)*, Turin, October 10–12, 2013.

a high school mathematics teacher, was called, in 1860, to a new chair of higher geometry at the University of Bologna (Brigaglia and Di Sieno 2011b). Bologna had just been liberated from the power of the Pope, a new step in the evolution that led to the unity of Italy: in fact, after a few months, in March 1861, the Parliament proclaimed Vittorio Emanuele II King of Italy. A new nation had achieved political unity and had impetuously entered nineteenth century Europe and the world stage in order to carve out a top-ranking position in the fields of industry and culture for itself.

In their letters, the two German colleagues show their strong appreciation for Cremona's early mathematical contributions to the field of geometry. Borchardt's letter, written in April, was still addressed to Cremona in Milan, as a teacher at the city's high school, Liceo Sant'Alessandro (from 1865 until now, Liceo Cesare Beccaria); he emphasized that Cremona was turning to pure (synthetic) geometry:

> comme votre nom ne m'est point inconnu après les mémoires que vous avez fait insérer dans le recueil italien et dans lesquels vous avez montré tant d'habileté dans les questions de géométrie analytique je ne doute pas que ce nouveau mémoire qui semble prendre plutôt la voie de la géométrie pure ne formera un heureux enrichissement du Journal allemand [Borchardt to Cremona, Berlin, April 2, 1860].

As for Clebsch, his first letter (Fig. 1) started an intense exchange that lasted twelve years, until Clebsch's death in 1872:

> Erlauben Sie Ihnen zunächst meinen herzlichsten Dank für die Uebersendung Ihrer ausgezeichneten Abhandlungen auszusprechen. Zugleich bin ich so frei, die beiden Abhandlungen beizulegen welche Sie die Güte hatten zu wünschen. Ich hoffe Ihnen bald Anderes über diese algebraischen Probleme zusenden zu können, welche nicht mit Unrecht die Mathematiker unserer Zeit so vielfach beschäftigen, und erlaube mir, Ihnen auch für die Zukunft einen Austausch unserer Arbeiten vorzuschlagen [Clebsch to Cremona, Carlsruhe, August 27, 1860].

Moreover, Cremona's intellectual and political-cultural figure as a whole marks the contents and meaning of the letters in the collection.[2] Cremona was a member of a generation of Italian scholars who shared a vision in which mathematics played a principal role in secular culture, because culture and education were intertwined with freedom and nationality; a generation of mathematicians with a radical attachment to national culture and progress. He had been a volunteer in the "Free Italy" battalion during the First Italian War of Independence (1848–1849), when he was only 18 years old. His introductory lecture to the higher geometry course in Bologna, held in November 1860, was an impassioned speech that reflects the strong feelings of a country, a large part of which had just been freed from the yoke of the foreigner (the "Austrian jailer," he said) and from the temporal power of the Church (the "livid Jesuit," in Cremona's words). In the face of oppression and obscurantism—this was Cremona's point of view—the new Italy offered reason and free thought, of which science was the model.

[2]For a biographical profile of Luigi Cremona, see Israel (2016) and the bibliography therein.

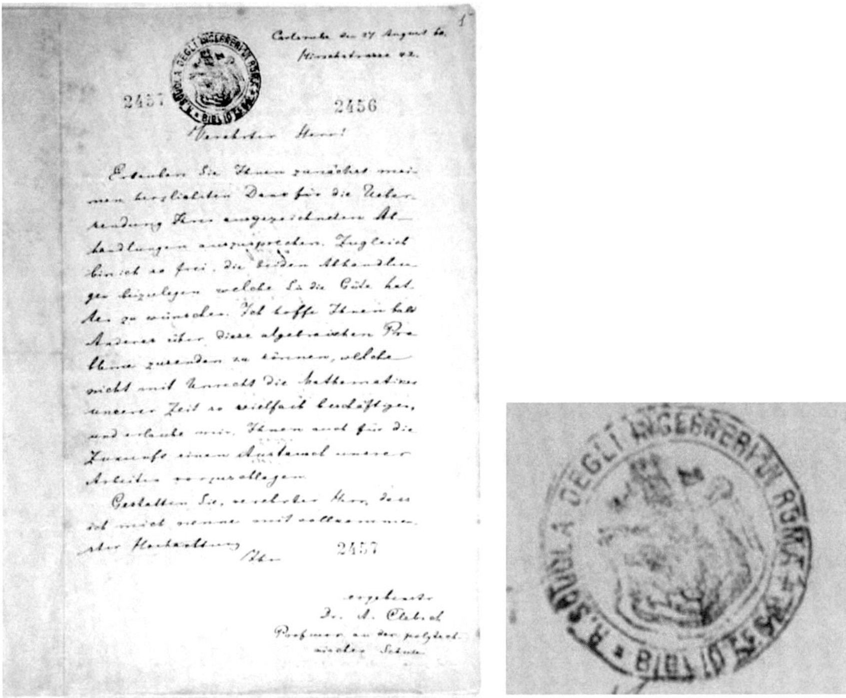

Fig. 1 Letter from Alfred Clebsch to Cremona, Carlsruhe, August 27, 1860. At the top, the stamp of the Library of the Royal School of Engineers of Rome and a catalogue number can be seen; the letters in the collection were stamped and numbered after Cremona's death, when his papers were acquired by the school library. Twenty-eight letters from Clebsch and two letters from his wife Minna Clebsch were found. The exchange and discussion of Clebsch's analytical methods and Cremona's synthetic methods is a *fil rouge* in the correspondence

The Bologna speech was very well received by his French colleagues. Eugene Prouhet (1817–1867), wrote, in May 1861, that Cremona's appeal to the patriotism of young people had moved him, as well as Terquem, Bonnet, Serret and Mannheim, and remembered at the same time the "common work shared by all the civilized nations" regarding modern sciences.[3] Olry Terquem appears to have been well aware of Cremona's involvement in the tension in Italian politics between moderate, monarchist positions and radical republican ideas: Terquem wrote to Cremona three months before his death in 1862, when he was eighty years old.[4] Political issues emerge in many letters, often entangled with tricky rivalries between nations and,

[3]Prouhet to Cremona, Paris, May 29th 1861; for further details, see Millán Gasca (2011, pp. 52 ff). Letters from Amédée Mannheim (1831–1906) can be found in the Genova Cremona Archive (see note 8).

[4]Terquem to Cremona, Paris, March 11, 1862. On Cremona's political evolution, see Brigaglia and Di Sieno (2009, 2010, 2017); see also Rossi 1984.

of course, even wars. Maximilian Curtze (1837–1903), a Gymnasium teacher in Thorn (now Torun, in Poland) and his main German translator, wrote, in February 1878, about the new German edition of Cremona's Introduction to a geometrical theory of plane curves (*Introduzione ad una teoria geometrica delle curve piane*, 1861), expressing his solidarity regarding King Vittorio Emanuele's death, and this is but one example of this kind of comment in letters from German correspondents (Knobloch 2013). As to the penetration of the pride of national identity into European intellectual and social life, Françosi Furet wrote:

> ... none of the 19th century wars—in any case few in number—presented the fearful nature of those of the 20th century. Even in Germany, where it displayed most intensely the blindness and the perils involved in it, the national idea remains incorporated into that of culture. It does not propose as sufficient per se its pure substance, the particular election of the Germans, their superiority as human beings. It strongly enhances Germany's contribution to ethics, the arts, philosophy, culture (Furet 1995, p. 45).

Several letters come from places in Europe that have since changed their position relative to national borders, such as Torun, or Bromberg (now Bydgoszcz), where Rudolf Sturm (1841–1919) lived until 1872, or Breslau, now Wroclaw, where Heinrich Schröter lived (1829–1892, born in Königsberg); many letters arrived from towns in the Austro-Hungarian or the Russian Empires, in areas such as Poland, Hungary and Bohemia, that were to become independent European countries—for example, Prague, where Cremona's research was translated into Czech.

Cremona's political and cultural evolution from his ardent mathematical and political youth to his moderate maturity was marked by the effort to develop mathematical and scientific institutions and culture in Italy. After 6 years in Bologna, starting in October 1867, he was professor of higher geometry at the Milan Royal Higher Technical Institute, directed by his mentor Francesco Brioschi (1824–1897), where he also taught graphical calculus and graphical statics, following the example of the Zurich ETH. In late 1873, he accepted an assignment to reorganize the old Pontifical School for Engineers so as to set up the third Italian polytechnic school in Rome (Fig. 2). Thus, he was appointed professor of graphical statics and Director of the Royal School of Engineers of Rome, and from that day—even though he received the Steiner prize for the second time in 1874—his institutional commitments made it difficult to carry through his scientific projects. Thus, his correspondence with Arthur Cayley (1821–1895), strictly regarding geometrical issues and characterized by a tone that is quite formal, stopped in 1872.

The new School of Engineers, where the letters comprising this collection were deposited and whose stamp they bear (see Fig. 1), was set up in the Roman convent of San Pietro in Vincoli, where the Faculty of Engineering of the "La Sapienza" University of Rome still has its premises.

In 1879, Cremona was nominated senator. His institutional work was intense in the field of educational reform, both in the schools and the university[5]; and

[5]He was offered the ministry of education twice: in 1881, by Quintino Sella, a request he turned down, and in 1898, this time accepting, although he only remained in office for the month of

Fig. 2 The venue of the Royal School for Engineers of Rome: the convent close to the church of San Pietro in Vincoli (with a Chiostro dating back to the sixteenth century); the library was set up in 1876 in the area of the former monks' refectory (Ippoliti 2012)

thus many of the letters include questions addressed to Cremona on educational organization, even on technical aspects of buildings.[6]

The ethos of research and national institutional commitment are both characteristics of Cremona's life and work; but to these should also be added his great attention to the European (mathematical) scene, as demonstrated by the numerous academies and foreign societies that selected him as one of their members, starting in 1867 with the Academy of Sciences of Lisbon[7]; and, of course, his correspondence with foreign colleagues is also noteworthy.

2 Cremona's Papers in Genoa and Rome

There are two main Italian archives containing Cremona's papers: the Rome Cremona archive at the Rome Sapienza University Department of Mathematics Library (Israel and Nurzia 1983) and the Genoa Cremona archive at the Istituto Mazziniano (Brigaglia and Di Sieno 2011a). Both of these are outstanding sources

June, owing to the political crisis within the government. In 1880, the minister Francesco De Sanctis appointed him government commissioner for the reorganization of the "Vittorio Emanuele" National Library in Rome housed in the Collegio Romano.

[6]For example, letters from Sturm initially addressed geometrical issues, but then turned towards institutional aspects; letters from German correspondents were edited in the CLC by Eberhard Knobloch and Karin Reich.

[7]Academy of Sciences of Lisbon (1867), Mathematical Society of London (1871), Society of Sciences of Bohemia (1872), Danish Academy of Sciences (1876), Cambridge Philosophical Society (1877), the Academy of Science of Munich (1878), the Royal Society of London (1879), the Society of Sciences of Göttingen (1880), the Dutch Academy of Sciences (1881), the Mathematical Society of Prague (1881), the Royal Society of Edinburgh (1883), the Prussian Academy of Berlin (1886), the Physico-Medical Society of Erlangen (1896), the Irish Academy of Dublin (1898), the Academy of Belgium (1899), the Institut de France (1899), the Swedish Academy (1901), and finally, the American Academy of Washington (1902).

on Cremona's scientific biography in the context of the evolution of science in Italy after the Risorgimento. The Rome collection came to its present location from the Library of the Rome School of Engineers, from which Cremona's library was acquired in 1909; the letters were transferred to the Sapienza University Mathematical Institute after it moved to its new premises (now Piazzale Aldo Moro) in 1935. Cremona's papers conserved in the Genoa Istituto Mazziniano—more than 6000 documents donated by Cremona's daughter Itala, probably in 1939—consist mainly of correspondences with Italian scientists and politicians or state officials, as well as correspondences with 34 foreign mathematicians.[8]

As Aldo Brigaglia and Simonetta Di Sieno have written, the Genoa archive is especially useful as a source regarding the history of science in Italy:

> It was also during his time in Bologna that Cremona became acquainted with a large number of Italian and European mathematicians. The Archive of the IMG [Mazzini Institute of Genoa] contains many new documents about the early stages of these contacts and their subsequent development. Cremona's correspondence with the Italian mathematicians (e.g., Eugenio Beltrami, Enrico Betti, Francesco Brioschi, Felice Casorati, Placido Tardy ...) is of particular importance, not only because of the clear description of Italian academic life and its problems, but also because of the discussions concerning two of the main organizational problems of the Italian scientific world: the problem of the development of the main Italian mathematical journal (the *Annali*) and the didactical problems relating to the programs and content of mathematical learning and teaching. The Archive of the IMG contains a large number of letters that shed light on these historical questions. Another important issue, strictly linked to the didactical problems (mainly at the university level), is the training of a new ruling group in Italy, a group no longer composed primarily of lawyers with a humanistic education, but one composed of engineers and technicians with sound scientific knowledge. The role played by Brioschi, Cremona, and many of the Italian mathematicians in this respect cannot be overstated (Brigaglia and Di Sieno 2011a, p. 101).

Moreover, letters in the Genoa archive from the period he spent in Milano, above all, the letters exchanged with Eugenio Beltrami (1835–1900), show "Cremona's efforts to keep pace with the rapidly changing face of modern mathematics. In

[8]In Genoa there are letters from 22 non-Italian correspondents included in the Sapienza Cremona Archive: Arthur Cayley (4 letters); Eugène Dewulf (2 letters); Lewis Carroll–Charles Dodgson (1 letter); James Glaisher (4 letters); Charles Hermite (1 letter); Thomas Archer Hirst (86 letters); Felix Klein (4 letters); Leopold Kronecker (1 letter); Ernst Eduard Kummer (1 letter); Sophus Lie (1 letter); Max Noether (1 letter); Emile Picard (1 letter); Eugène Prouhet (1 letter); Theodor Reye (1 letter); George Salmon (3 letters); Ludwig Schläfli (1 letter); Kyparissos Stephanos (1 letter); Rudolf Sturm (4 letters); James Sylvester (1 letter); Peter Tait (5 letters); Emil Weyr (1 letter); and Hieronymus Georg Zeuthen (1 letter). There are also letters from the following additional 12 foreign correspondents: James Booth (1806–1878) (4 letters); Maurice D'Ocagne (1862–1938) (1 letter); Morgan Jenkins (1 letter); Seligmann Kantor (1857–1902) (8 letters); Jacob Lüroth (1844–1910) (7 letters); Gösta Mittag Leffler (1846–1927) (2 letters); Amédée Mannheim (1831–1906) (55 letters); Henri Poincaré (1854–1912) (3 letters); Henry Smith (1826–1883) (9 letters); William Spottiswoode (1825–1883) (16 letters); J. Vanecek (7 letters); and Gustav Wolff (1834–1913) (20 letters, 1883). Data from Brigaglia and Di Sieno (2011a); the study of this collection is ongoing (see website www.luigi-cremona.it).

particular, he tried hard to fully understand Riemann's theory and to translate it into a more geometric language" (Ibid).

Other papers are conserved in Italy, for example, the correspondence with Domenico Chelini in the Rome Archive of the Piarist Order (*Archivio Generale delle Scuole Pie*). Some of this material has already been published.[9] Further research will lead to the discovery of letters written by Cremona to his correspondents in various archives, mainly in Europe. Thus, the complete publication of Cremona's correspondence was not and is not a pursuable aim. Letters written to Hirst were published in Nurzia 1999; letters written to Darboux and Klein have been included in the CLC edition discussed in our paper. But apart from the few exceptions mentioned, the CLC contains only the letters sent to Cremona.[10]

3 The Letters in the Cremona Rome Archive and Their Authors

The bulk of Cremona's international correspondence was found in 1982 at the Mathematical Institute of Sapienza, the University of Rome, inside 28 envelopes, during a search organized by one of the authors of this paper, Giorgio Israel. Let us recall how he narrated this finding:

> In November 1982, I was associate professor at the Istituto Matematico "Guido Castelnuovo" of Rome University. With the conviction that the Institute, which already had a rich library, probably contained other documents of historical interest, I planned a search of the building. The places to be explored consisted of the library and a local storeroom containing the duplicates of books, papers to be disposed of and many other kinds of objects, including broken furniture and an old bicycle used by the janitor in the 1950s. The search was carried out with the help of Laura Nurzia, then researcher at the same Institute. After a few days exploring this room, among the jumble of material, on the floor in one corner, under a pile of documents, I found twenty-eight envelopes containing letters that, at first view, had obviously been sent to the Italian mathematician Luigi Cremona. The many correspondents, more than 170, included the names of the most eminent 19th century European mathematicians (Israel 2016, p. 17).

The envelopes also contained two important archival documents. The first one was an autograph by Gauss, a small but accurate sheet of paper donated to Cremona by Alfred Enneper (1830–1885) in 1881, which Enneper had received in 1852 while attending a lesson by Gauss on the method of least squares. The second one was a group of four sheets of paper handwritten by Jean Victor Poncelet; on the envelope,

[9]See Carbone et al. (2001, 2002), Cerroni (2014), Cerroni and Fenaroli (2007), Enea and Gatto (2009) and Palladino et al. (2009). All of the publications are included on the web site www.luigi-cremona.it

[10]In fact, the overall research into Cremona's letters to every correspondent was so immense that it soon became obvious that the project would inevitably remain incomplete and that failure to acknowledge this fact would postpone the conclusion of the book indefinitely.

Cremona wrote that it was donated to him by Poncelet's widow[11] when he visited her on May 4th, 1884. Three drafts of mathematical notes by Cremona were also found, attached to the letters sent by Max Noether.

The envelopes contained letters from 176 mathematicians addressed to Cremona (among them, only three Italian colleagues),[12] and from representatives of three scientific societies (the British Association for the Advancement of Science, the Göttingen Königliche Gesellschaft der Wissenschaften, and the Société Mathématique de France).[13] The *size* of the single correspondences varies. Fifty colleagues sent only one letter to Cremona (one of them a postcard). The longest correspondence was with his French translator Eugène Dewulf (1831–1896); the heftiest correspondences were those with other translators, with Thomas Hirst, Cayley, and George Salmon, and with many German-speaking correspondents: Clebsch, Sturm, Elwin Bruno Christoffel, Wilhelm Fiedler, Johann Nicolaus Bischoff, and Theodor Reye, as well as Carl Friedrich Geiser, Ludwig Schläfli, and Emil Weyr. In addition, a few letters from four correspondents were found that were not addressed to Cremona.[14]

The letters were sent mainly from European *places*, in Germany, France, and Great Britain. There are also letters from towns in the United States and Canada, and from British scholars in Calcutta (James B. Chalmers) and Adelaide (Horace Lamb (1849–1934), who, in 1885, returned to Manchester). The letters are written in seven different *languages*: German, French, English, Italian, Spanish, Portuguese, and Latin. In several cases, letters from a single correspondent are written in more than one language: for example, the first letter from Emil Weyr is written in French, followed by three letters in Italian, and the rest of the correspondence is written in German (Bečvář and Bečvářová 2006). A surprising inter-European linguistic facility emerges from the correspondences, if compared with current scientific exchanges among university scholars who use a single language—English—while only a few of them are able to read languages other than English and their own. Seven correspondences in the collections mainly regard the translation of mathematical works. First, the correspondence with Richard Baltzer (1818–1898) regards Cremona's Italian translation of Baltzer's second edition of *Elemente der Mathematik* (first published in Leipzig in two volumes in 1860 and 1862). Secondly, several research essays and three textbooks written by Cremona in 1872–1874 were translated into German, French, English, and Czech (see Millán Gasca 2011) by six different translators whose letters are included in the collection. The distinct

[11]Louise Palmyre Gaudin (1813–1889).

[12]Ettore Caporali (2 letters), Valentino Cerruti (1 letter), and Carlo Saviotti (6 letters, referring to Louis Bossut's French translation of his 1872 essay *Le figure reciproche nella statica grafica*).

[13]See Table 1.

[14]Four letters from Martin Krause (1851–1920) to Eugenio Beltrami (written in 1898–1899); a letter from Rudolf Clausius (1822–1888) and 5 letters from a certain Heinrich Schramm to Francesco Brioschi dating back to the years 1867–1869; and three letters from Édouard Combescure (1824–1889) to an unidentified member of the editorial office of the journal *Annali di matematica*, written in 1871–1872.

impression that arises is one of intellectual richness: a widespread facility of exchange, together with great attention to the national language.

The scholars who wrote to Cremona belonged to different *generations*: the oldest, like the already mentioned Terquem, Borchardt, Prouhet and Baltzer, all of whom were born before 1820; his peers, born around 1830, like Clebsch, Fiedler and Dewulf; and younger scholars like Emil Weyr from Prague (born in 1848, a graduate student in Rome in 1870–1871), the Greek Kyparissos Stephanos (born in 1857, a graduate student writing from Paris in 1881), the English Carslaw (born in 1870, a graduate student in Rome in the late 1890s) and the American Julian Coolidge (born in 1873, a student in Hessen in the same period). Some of these mathematicians were at the forefront of research, some have not made many original contributions to mathematics, and among the latter were the scholars who took charge of translating Cremona's books, as well as scholars who were committed to developing mathematical culture in their own country (like Zoel García de Galdeano (1846–1924) from Zaragoza, in Spain).

Thus, the style of the letters (formal, friendly, deferential, and so on . . .) depends on the kind of correspondent; in each case, they wrote to Cremona with the confidence that they would be shown consideration, interest and cordiality. In correspondences spanning longer periods of time, a respectful, polite tone often evolves into a more informal or familiar one, as Eberhard Knobloch (2013) has underlined. There are letters dealing with problems regarding academia, especially problems of priority or lack of acknowledgement of results published or privately communicated, such as Dewulf versus Mannheim in a letter dated March 18, 1886. Others are light-hearted and even jocular. For example, the 22-year-old Coolidge wrote to Cremona about his difficulties in finding and buying his *Introduzione a una teoria matematica delle curve piane*:

> Unfortunately, my knowledge of Italian is to be reckoned among the imaginary quantities, so I must have the work in an English, French or German translation [Coolidge to Cremona, Hessen, August 28, 1896].

But perhaps Cremona did not appreciate this very informal tone, if one were to judge from the second, more humble letter (in French):

> Je vous prie de m'excuser si je n'ai pas très bien exprimé ce que j'ai voulu dire. Je suis depuis trois mois en Allemagne, et ainsi je le trouve plus difficile qu'ordinairement d'ecrire bien le Français [Coolidge to Cremona, Hessen, September 27, 1896].

The few correspondents who were not mathematicians deserve some brief comment. Special mention should be made of Jean-Albert Gauthier Villars (1828–1898), the well-known French mathematics and science publisher, who was educated at the École Polytechnique and graduated as a telegraph engineer (Paul 1985). His letters were found together with the 85 letters from Dewulf, who translated Cremona's textbook on projective geometry; many comments on his industrial venture, both from the economic point of view—including competition with other European editors—and as a cultural project are included in his letters to both Cremona and

Dewulf. For example, in exchanges regarding the second edition of the above-mentioned textbook, he wrote:

> La Géométrie est absolument délaissée en France ; elle n'est plus représentée à l'Académie et n'a pas une seule chaire où on la professe. J'ai pensé qu'un des meilleurs moyens de raviver, dans la limite du possible, le goût de cette Science, était de réimprimer l'ouvrage d'un maître, comme celui de M. Cremona.
>
> Si j'agissais come un Editeur, uniquement préoccupé du côte industriel, je réimprimerais d'abord des ouvrages épuisés, s'adressant à un nombreux public, comme mes traductions de Tyndall. Mais, forcé de faire un choix, à cause de l'encombrement de mon Imprimerie, j'ai préféré la Géométrie de M. Cremona, qu'est appelée de rendre plus des services.
>
> Je cherche toujours, dans la limite de mes moyens, publier des traductions pouvant développer certains courants d'études dans notre pauvre pays, qui ne lit rien de ce qui se fait à l'Etranger et qui a si grand besoin d'être tenu au courant des productions nouvelles. C'est ainsi qu'au lieu de publier des ouvrages à succès, j'imprime en ce moment les Quaternions de Tait, parce qu'on ne veut pas jusqu'à ce jour introduire ces nouveaux symboles dans notre enseignement ; c'est ainsi que je prépare des traductions d'ouvrages sur l'Electricité (Maxwell, Jenkins, Kempe etc.) parce que la dernière Exposition a montré notre infériorité dans cette branche de la Physique [Gauthier-Villars to Dewulf, December 27, 1882].

The Scottish industrialist Walter Macfarlane (1817–1885), an ironwork manufacturer, was also a correspondent: Cremona met him during his first visit to the UK when he attended the 1876 Glasgow meeting of the British Association for the Advancement of Science. There are no letters from women mathematicians in the archive, but only from relatives, such as Clebsch's second wife Minna Rays Clebsch, after the early death of her husband, and Clara, Karl Weierstrass' sister, who was in contact with Cremona during a stay in Italy in 1874. Nevertheless, the letters mention Cremona's and other mathematicians' support for Sofya Kovaleskaya's interest (as Charles Hermite put it, this meant defending "the interest of Science"); and Zeuthen recommended a pupil of his to Cremona, Miss Ellie (see Millán Gasca 2011, pp. 62–63).

We offer now a description of the distribution of the letters, which we obtained after the completion of the work. In Fig. 3, a chart of the chronological evolution over the years 1860–1903 is shown, elaborated from the chronological index included in the edition.

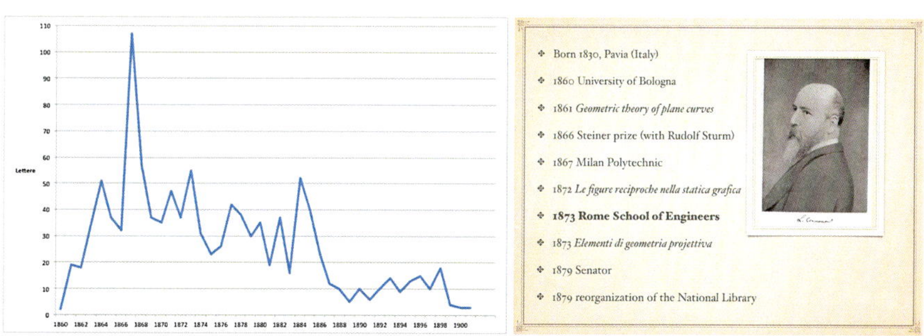

Fig. 3 The intensity of the exchange with foreign colleagues, as reflected by the number of letters in Cremona's Archive at the Dipartimento di Matematica, Sapienza University of Rome

Table 1 The geographical/national distribution of Cremona's correspondents outside Italy, as represented in Cremona's Archive at the Dipartimento di Matematica, Sapienza University of Rome

Nation	No.	Names of correspondents									
Germany	69	Baltzer	Bauernfeind	Becker	Bischoff	Boehm	Borchardt	Bremiker	Brill	Christoffel	Clausius~
		Clebsch	Curtze	du Bois-Reymond	Enneper	Fuchs	Gordan	Grosmann	Grunert	Gugler	Gundelfinger
		Günther	Hankel	Harnack	Helmholtz	Hess	Hesse	Klein	Koenigsberger	Krause~	Krimphoff
		Kronecker	Kummer	Lampe	Lipschitz	Maser	Matthiessen	Mayer	Meyer	Neumann	Noether
		Ohrtmann	Pasch	Plücker	Prÿm	Reye	Rosanes	Röthig	Schell	Schering	Schilling
		Schlegel	Schmidt A	Schmitz-Dumont	Schoenflies	Schröter	Schwarz	Siebeck	Stammer	Stern	Sturm
		Thomae	Timerding	Vahlen	von Lindemann	Weierstrass	Weissenborn	Wiedemann	Wöhler (K. Gesell.)	Zimmermann	
France	36	Abakanowicz#	Aoust	Autonne	Bossut	Boucher	Brisse	Chasles	Combescure~	Darboux	de Ficquelmont
		de Jonquieres	de la Gournerie	de Longchamps	de Marsilly	Debacq	Dewulf	Fouret	Gauthier-Villars	Gerono	Grouard
		Halphen	H. de la Goupillerie	Hermite	Hoüel	Jordan	Lemoine	Levy	Lucas	Painvin	Picard
		Picquet	Poudra	Prouhet	Saltel	Terquem	Weill (S. M. de France)				
United Kingdom	20	Beare	Cayley	Chalmers	Chrystal	Cotterill	Dogson (Carroll)		Esson	Glaisher	Henrici
		Hirst	Lamb	Leudesdorf	Macfarlane	Miller	Price	Sylvester	Tait	Thomson	Tucker
		Crompton, Carpenter and Clarke (BAAS)									
Ireland	7	Anglin	Casey	Haughton	Malet	Roberts	Salmon	Townsend			
Belgium	4	Catalan	de Cuyper	Folie	Le Paige						

Denmark	2	Petersen	Zeuthen								
Finland	1	Lindelöf									
Greece	2	Stephanos	Vitalis								
The Netherlands	2	Baehr	Schoute								
Norway	2	Bjerknes	Lie								
Portugal	2	Albuquerque	de Ponte Horta								
Romania	1	Urechia									
Russia	3	Boguslavsky	de Khanikof	Vassilief							
Spain	3	G. de Galdeano	de Moÿ	Vázquez Illá							
Switzerland	5	Affolter	Culmann#	Fiedler#	Geiser	Schläfli					
Austria*	2	Härtenberger	Schramm~								
Czech Territories*	4	Durege	Houdek	Weyr Ed	Weyr Em						
Hungary*	1	Schmidt F									
Poland*	1	Czarnowski									
Australia	1	Carslaw#									
Canada	1	Young									
United States	5	Coolidge	Dickson	Eddy	Elliott	Schwatt#					

National areas (Czech territories, Hungary, Poland) with national languages that were to become independent countries are marked with an asterisk. For mathematicians writing on behalf of a scientific society, the name of the society is included. Mathematicians who are authors of letters not addressed to Cremona are included with a ~. The symbol # indicates mathematicians working abroad.

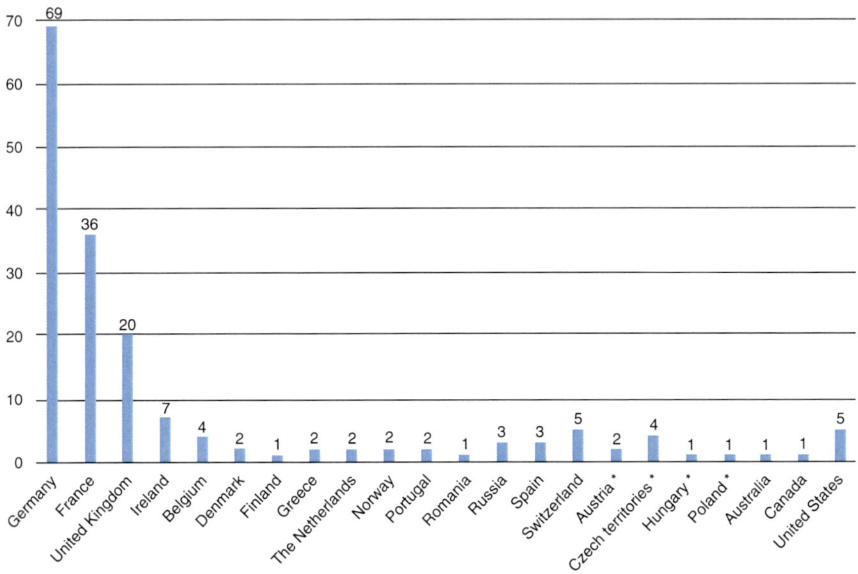

Fig. 4 Numbers of correspondents in European countries or areas, Australia, Canada, and the United States. It should be taken into account that subjects from the Austro-Hungarian and Russian Empires lived in areas such as Poland, Hungary or Bohemia, which were to become independent European countries

In Table 1, the correspondents are listed by country or European national area (Czech territories, Hungary, Poland, Romania, with national languages that would eventually become independent countries). The country where each mathematician mainly developed his professional activity has been chosen: thus, Isaac Joachim Schwatt (1863–1934) was born in Latvia, but emigrated to the United States; the Briton Horatio Scott Carslaw (1870–1954) had an important role in Australian mathematics [or in "taking mathematics to Ultima Thule," as Michael Deakin has put it (see Deakin 1997)]; the Polish-Lithuanian Bruno Abdank-Abakanowicz (1852–1900) was in exile in Paris from 1881; Heinrich Durege (1821–1893), born in Danzig (now Gdansk), worked in Prague from 1869 on; the German Karl Culmann (1821–1881) and Fiedler were important figures in Zurich.

Figure 4 offers a chart comparing the numbers of correspondents in each area, showing the prevalence of Germany, and the presence of 1–3 correspondents in many areas with a developing mathematical community.

4 The Editing

Two main aspects had to be considered when planning the editing of this material: (a) the organization of the letters and (b) the critical apparatus.

The numbering of the letters marked with the stamp of the Royal School of Engineers goes from number 1742 to 2882, with some of the letters bearing the same number and others not being numbered at all. The first 1741 letters are missing, so a large part of the collection of Cremona's letters has been lost, perhaps becoming hopelessly disintegrated[?] in the storeroom over the years. Owing to its faulty and incomplete nature, this numbering was of no use, and was thus disregarded.

As to the organization of the edition, a first possibility was to publish the letters in chronological order. This criterion would perhaps have been useful if the main focus of the archival source had been either Cremona's scientific biography or the evolution of nineteenth century geometry. In fact, the collection offers insights into the evolution of nineteenth century geometry, because it contains a dialogue among a group of distinguished geometers.[15]

The second possibility was to order the letters by correspondent. In fact, the letters had been found divided into envelopes according to correspondent, so this organization was based on archival considerations; partial publications of letters in several booklets, starting in 1992, had already followed this mode of organization.[16] Moreover, the contents of many of the single correspondences led us to consider the correspondents as individual scholars within their own national context, in parallel with the activity of Cremona himself. Besides their mathematical content, the letters also address political aspects and show that the participation of science in the process of modernisation was experienced by mathematicians all over Europe as a patriotic commitment (Millán Gasca 2011). The general editor decided—in agreement with the director of the series—to publish the letters in alphabetical order according to correspondent, and to include, at the end of the volume, a chronological index. A two-volume book has been produced. It is introduced by a foreword and an essay on Luigi Cremona by G. Israel, with a bibliography of Cremona's works edited by G. Israel and L. Regoliosi.

The critical apparatus was intended to be neutral, but capable of directing the reading; the main goal was to offer the letters for further studies regarding various historical problems. Each chapter is equipped with: (a) a short biographical note accompanied by biobibliographic references; and (b) a short introduction regarding the main topics and the meaning of the correspondence. The letters, published in the original language, are annotated, with cross-references to Cremona's works, which are listed in the bibliography, as well as full references to the books and papers mentioned in single letters and other helpful information. The critical apparatus in English has been translated or revised by Ian McGilvray.

A team of nearly 20 researchers from 6 European countries (Czech Republic, Germany, Greece, Italy, Portugal and Spain) have contributed to the editing of the collection: Martina Bečvářová, Aldo Brigaglia, Luca Dell'Aglio, Simonetta Di

[15]See Israel (2016).

[16]These booklets have been useful for obtaining a better understanding of the general historical meaning of this archival material and establishing the final criteria for the edition (Millán Gasca 1992a, b; Menghini 1994, 1996; Nurzia 1999).

Sieno, Paola Gario, Livia Giacardi, Angelo Guerraggio, Eberhard Knobloch, Marta Menghini, Ana Millán Gasca, Mara Monaldi, Pietro Nastasi, Efthymios Nicolaidis, Luigi Regoliosi, Karin Reich, Enrico Rogora, Luís Ribero Saraíva, Paola Testi Saltini, and Claudia Umani. One or two scholars in the group are the editors of each single correspondence.

The second volume includes an index of names and a chronological index.[17] The names cited number more than one thousand. It was not always possible to identify them and obtain their date of birth and death; but the information available on the Internet was very helpful,[18] if we compare it with the resources available in the 1990s. In some cases, identification was impossible, as they were minor figures of which all memory has been lost; in all cases in which only the family name was known, it was omitted from the index of names. In any case, it cannot be ruled out that more thorough research will lead to further information.

5 A European Network of Scientists

The letters encourage us to consider the set of correspondents as a *European network*, an evidence of the "European space," the Europe of sciences (Blay and Nicolaidis 2001, Goldstein et al. 1996, Pepe 2013). The patriotic commitment returns in letters from every corner of Europe, so it may then seem a paradox to speak about a European network: the fresh desire that was spreading among European mathematicians to develop autonomous mathematical research in their own national languages might well have been detrimental to the universalistic ideal of mathematics and acted as a concrete obstacle to communication. This was not the case: one meaningful example is a letter from Weierstrass to an unknown mathematician, written in 1867, in which "he underlines the willingness of the German mathematicians to continue the fruitful cooperation with their Italian colleagues, just as the political alliance between their two countries had led to good results. The German mathematical achievements, he writes, are better understood and appreciated in Italy than in France or England."[19]

In the preface to the essay on the "Europe of science" as a scientific space (*L'Europe des sciences. La constitution d'un espace scientifique*, 2001), Michel Blay and Efthymios Nicolaidis highlight the interest in approaching "the develop-

[17]Edited by G. Israel and L. Regoliosi.

[18]For example: at bbf.dipf.de (German Institute for International Educational Research (DIPF), Bibliothek für Bildungsgeschichte Forschung, Germany), the BBF/DIPF/Archiv, Gutachterstelle des BIL—Personalbögen der Lehrer höherer Schulen Preußens; at www.culture.gouv.fr/public/mistral/leonore_fr. The database Léonore (Légion d'honneur), Archives Nationales (France).

[19]Knobloch, Reich, in CLC, pp. 1651–1652. This letter, written in Italian and dated March 25, 1867, in Berlin, was found together with two letters from Weierstrass to Cremona dated 1874 (and two letters from Weierstrass's sister Clara). See Casorati's letter published in Neuenschwander (1978, p. 72 ff). We have already mentioned the letters from Prouhet, showing shared political feelings; also Neuenschwander (1986).

ment of scientific knowledge as such along with its relations with the space in which it developed, as well as with the dialogue or conflicts those relations aroused." In fact, they add that, although there is a rich bibliography on the development of European scientific knowledge, the problem of "scientific Europe as an intellectual unit throughout the centuries" is an issue that deserves attention: the interest in that approach stems from the fact that it would be able to offer "[...] a global grasp of the origin and the development of scientific knowledge in its original space, as well as of the influence this knowledge had on the *homogenization of the societies that occupy this space*."

The evolution of the modern mathematical profession came about as a result of the growth in the number and size of national communities: societies and journals in the national languages were launched, thanks to the widening of the social platform of mathematics and the emergence of a national leadership; the deployment of the state school systems increased mathematical information; and mathematics played a role and received encouragement from the processes of social and economic modernization and development of state institutions. We know that many common traits were shared by national mathematical communities that were far apart geographically (from the Czech lands to Japan), culturally (from northern to southern Europe) or in regard to the dynamism of the original research (from Germany to the United States) (Grattan Guinness 1994, pp. 1427 ff.). The letters to Cremona offer a point of view on the backstage developments of this evolution that can explain their common traits: they show the interplay between national leaders and the circles in the capitals and mathematicians working in isolation (even in Germany and France); they show a variety of connected activities—research, institutional commitments, and cultural fostering, including translations and textbooks. International dialogue grew out of this nebula of initiatives, driven by national passion and philosophical and political convictions.

A new kind of communication developed in that period, communication stimulated by competition—typical of economical liberalism—which led the single nations to observe and imitate the successes, or the best practices, as we would say today, of other countries. These contacts driven by competition, combined with the traditional universal spirit of mathematics scholarship, helped to establish a new kind of international contact that contributed to the diffusion of ideas and the homogenization of the European scientific space.

6 Democratization of Mathematics and Science as a Secular Religion in the Nineteenth Century

As we have noted, Cremona was a member of a generation of Italian mathematicians with a profound attachment to the national ideal and the national secular religion, who also shared the view of the key role of science and mathematics in a liberal democracy. Science was viewed not only as a fundamental tool for the development of technology and industry at the national level, but also as a force that could liberate thinking from all dogmatic constraints and from the chains of backwardness.

The Italian model meant that mathematics was a patriotic activity, an element of technological and industrial modernization, but also a democratic activity, a universal element of culture inherited from the Greek world: the mathematical professions (math teachers, engineers, actuaries) were potentially open to everyone and were needed for both modernization and progress in regard to political and economic liberty.[20]

In the years that have elapsed since the collapse of the Soviet political project and empire, Late Modern Age historians have investigated the creation of modern society in depth, spurred, above all, by the desire to identify the symptoms of its political malaise, the symptoms that could account for the First World War and the catastrophes and massacres caused by the ideologies of fascism and communism during the twentieth century.[21] One of the greatest experts on the French Revolution, François Furet, with reference to what he calls the "revolutionary passion" that marked nineteenth century European society, wrote that "as the century advances, the Europeans no longer conceive of the political scene but through the death of God, as a pure creation of human will, intended at last to assure the liberty of all and the equality of each of us with the other" (Furet 1995, 44). A decisive contribution to the profound confidence in human will was made by the scientific revolution, which affirmed a human omnipotence that replaced that previously reserved for God in the mediaeval history of Christian Europe (Israel 2001); this substitution actually took place through a long process that has continued down to our times, to the era of biotechnology, far outrunning the intentions and convictions of the fathers of modern science. The Enlightenment disseminated this acquisition of modern science and introduced it into the eighteenth century political philosophy debate, which challenged the religious basis of society and opened the way to the revolution (Cassirer 1931). The prodigious development of science in the nineteenth century continued to feed the slow but gradual departure of Europeans from their traditional spiritual vision of society and its substitution for a materialistic vision of interactions among individuals.

Together with this philosophic contribution, science offered itself as a concrete and increasingly effective tool for the construction of this modern society, through the boost it gave to technological innovation and also thanks to its role in the democratic transformation of education. This philosophic and concrete contribution by science to the new bourgeois society inevitably led to the democratization of science itself. Indeed, the transformation of the network of European universities following the model of Berlin University, inspired by the ideas of the reformer Wilhelm von Humboldt, actually turned an ancient mediaeval European institution into a typically modern one based on the intellectual and teaching freedom of individuals emancipated from political and religious powers, as well as from

[20]For its influence in Spain, see Millán Gasca (2012).

[21]The break-up of the Soviet Union brought to a close a cycle of development of modernity that began with the French Revolution and had as its guiding principle the development of democracy, enveloped as this was in the tension between universal aspiration and national dimension.

all other utilitarian servitudes (Turner 1971). A university professor was thus transformed into a researcher, where research was considered less as a form of study and the transmission of knowledge and more as an enterprise conducted within an intellectual environment, with the same drive towards originality and innovation as were present in other environments in European society. In this way, as the nineteenth century drew to a close, the scientific researcher became a professional figure who was no less important than the engineer in the industrial and economic development of the nation. Engineers, university professors, secondary school teachers, and later, other figures with a scientific background, such as actuaries and medical doctors: in every country, the number of "scientists" grew as the new bourgeois society advanced and, as is typical of a liberal society, scientists also organized themselves into numerous associations, publishing periodicals, organizing meetings and congresses, and interacting with the other economic and political organizations.

As András Gerö emphasized when examining the case of national sentiment in an area on the "periphery" of Europe, such as Hungary, national identity became a factor of social cohesion that replaced religion, and the traditional factors were gradually rejected to make way for the ideals of 1789 (Gerö 2006, p. 2):

> Feudal Europe thought of itself primarily as a community of estates intent on safe-guarding the general value of Christianity. Identity was provided by the divergent legal status, the presence or lack of privilege, the commonality was provided by the religious culture. Therefore, the Middle Ages were the triumphal march not of the vernaculars, but of Latin. The marriage strategies of the ruling houses gave no consideration to the "nationalist principle," and the same may be said of European aristocracy in general. This kind of universalism was seriously challenged by the Protestant Reformation, for the schism within Western Christendom created almost irreconcilable identities. Yet, the national dimension was far from dominant as yet; it remained without significance relative to the differences in religion. Nevertheless, feudal universalism had suffered its first setback, and it could not be mended or covered up by any religious peace.
>
> The agony of universalism began with a process with roots in the eighteenth century. This was the tendency to contest the priority of the estates and of religion, two processes along parallel lines, although there were differences in pace: namely, the process of secularization and the development of national consciousness.

However, the association, science and the nation could not be too exclusive. Modern science was born as a universal intellectual undertaking. Indeed, the development of modern science followed on from the erudite mediaeval debate (in its themes, in the constant comparisons with the Greek classics) that took place across the frontiers, among educated men—many of whom were members of the clergy—and in the universal language represented by Latin. If anything, in modern science, even greater emphasis was laid on the universal character, since, in addition to the *lingua franca*—still Latin, and later French—the universal language of mathematics had been added. Mathematical universalism had its roots in the assimilation of Euclid's *Elements* and of the Greek mathematical corpus among European scholars in the sixteenth to seventeenth centuries[22] and has as its exemplars the correspondence of Father Marin Mersenne (1588–1648)[23] and that of Leonhard Euler (1707–1783),[24] as well the *Acta eruditorum* founded in 1682, in which Euler published his first works addressed to the "community of men of letters" so that they could be "subjected to careful scrutiny." We have noted above the astonishing linguistic knowledge and flexibility shown by the nineteenth century Cremona and his correspondents. Two letters written in Latin by Hermann Weissenborn (1830–1896) are survivors of the classical universal tradition; and, as Zeuthen wrote, apologizing for the dispatch of a note in Danish, "[. . .] in the most essential part, the table, I use the universal language of Mathematicians" [Zeuthen to Cremona, Copenhagen, August 25, 1866].

The link-up between universalism and nationalism in the political and cultural worldview of nineteenth century mathematicians certainly represented a strong point that contributed to the development of the singular national communities. However, this interaction was only one aspect of the more general process of transformation that the world of mathematics was undergoing as a result of the rise of the modern bourgeois society. Cremona's letters give us a picture of the world of mathematics in the second half of the nineteenth century, marked as it was by a strong dynamism, which was successful in coping with a harsh political and cultural challenge. Mathematics actually succeeded in passing through the deep cleft opened by 1789—as well as that of Jacobinism—without diminishing the value assigned to it by the preceding European tradition as an essential and universal element of culture inherited from the Greek world and reserved for the European intellectual aristocracy under the rigorous control of the Church. Indeed, the "progressive" sectors, ranging from the moderate liberal positions to those of the socialists, shared the same view of the role of mathematics in a modern society, a fresh and hybrid view stemming from both traditional and modern ideas. In this view, mathematics was to remain the main focus of education, to which every citizen had a right. Moreover, mathematics was to provide the intellectual platform on which to build technological innovation, as well as the future ruling classes required for the development of industrialization and the running of the State. Lastly, mathematics—and this was perhaps the idea most strongly resented by conservative thinkers—would be able to provide useful tools for the rational management of society. This view became widespread during the nineteenth century in all countries, East and West, even those lacking any democratic institutions but engaged in a

[22]"The boundless number of editions, translations and reprints that followed each other throughout the sixteenth century bears witness to the circulation at all levels of Euclid's works, the assimilation of which was to make a substantial contribution to a unitary mathematical culture, and thus to the formation of a universal scientific community." (Giusti 1993, p. 2). Euclid's Book V theory of proportions became the universal language of natural philosophy, "almost a metageometry, or better *a mathesis universalis*" (ibid.).

[23]Fletcher (1996).

[24]Euler, L. 1975. *Opera Omnia. Series quarta A, Commerciumepistolicum*, vol. 1. Basel: Birkhäuser.

process of modernization, and also by virtue of the strong commitment made by professional mathematicians and their combined international efforts.

The role of mathematics in the construction of a modern society, the offspring of a liberal democratic view, is widely accepted and proposed in all latitudes as a pathway to development, with a much greater sense of conviction than the actual political democratization.[25] Perhaps even more noteworthy is the fact that, although mathematics was wholly an heir to the European tradition, the "importance" of mathematics was also accepted by those who continued to be inflamed by revolutionary and palingenetic passion, particularly in communist countries. Perhaps the principal explanation is linked to the scientism of Marxism and Marx's personal interest in both mathematics and its applications within economics. A more in-depth analysis of the penetration of mathematics into individual national cultures and, above all, of the political aspects that played such an important role among the men of the nineteenth century up until World War I is a task that essentially still remains to be carried out. As we have just seen, it would provide a deeper insight into issues that continue to be extremely topical today. Throughout the nineteenth century, in the case of several professional groups with a technical-scientific educational background, national passion—and democratic convictions themselves—found an outlet precisely through the establishment of national mathematical communities.

In this phase, the national spirit, according to Cremona's letters, did not run counter to the universal spirit.

References

Bečvář, Jindrich, and Martina Bečvářová. 2006. Emil Weyr e Luigi Cremona. *Bollettino di Storia delle Scienze Matematiche* 26: 245–261.

Blay, Michel, and Efthymios Nicolaidis, eds. 2001. *L'Europe des Sciences. Constitution d'un espace scientifique*. Paris: Seuil.

Brigaglia, Aldo, and Simonetta Di Sieno. 2009. L'opera politica di Luigi Cremona attraverso la sua corrispondenza. Prima parte: Gli anni dell'entusiasmo e della creatività. *Rivista dell'Unione Matematica Italiana* 2: 353–388.

———. 2010. L'opera politica di Luigi Cremona attraverso la sua corrispondenza. Prima parte: Gli anni dell'entusiasmo e della creatività. *Rivista dell'Unione Matematica Italiana* 3: 137–179.

———. 2011a. The Luigi Cremona Archive of the Mazzini Institute of Genoa. *Historia Mathematica* 38: 96–110.

———. 2011b. Luigi Cremona's years in Bologna: From research to social commitment. In *Mathematicians in Bologna 1861–1960*, ed. Salvatore Coen, 73–104. Milano: Springer.

———, eds. 2017. *La corrispondenza massonica di Luigi Cremona con Giosuè Carducci e Francesco Magni*. Milano: Mimesis.

[25]This is accurately summed up in the declaration of November 11, 1997, of UNESCO support for the IMU's decision to declare 2000 the International Year of Mathematics on the basis of the role of mathematics and its current applications in science, technology, communications and economy; of its ancient roots and universal character; and of the importance of a mathematical education for the development of rational thinking.

Carbone, L., R. Gatto, and F. Palladino. 2001. *L'epistolario Cremona-Genocchi 1860–1886: la costituzione di una nuova figura di matematico nell'Italia unificata*. Firenze: Olschki.
———. 2002. *Una comunità e un caso di frontiera: l'epistolario Cremona-Cesàro e i materiali correlati*. Napoli: Liguori.
Cassirer, Ernst. 1931. Enlightenment. In *Enciclopedia of Social Sciences*, 547–552. New York: Macmillan.
Cerroni, Cinzia, ed. 2014. *Il carteggio Cremona-Guccia (1878–1900)*. Milano: Mimesis.
Cerroni, Cinzia, and Giuseppina Fenaroli. 2007. *Il carteggio Cremon-Tardy*. Milano: Mimesis.
Deakin, Michael A.B. 1997. Taking mathematics to Ultima Thule: Horatio Scott Carslaw – his life and mathematics. *Gazette of the Australian Mathematical Society* 24: 4–16.
Enea, M. Rosaria, and Romano Gatto. 2009. *Le carte di Domenico Chelini dell'Archivio Generale delle Scuole Pie e la corrispondenza Chelini-Cremona (1863–1878)*. Milano: Mimesis.
Euler, Leonhard. 1975. *Opera Omnia. Series quarta A, Commercium epistolicum*. Vol. 1. Basel: Birkhäuser.
Fletcher, Colin. 1996. Mersenne: sa correspondance et l'academia parisiensis. In eds. Goldstein, Gray, and J. Ritter, 143–153.
Furet, François. 1995. *Le passé d'une illusion. Essai sur l'idèe comuniste au XXe siècle*. Paris: Robert Laffont/Calmann-Lévy.
Gerö, András. 2006. *Imagined History: Chapters from 19th and 20th Century Hungarian Symbolic Politics*. Wayne/Budapest: Center for Hungarian Studies and Publications/Institute of Habsburg History.
Giusti, Enrico. 1993. *Euclides reformatus. La teoria delle proporzioni nella scuola galileiana*. Torino: Bollati Boringhieri.
Goldstein, Catherine, Jeremy Gray, and J. Ritter. 1996. *L'Europe mathématique/Mathematical Europe*. Paris: Éditions de la Maison des sciences de l'homme.
Grattan-Guinness, Ivor. 1994. *Companion Encyclopedia of the History and Philosophy of the Mathematical Sciences*. London: Routledge.
Ippoliti, Alessandro. 2012. *Il progetto di restauro del palazzo di San Pietro in Vincoli a Roma*. Roma: Gangemi Editore.
Israel, Giorgio. 2001. L'idéologie de la toute puissance de la science. La constitution des champs disciplinaires, in *L'Europe des sciences. Constitution d'un espace scientifique*, ed. M. Blay, E. Nicholaidis, 135–162. Paris: Éditions du Seuil.
———. 2016. Luigi Cremona. In Israel 2017, 33–72.
———. 2017. *Correspondence of Luigi Cremona (1839–1903) conserved in the Department of Mathematics, "Sapienza" Università di Roma*. De diversis artibus, vol. 97 (N. S.: 60). Turnhout: Brepols. Abbreviation in this paper: CLC.
Israel, Giorgio, and Laura Nurzia. 1983. Correspondence and manuscripts recovered at the IstitutoMatematico "Guido Castelnuovo" of the University of Rome. *Historia Mathematica* 10: 93–97.
Knobloch, Eberhard. 2013. Luigi Cremona and his German correspondents. Unpublished lecture, International Conference Mathematical Schools and National Identity (16th–20th centuries), Turin, October 10, 2013.
Menghini, Marta, ed. 1994. *La corrispondenza di Luigi Cremona (1830–1903)*, Quaderni della Rivista di Storia della Scienza. Vol. II, 3. Roma: Università La Sapienza.
———, ed. 1996. *La corrispondenza di Luigi Cremona (1830–1903)*, Per l'Archivio della corrispondenza dei Matematici italiani. Vol. III. Palermo: P.RI.ST.EM. – Universita Bocconi.
Millán Gasca, Ana, ed. 1992a. *La corrispondenza di Luigi Cremona (1830–1903), vol. I*. Quaderni della Rivista di Storia della Scienza, 1. Roma: Università La Sapienza.
———, ed. 1992b. *I corrispondenti spagnoli di Luigi Cremona*, in *La corrispondenza di Luigi Cremona (1830–1903), vol. I*, ed. A. Millán Gasca,165–172. Roma: Università La Sapienza.
———. 2011. Mathematicians and the nation in the second half of the nineteenth century as reflected in the Luigi Cremona correspondence. *Science in Context* 24 (1): 43–72.

———. 2012. La matematica nella "sfida della modernità" della Spagna liberale e il ruolo del modello italiano. In *Europa matematica e Risorgimento Italiano*, ed. L. Pepe, 81–90. Bologna: CLUEB.

Neuenschwander, Erwin. 1978. Der Nachlaβ von Casorati (1835–1890) in Pavia. *Archive for History of Exact Sciences* 19 (1): 1–89.

———. 1986. Der Aufschwung der italienischen Mathematik zur Zeit der politischen Einigung Italiens und seine Auswirkungen auf Deutschland. *Symposia Mathematica* 27: 213–237.

Nurzia, Laura, ed. 1999. La corrispondenza tra Luigi Cremona e Thomas Archer Hirst (1864–1892), in *La corrispondenza di Luigi Cremona (1830–1903)*, vol. 4, ed. L. Nurzia, 1–224. Palermo: Università Bocconi.

Palladino, N., A.M. Mercurio, and F. Palladino. 2009. *Per la costruzione dell'unità d'Italia: la corrispondenza epistolare Brioschi-Cremona e Betti-Genocchi*. Firenze: Olschki.

Paul, Harry W. 1985. *From Knowledge to Power: The Rise of the Science Empire in France, 1860–1939*. Cambridge: Cambridge University Press.

Pepe, Luigi, ed. 2013. *Europa matematica e Risorgimento italiano*. Bologna: CLUEB.

Rossi, Arcangelo. 1984 Luigi Cremona (2nd part). *Dizionario Biografico degli Italiani* 30, ad vocem.

Turner, R. Steven. 1971. The growth of professional research in Prussia, 1818 to 1848: Causes and context. *Historical Studies in the Physical Sciences* 3: 137–182.

On Giusto Bellavitis's Correspondence

Paolo Freguglia, Giuseppina Fenaroli, and Giuseppe Canepa

Abstract

Some of the aspects of Giusto Bellavitis's life dealt with in this paper provide an opportunity to contribute to a better knowledge of both the Venetian mathematician's background and the Italian mathematical culture of that period, particularly the first part of the nineteenth century. Apart from mathematical topics (the calculus of the equipollence), we will discuss Bellavitis's political affiliation and the shared or diverse attitude of his partners, i.e., colleagues and friends, concerning the political and military events that occurred during the Third Italian War of Independence (June 19th, 1866–October 3rd, 1866).

1 Preliminaries

The study of letters exchanged by scientists and scholars is fundamental to the development of the history of mathematics (see Borgato 2012): in fact, for some years now in Italy, research works on unpublished correspondence have increased the publication of letters that Italian scientists wrote to one another and to colleagues from other countries. Giusto Bellavitis is one of these figures, a considerable mathematician of the "Studio Padovano" (Baldassarri 2008; Canepa 2012; Canepa et al. 2012a, b) and founder of the calculus of equipollence (Canepa 1994; Freguglia 1998), who maintained contact with the majority of Italian and foreign mathematicians around the middle of the nineteenth century.

Most of Bellavitis's correspondence (about 1270 letters) is housed at the Istituto Veneto in "Archivio Bellavitis" in Venice.

P. Freguglia (✉)
University of L'Aquila, L'Aquila, Italy

G. Fenaroli · G. Canepa
University of Genoa, Genova, Italy

M. T. Borgato et al. (eds.), *Mathematical Correspondences and Critical Editions*,
Trends in the History of Science, https://doi.org/10.1007/978-3-319-73577-1_14

Others letters are housed in Genoa, Rome and Piacenza, as follows:

- Letters to Placido Tardy at the *Biblioteca universitaria*, in Cassetta Loria, Genoa[1] (109 documents);
- Letters to Luigi Cremona at the *Museo Mazziniano*, in Fondo Cremona-Cozzolino, Genoa[2](61 documents);
- Letters to Domenico Chelini at the *Archivio Generale scuole Pie*, in Rome (5 documents);
- Letters to Angelo Genocchi at the *Biblioteca Passerini Landi*, in Fondo Genocchi, Piacenza (100 documents).[3]

Some of these letters have been edited in part.

According to C. Alasia (Alasia 1906), in spite of the lack of proof, many letters to European mathematicians most likely exist. In fact, Bellavitis himself wrote (Legnazzi 1881) that he had indeed maintained an epistolary intercourse with several European mathematicians.

2 The Letters

Giusto Bellavitis was born in Bassano del Grappa (Vicenza, Italy) on November 22nd, 1803, and he died in Tezze sul Brenta (Vicenza, Italy) on November 6th, 1880. He began his studies under the guidance of his father Ernesto, but he was essentially a self-taught man (Legnazzi 1881).

In 1843, Bellavitis obtained the teaching post of mathematics and mechanics at Vicenza High School. In 1845, he was awarded an honorary degree in Philosophy and Mathematics (*Laurea honoris causa*).

He held courses at the University of Padua from 1845 to 1880.

Descriptive Geometry with drawings	1845–67
Advanced Geometry	1857–59; 1871–73
Theory of Probability	1857–67
Advanced Geometry and probability calculus	1863–67
Co-ordinate Geometry	1867–80
Complementary Algebra	1867–80
Algebra with exercises	1879–80

[1]This group of letters was published in Canepa and Freguglia (2009).

[2]This group of letters was the subject of the report "La corrispondenza G. Bellavitis-L. Cremona", a summary of which is accessible on the site: http//www.dm.unito.it/sism/. The report was presented to the XI Congresso della Societa' Italiana di Storia delle Matematiche", Genova, November 17–19, 2011.

[3]This group of letters was the subject of the work: Canepa and Freguglia (1991), when the 28 letters from Genocchi to Bellavitis were still unknown.

He wrote 223 works and edited the *Rivista di Giornali* (Canepa et al. 2012a, b, 2014).

In 1840, he became a fellow of the Istituto Veneto di Scienze, Lettere e Arti, keeping the position of a retired member later on. In 1850, he was one of the members of the Italian Society of Sciences (called of the ***XL***), and in 1854, he became a correspondent of the Accademia dei Lincei. Other positions he held were Inspector of Scuola Reale Superiore di Venezia (1855–1857), Senator of the Italian Kingdom (1866) and Rector of Padua University (1866–1867).

His main scientific contributions concern analysis, the resolution of equations, the classification of curves, and the algebra of complex numbers; his studies involved physics, chemistry and linguistics, among other topics (Canepa 2010).

In 1832, he proposed "equipollence calculus" for the purpose of studying the "nature" of complex numbers, i.e., their geometrical foundation (Freguglia 1992; Caparrini 2003; Dell'Aglio 2008). It should be remembered that equipollence calculus arises from the same concepts as those in Lazare Carnot's *Géométrie de Position* (Carnot 1803) and it represents the first kind of geometrical plane calculus. Of course, we must also consider the work of A. F. Möbius (*Der barycentrische Calcül*), (Möbius 1827). Afterwards, in 1844, the grand constructions of geometrical calculus by W.R. Hamilton (*Quaternions*) (Hamilton 1853) and H.G. Grassmann (*Ausdehnungslehre*) (Grassmann 1844) were conceived.

Bellavitis's cultural path was particular, and his correspondence covered a significant cross-section of Italian scientific and political situations in the middle of the nineteenth century.

Together with Placido Tardy, Angelo Genocchi and Domenico Chelini (although the latter had a different approach to life, owing to his involvement in particular political events), Bellavitis belonged to a generation of mathematicians who had the important role of becoming the link between the "Ancien Régime" and the Risorgimento, that is, between the great masters like Bordoni, Piola, Chiò, and Mossotti, and the new group of mathematicians allured by the European research, such as Betti, Brioschi, Casorati (Volterra 1902; Guerraggio and Nastasi 2010), and Cremona. Bellavitis was witness to the passage from a split country to a unified one (Bottazzini 1989, 1994, 2003; Giacardi 2012). His several letters to Italian and foreign mathematicians testify to these important events.

The epistolary intercourse between Bellavitis and his correspondents took place from the beginning of the 1830s to the end of the 1870s, 50 years of total transformation in the political, economic and social orders of a country, Italy, that was resisting the political fragmentation set by the Congress of Vienna in order to accomplish national unification, through the expression, over the decades, of higher ideals representing an extension of Piedmontese provincialism. In this period, one of the greatest developments in transport of all time occurred: the standard use of old carriages began to give way to a new railway connecting all of the main Italian cities; among Bellavitis's friends and letter recipients were the engineers who studied the development of the railway system or supervised the works (see the studies by Carlo Conti and the activities by Gustavo Bucchia). In a different field, great hydraulic works were designed (one example is D. Turazza's projects). While the first telecommunication systems were spreading (telegraph), the question of the literacy of the rural population was gaining more and more attention, eventually leading to the transformation of the peasantry into the middle class. Step by step, this change is testified to by the epistolary correspondence, sometimes enriched with a new kind of portrait: photographs.

Since Bellavitis's interlocutors were mostly mathematicians, the focus was on mathematics: the debates on new ideas produced new mathematical entities, shown in issues and publications that rapidly spread all over Europe, thanks to the new means of communication, and often in *leaflets* or *abstracts* that travelled along with the scientists' correspondence.

Most of the letters currently belong to the Istituto Veneto di Scienze, Lettere e Arti. They were a gift from Claudio Bellavitis on July 11th, 1991.

Below is the list of Bellavitis's correspondents who participated in the above-mentioned letters taken from the article (Canepa 1994)[4] :

Correspondent	No.	Kind of document		No. (Bellavitis's draft to Correspondents)	Kind of document
Giovanni Bizio	119	Letters	Bellavitis to Bizio	7	Letters
Carlo Conti	105	Letters	Bellavitis to Conti	120	Letters
Vincenzo Gallo	72	Letters	Bellavitis to Gallo	10	Letters
Ambrogio Fusinieri	72	Letters	Bellavitis to Fusinieri	76	Letters
Raffaele Rubini	64	Letters	Bellavitis to Rubini	8	Letters
Gaetano Barbieri	59	Letters	Bellavitis to Barbieri	42	Letters
Domenico Turazza	50	Letters	Bellavitis to Turazza	29	Letters
Lorenzo Casari	34	Letters	Bellavitis to Casari	14	Letters
Gaspare Mainardi	29	Letters	Bellavitis to Mainardi	24	Letters
Angelo Genocchi	28	Letters	Bellavitis to Genocchi	*	Letters
Paolo Volpicelli	21	Letters	Bellavitis to Volpicelli	0	Letters
Domenico Chelini	16	Letters	Bellavitis to Chelini	*	Letters
Giacinto Namias	14	Letters	Bellavitis to Namias	13	Letters
Pietro Dalla Balla	13	Letters	Bellavitis to Dalla Balla	12	Letters
Ludovico Pasini	12	Letters	Bellavitis to Pasini	14	Letters
Gio Batta Berti	9	Letters	Bellavitis to Berti	7	Letters
Rafaele Minich	8	Letters	Bellavitis to Minich	13	Letters
Girolamo Stecchini	8	Letters	Bellavitis to Stecchini	4	Letters
Paolo Frisiani	7	Letters	Bellavitis to Frisiani	5	Letters
Giovanni Codazza	7	Letters	Bellavitis to Codazza	3	Letters
Angelo Balestra	7	Letters	Bellavitis to Balestra	9	Letters

(continued)

[4]http://www.istitutoveneto.it/flex/cm/pages/ServeBLOB.php/L/IT/IDPagina/114

Correspondent	No.	Kind of document		No. (Bellavitis's draft to Correspondents)	Kind of document
Girolamo Resti Ferrari	6	Letters	Bellavitis to Resti Ferrari	5	Letters
Antonietta Parolini	6	Letters	Bellavitis to Parolini	2	Letters
Salvatore Dal Negro	5	Letters	Bellavitis to Dal Negro	9	Letters
....Dal Vecchio	4	Letters	Bellavitis to Dal Vecchio	0	Letters
Ludovico Alberti	4	Letters	Bellavitis to Alberti	3	Letters
Luigi Zerlotti	8	Letters	Bellavitis to Zerlotti	3	Letters
Pietro Configliachi	3	Letters	Bellavitis to Configliachi	4	Letters
Marco Santini	3	Letters	Bellavitis to Santini	1	Letters
Bartolomeo Iattara	3	Letters	Bellavitis to Iattara	6	Letters
Bartolomeo Gamba	2	Letters	Bellavitis to Gamba	2	Letters
Baldassarre Poli	2	Letters	Bellavitis to Poli	1	Letters
Antonio de Luca	2	Letters	Bellavitis to de Luca	2	Letters
Gabrio Piola	2	Letters	Bellavitis to Piola	3	Letters
Andrea Meneghini	2	Letters	Bellavitis to Meneghini	2	Letters

The symbol "*" signifies that the letters stay in the collection of the considered correspondent

There are also correspondents for whom only one letter is known to exist, such as Barilari, Tardy, Rizzo, Gregoretti, De La Casa, Sereni, Zerbinatti, Spitzer, and Toffoli.

Besides these, there are copies or drafts from Bellavitis to Zantedeschi, Libri, Möbius, and Toblini.

The Bellavitis Archive of IVLSA includes a huge quantity of documents concerning education, the reporting of publications and his relationships with universities. It also includes a donation made by his descendant Paolo Bellavitis, composed mainly of booklets, abstracts from articles that Bellavitis received from his correspondents, and his *Repertori* (Belcastro et al. 2003; Canepa et al. 2014).

The topics in the Bellavitis correspondence mostly concern private subjects, the XL society (Penso 1978), social and political situations, Risorgimento events, opinions on scientific papers and their mutual exchange, the calculus of equipollence, various mathematical and scientific items, programmes and teaching matters, and the review *Rivista di Giornali*.

Given the difficulty of dealing with all of the epistolary correspondence that still remains unpublished, we have chosen to take into account only certain letters, partly copied down, exchanged between Bellavitis and three notable authors, that provide information on the above-mentioned topics. The recipients are Domenico Chelini, Luigi Cremona and Angelo Genocchi.

Domenico Chelini studied geometrical calculus and its applications to Mechanics (see *Saggio di geometria analitica trattata con nuovo metodo*, and *Elementi di meccanica razionale con appendice sui principi fondamentali delle matematiche*) (Chelini 1838, 1860). Besides Luigi Cremona's interest in geometrical calculus and his scientific stature (he was the father of *Algebraic Geometry*), his political commitment had always been evident, in particular regarding the events of the

Italian Risorgimento. Angelo Genocchi had great intellectual as well as cultural and political affinities with Bellavitis.

3 Some Mathematical Contents in the Letters

Firstly, we will examine the Bellavitis—Chelini correspondence. There are 16 letters sent to Bellavitis by Chelini (1802–1878), a Bellavitis-Chelini draft (Istituto Veneto) and 5 letters to Chelini written by Bellavitis (Archivio Generale Scuole Pie in Rome).[5]

Domenico Chelini was born in Gragnano (Lucca) on October 18th, 1802, and he died in Rome on November 16th, 1878. He was ordained a priest in 1827, in the *Scolopi* congregation. He was a self-taught man and, in particular, he studied mathematics. His research would be devoted to Pure Mechanics and Geometry. He became Professor of Mechanics and Hydraulics at the Papal University of Bologna in 1851, transferring to Rome in 1871, but he ultimately suffered the consequences of his loyalty to the Catholic Church.

The relationship between spiritual and temporal powers are often considered in abstract terms, and there is a tendency to neglect the consequences it may have on individuals, especially during moments of transition. Chelini's case is emblematic, because it brings out the moral and civil struggle imposed on the conscience of a man of the church in his decision-making. During the first half of the 1860s, the letters exchanged by Cremona, Genocchi, Tardy (Cerroni and Fenaroli 2007), and Chelini himself dealt with such topics; they were collected by M. R. Enea and R. Gatto (Enea and Gatto 2009), who explained the reason why it was not easy for many mathematicians and politicians of that period to neglect the "Chelini question."

Let us now look at some mathematical details mentioned in some of the letters.

The correspondence we have analysed starts with a letter dating back to 1852: Chelini thanked Bellavitis for the copy of *Saggio sull'algebra degl'immaginarii* (Bellavitis 1851), which he had received as a gift. He praised some "*right and profound* [...] *views*" by citing some of his articles.

In the answer dated September 2nd, 1852, Bellavitis declared his uniformity of opinion with some works by his colleague when he wrote:

> I am finding pure delight in studying them, also because of a certain accordance that, if not deceived by vainglory, I seem to descry in your way of treating Geometry. I myself considered the composition of straight lines, of areas, etc., as geometric topics, and I collected several theorems concerning them.[6]

---

[5]See the Beltrami-Chelini correspondence (1863–1873) (Enea 2009) on these themes; the letters exchanged in Chelini-Cremona (1863–1878) (Enea and Gatto 2009) and Cremona-Genocchi (1860–1886) (Carbone et al. 2001) are very interesting as well.

[6]See Bellavitis (1832) a work that Bellavitis himself reported he sent to the Istituto delle Scienze di Bologna in 1844.

He recalled his definitions of *poligonoide (polygon)* or *multilatero (multilateral figure)* (Bellavitis 1837), (Bellavitis 1838), a system of straight lines, whose composition is null, of *poliedroide (polyhedron)* or *multifacce (multisides)*, a system of areas whose composition is null, and of *pseudocentro (pseudocentre)*. Furthermore, he states that (Bellavitis 1847; Freguglia 1994):

> I also remarked the necessity of the statement
>
> $$\mathrm{AB} + \mathrm{BC} + \mathrm{CA} \simeq 0;$$
>
> hence, we have
>
> $$\mathrm{AB} + \mathrm{BC} + \mathrm{CA} \simeq 0.$$
>
> I point out that such equipollence exists not only for the points of a straight line, but also for the points of a space, as long as the sign + does not have the meaning of addition, but that of composition [...]. For the segments that belong to the same plane, we can consider equipollence of a greater degree than the first, such as, for instance:
>
> $$\mathrm{AE.BC} \simeq \mathrm{AB.BD} - \mathrm{AC.CD}$$
>
> which is evident when five points belong to the same straight line; for five points in a plane, this equipollence includes a lot of elementary geometry theorems. Hence, by virtue of these principles, my method of equipollence includes a lot of elementary geometry theorems.

This passage of Bellavitis's letter to Chelini expresses Bellavitis's *fundamental theorem* (or *canon*) of the calculus of equipollences.

Its important corollary states that:

If any relationship concerning distances of points on a straight line is determined and expressed by an equation, then it is possible to establish a corresponding relationship among the points on the plane transforming the equation into an equipollence.

Here is an example of an application of the corollary of the fundamental equipollence theorem (Fig. 1):

$$AE \cdot BC = AB \cdot BD - AC \cdot CD$$

$$x \cdot y = (x + a)(y + a) - (x + y + a)a = x \cdot y$$

$$CD \simeq EB$$

$$AE.BC \simeq AB.BD - AC.CD$$

The algebraic importance of this corollary consists of the fact that it is possible to extend the calculus among oriented segments in a straight line to the calculus among oriented segments in a plane. That is, the algebraic properties of the calculus with segments in a straight line (real numbers) are the same as the segments in a plane (complex numbers). This result is linked to elementary geometry bases.

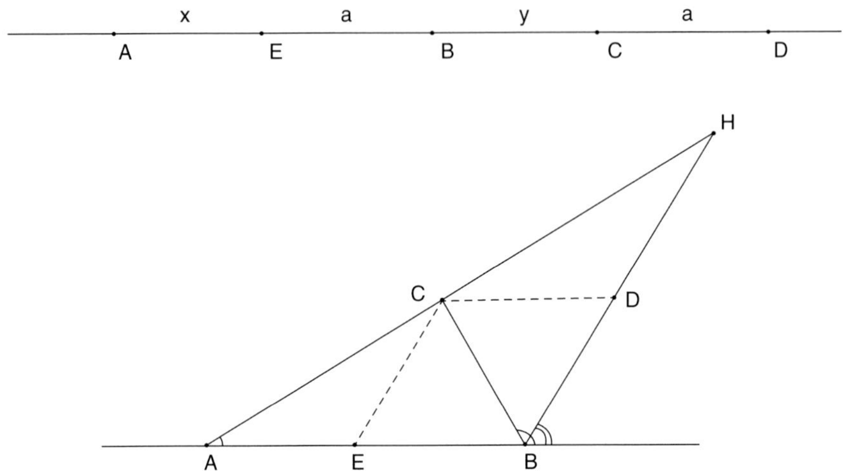

Fig. 1 The straight line case (up) and the plane case (down) of the same expression

Moreover, it is interesting to notice that, in the letter dated September 2nd, Bellavitis hinted at a topic of disagreement with Chelini:

> Following the same principle of scientific correspondence, rather than singing praises and congratulations for the beautiful things you are enriching science with (praises of little value, considering my meanness), it would be better to mention that which does not find me in complete agreement with your opinion, that is, the composition of rotating finite motions, since it seems to me that too many explanations are needed in order to make it valid.[7]

Chelini's answer, dated October 16th, 1852, is very interesting. After thanking Bellavitis for sending him some works on Descriptive Geometry, he wrote:

> Whenever I analyse and study your work, I can find and admire many cunning and original concepts, besides several new propositions; nevertheless, I will tell you sincerely that not all of it satisfies me as far as the metaphysics of science is concerned, but, to be perfectly clear, by expressing myself in this way, I do not assume to be pedantic towards such a distinguished Scientist as you are; I do nothing but frankly expose my opinion following your kind question.[8]

[7]Per lo stesso motivo di corrispondenza scientifica, meglio che presentarle i miei elogi e congratulazioni per le belle cose di cui Ella arricchisce la scienza (elogi di poco valore per la mia pochezza) potrà giovare che io accenni ciò di che non mi trova pienamente della sua opinione, e ciò sarebbe sulla composizione dei moti rotatorii finiti, sembrandomi che per renderlo esatto occorrerebbero troppe dilucidazioni.

[8] A misura che vado percorrendo e studiando i suoi lavori vi trovo e vi ammiro, oltre ad un gran dovizia di nuove proposizioni, molti concetti ingegnosi ed originali; nondimeno le dirò francamente che non tutto mi soddisfa dal lato di ciò che si chiama la metafisica della scienza, ma, intendiamoci bene, nell'esprimermi in questo modo non presumo già di fare il saccente ad uno Scienziato così eminente siccome Ella è, io non fo altro che esporre schiettamente il mio parere in conformità della sua gentile dimanda.

And he immediately exposed one. He argued that, in his treaty on equipollence, Bellavitis did not use the symbology of the imaginary numbers:

> For example, I cannot understand the reason why we should withdraw from the common concept of imaginary quantities. On the contrary, by taking into account the common idea of imaginary numbers, I believe that your theory on equipollence could become a lot more apparent and would be acclaimed more favourably by everyone.[9]

Chelini explains his idea in the following way:

> Given the imaginary quantities:
>
> $$a\left(\cos\alpha+\sin\alpha\sqrt{-1}\right),b\left(\cos\beta+\sin\beta\sqrt{-1}\right),c\left(\cos\gamma+\sin\gamma\sqrt{-1}\right),$$
>
> let $e^{\sqrt{-1}}=\varepsilon$, therefore
>
> $$a\varepsilon^{\alpha},b\varepsilon^{\beta},c\varepsilon^{\gamma}.$$
>
> If the moduli $\underline{a}$, $\underline{b}$, $\underline{c}$ are considered as the first sides *AB, BC, CD* of a polygon inclining on an $\underline{x}$ axis with the angles α, β, γ, and if δ is the angle that the straight line *AD* creates with the same axis, the sum of those imaginary numbers will produce the equation
>
> $$\varepsilon^{\alpha}AB+\varepsilon^{\beta}BC+\varepsilon^{\gamma}CD=AD.\varepsilon^{\delta},\qquad(1)$$
>
> where the straight line *AD* is the one that can be called the resultant from the straight lines *AB, BC, CD,* by the property stating that its projection on any axle is always equal to the sum of the corresponding projections of the other straight lines, which are its components.
>
> Equation (1), which essentially is equal to two, written in brief, like you do, as
>
> $$AB+BC+CD\simeq AD$$
>
> is the fundamental one of your theory on equipollence. Now, I cannot understand why the terms of (1) should not be considered as imaginary numbers, given the result, although by their moduli and arguments, they reveal sides and angles of a real polygon. And if the term $\varepsilon^{\alpha}.\overline{AB}$ is multiplied by $\varepsilon^{\frac{\Pi}{2}}=\sqrt{-1}$, the module of the new term $\varepsilon^{\alpha+\frac{\Pi}{2}}.\overline{AB}$ will absolutely be perpendicular to the module of the proposed term, provided that all of the arguments α β γ are counted on the same axle. This being true, why should we recoil from the ideas on imaginary numbers, which we obtained as a necessary consequence of the solution of algebraic equations? In my opinion, there are two main reasons why (1) is so fruitful: the former deals with the several good properties of the resultant straight line, the latter concerns the properties of the exponential ε^{α}, ε^{β}, etc., which are not worth less.[10]

---

[9] Per esempio, io non vedo né la ragione né la necessità di scostarsi dall'idea comunemente ricevuta delle quantità immaginarie. Anzi partendo dall'idea comune degl'immaginarii, son d'opinione che la sua teoria dell'equipollenze acquisterebbe maggior chiarezza e sarebbe accolta dall'universale con maggior favore.

[10] Date le quantità immaginarie $a\left(\cos\alpha+\sin\alpha\sqrt{-1}\right),b\left(\cos\beta+\sin\beta\sqrt{-1}\right)$, $c\left(\cos\chi+\sin\chi\sqrt{-1}\right)$, Le quali, posto $e^{\sqrt{-1}}=\varepsilon$, prendon la forma $a\varepsilon^{\alpha}$, $b\varepsilon^{\beta}$, $c\varepsilon^{\chi}$ Se i

Chelini noted that the idea of composing straight lines and areas had been suggested to him "*directly from mechanics.*" In 1834, he used this idea when he wrote his "Saggio di geometria analitica"[11] without any reference to the work of Chasles (Chasles 1830), who had a similar point of view, nor Bellavitis's memoirs on equipollence.

Chelini's aim was to establish the teaching of analytic geometry for the systematic use of the resultant straight line properties by homologating, as far as *the metric relationships of the extension* are concerned, what the *principle of virtual speeds* applies in mechanics concerning *the metric relationships of forces.*[12]

In the letter, in accordance with the above-mentioned method, Chelini demonstrated Carnot's theorem represented by the equation

$$2\overline{MM'}.\overline{NN'}\cos\left(MM', NN'\right) = \overline{MN\prime}^2 + \overline{M\prime N}^2 - \overline{MN}^2 - \overline{M\prime N\prime}^2,$$

which, in the author's opinion,

> [...] associated with the fundamental terms of trigonometry, is the principle that demonstrates the important propositions on polygons and polyhedrons shown in your gracious [letter].[13]

Bellavitis answered on November 13th, 1852, by writing that he would have been very glad to discuss the metaphysics of imaginary numbers with him, but he set out that:

---

moduli a, b, c si riguardano come i primi lati *AB, BC, CD* di un poligono inclinanti sopra un asse x cogli angoli α, β, γ, *e se* δ *è* l'angolo che la retta *AD* fa collo stesso asse, la somma di quegl'immaginarii darà luogo all'equazione $\varepsilon^{\alpha}AB + \varepsilon^{\beta}BC + \varepsilon^{\chi}CD = AD.\ \varepsilon^{\delta}$ (1) Dove la retta *AD* è quella che si può chiamare la risultante delle rette *AB, BC, CD,* definendosi per la proprietà che la sua projezione sopra un asse qualunque è sempre uguale alla somma delle projezioni omologhe delle altre rette, sue componenti. L'equazione (1), che in sostanza equivale a due, e che per brevità si può scrivere com'Ella fa

$$AB + BC + CD \simeq AD$$

è la fondamentale della sua teoria dell'equipollenze. Ora io non so comprendere perché i termini della (1) non si debbano riguardare come immaginarii giusta il significato ricevuto, sebbene coi loro moduli ed argomenti mettano in evidenza i lati e gli angoli di un poligono reale. E se il termine $\varepsilon^{\alpha}.\overline{AB}$ si moltiplica per $\varepsilon^{\frac{\Pi}{2}} = \sqrt{-1}$, il modulo del nuovo termine $\varepsilon^{\alpha+\frac{\Pi}{2}}.\overline{AB}$ sarà certamente perpendicolare al modulo del termine proposto purché siasi fatta la convenzione che tutti gli argomenti α, β, γ si contano a partire da un medesimo asse. Ciò essendo, che bisogno v'è di allontanarsi dalle idee ricevute sugl'immaginarii , le quali sono una conseguenza necessaria della risoluzione dell'equazioni algebriche? Due sono le cagioni, secondochè a me pare della fecondità della (1), delle quali la prima sta nelle belle e numerose proprietà della retta risultante, e la seconda sta nelle non men belle e numerose proprietà degli esponenziali ε^{α}, ε^{β}, etc.

[11]The issue mentioned here is clarified in Chelini (1863) Note (*) p. 4

[12]Chelini had pointed out his authorship in the Memoir: Chelini (1849).

[13] associato ai termini fondamentali della trigonometria, è il principio onde si dimostrano le importanti proposizioni sui poligoni ed i poliedri indicatemi nella sua gratissima [lettera].

> [. . .] the speed of oral conversation would be appreciated; even though I know how difficult such an arrangement would be, given some previous experience with my dearest friends; and I noticed that the most frequent (nonetheless useful) result of such discussions is to reassure everyone of his own opinion, by letting him study the topic in-depth, but following his point of view. To reach an arrangement, one should get back to the origins of so many different opinions, perhaps even more than prudence allows us to do.[14]

Bellavitis suggested that Chelini should read the introduction to the *Saggio sull'Algebra degli immaginarii*, in which the former had dealt ironically with this very topic.

Bellavitis, however, admitted that his style in regard to the definitions contained in the "*Geometria pura*" could be annoying to some. Moreover, he thought that his symbology had a sort of agility and ease of use that was very close to imaginary numbers, and he added:

> Perhaps by establishing the method of equipollence, unlike what happened in other greater discoveries, I did nothing but express, synthetically, some ideas that had already been developing and gathering on their own; and, even without this synthetic principle, studies will get the same results under another form; in fact, Geometricians made some progress without (I believe) any knowledge at all of what I had already published.[15]

He also pointed out that his dear friend Carlo Conti had tried to modify the language of the method of equipollence, solely for the purpose of making it more suitable in regard to settled habits.

Through their correspondence, it can be seen that the two mathematicians continued to exchange their works. When Chelini wrote the *Elementi di Meccanica razionale*, principally for educational purposes, he asked Bellavitis to read them and propose improvements.

They also discussed matters regarding the Accademia dei XL, whose possible unification with the Accademia Nazionale dei Lincei was one of the main debates in which Bellavitis took part, expressing strong dissent.

In March 1863, just before his nomination as a member, appointed by Bellavitis, Chelini wrote on the subject:

> I totally agree with what you said about the project, which, rather than reforming, is trying to decompose and destroy this Society, one of our purest glories, clearly founded under liberal

[14] ...a ciò occorrerebbe la rapidità della conversazione vocale; d'altronde so, per esperienza con perspicaci e carissimi amici, quanto è difficile porsi d'accordo; ed osservai che il più frequente (e d'altronde tutt'altro che inutile) risultamento di tali discussioni si è di confermare ciascuno nella propria opinione, dandogli occasione di esaminare più profondamente, ma secondo le proprie vedute, l'argomento. Per andare d'accordo bisognerebbe risalire ai principi di tante opinioni, forse più di quanto sia prudente il farlo.

[15] Forse nello stabilire il metodo delle equipollenze io non ho fatto, come avvenne in altre maggiori scoperte, che esporre sotto un principio sintetico idee, che già si andavano di per sé maturando e riunendo; ed anche senza l'espressione di questo principio sintetico gli studii giungeranno agli stessi risultamenti sotto altro aspetto; ed infatti i Geometri progredirono senza avere (io credo) la menoma certezza di quanto fu da me pubblicato.

aims, a Society that, having gathered lively minds from everywhere in a unique common concept of scientific progress, has formed and still forms the intellectual unity of Italy, being the principle and basis of a more complete unification.[16]

And, in 1866:

I completely agree with your observations on the modifications to be introduced in the Accademia or Società Italiana delle Scienze, and it seems to me that no objections could be presented. As for me, before leaving for Lucca, I had already sent to Marianini in Modena, all of the forms with the names of Prof. Padula and Prof. Marianini, noting that there was no name among those of the 5 Counsellors, and relying on the President's wisdom, since it seemed very strange to get such harmony in the voting of so many people.[17]

Many years later, on December 28th, 1874, Bellavitis wrote about it once more:

Our Società dei XL is declining, moreover, it seems that it is going to be destroyed by its unification with the R. Accademia dei Lincei. I believe that no Member should agree with such a proposal, especially if made by a President whose seven year mandate has already expired, a proposal that, in any form, would destroy such an excellent institution, which has been honouring Italian Dotti for a century, with no exception or distinction, thanks to the free vote by Dotti all over the Country. It seems obvious to me that we should nominate a new President who wishes to preserve our Society.[18]

In 1875, Chelini stated, with ill-concealed resentment:

As for the Società Italiana dé XL, I feel that new radical innovations will be introduced, up to the mark of our times. I am taking into account that I will be left out, as I so wish, given my condition as a septuagenarian clergyman, or a perfect scientific nullity. So, I do not care what the others think about.[19]

---

[16] Trovo poi giustissimo quanto Ella dice intorno al progetto il quale, più che a riformare, tenderebbe a decomporre e a distruggere questa Società, una delle nostre glorie più pure, fondata con fini manifestamente liberali, Società che, riunendo da ogni parte in un comune pensiero di progresso scientifico i varii ingegni che più si distinguono nella nostra nazione, ha formato e forma in qualche modo l'unità intellettuale dell'Italia, principio e base di un'unità più completa.

[17] Trovo giustissime le riflessioni che mi fa intorno alle modificazioni che si vogliono introdurre nell'Accademia o Società italiana delle Scienze, e parmi che non si possano fare obbiezioni ragionevoli in contrario. Quanto a me avevo già rimandato al Marianini in Modena, prima di partire per Lucca, tutte le schede ricevute coi nomi del prof. Padula e del prof. Marianini, e senza notare alcun nome in quella de' 5 Consiglieri rimettendo per altro la cosa alla saviezza del Presidente, perché mi era parso assai strano di ottenere un po' di concordia nella votazione per tanta gente.

[18] La nostra Società dei XL va deperendo, inoltre mi si fa credere che si abbia in animo di distruggerla rinserrandola alla R. Accademia dei Lincei. Mi pare che nessun Socio dovrebbe acconsentire a tale proposta, se fosse fatta da uno il cui settenario di Presidente è già spirato, proposta che qualunque fossero le sue forme distruggerebbe una sì bella istituzione che da un secolo conferisce ai Dotti Italiani senza alcuna eccezione né distinzione un'onorificenza tanto più apprezzata quanto viene dal libero voto de' Dotti sparsi per tutta la penisola. Parmi che la sola cosa da farsi sia nominare un nuovo Presidente che abbia in animo di conservare la Società.

[19] Quanto alla Società italiana dé XL, sento anch'io che si vogliano fare innovazioni veramente radicali, che stiano all'altezza della sapienza dé nostri tempi. Tengo per fermo che io sarò lasciato

Chelini approved Giacinto Namias, Arcangelo Scacchi and Domenico Turazza's membership in the Società.[20]

Chelini and Bellavitis had many mutual friends among mathematicians, for instance: Battaglini, Beltrami,[21] Cremona, Gherardi, Piuma, Tortolini, etc., with whom they clearly had a remarkable correspondence and a profitable exchange of their works.

In addition, the Rivista di Giornali was sent regularly to the Istituto delle Scienze in Bologna.

On May 7th, 1863, Chelini wrote, in reference to this journal:

> I read hereby, it is not much, a Memoir about the different systems (that I call simple) of coordinates, and about the discussion of the general equation of a second degree expressed first by triangular coordinates, and then by tetrahedral coordinates, both considering the line and the area represented as a location of points, and as an envelope of straight lines or planes. I know nothing about a complete and methodical discussion on the subject. Prof. G. Battaglini dealt with it, it is true, (and very smartly, as you reported in your Rivista), but only partially. The way I have chosen is completely different, much plainer, clearer and wider, without need for circular points or straight lines or infinite plains. When it is issued,[22] you will judge it, if you have time to consider it with attention; I could, however, be wrong.[23]

Furthermore, the letters reflect the difficult moments Chelini had to cope with because of his loyalty to his moral principles and his condition as a clergyman. For instance, on July 26th, 1863, he wrote:

> I hope you will forgive me if I have not written to you for a long time, as I should have done, but I had to face many troubles that have caused me affliction and despair.[24]

---

in disparte, come appunto desidero e voglio, e come lo richiede il mio stato di religioso più che settuagenario, o di una perfetta nullità scientifica. Non mi curo adunque di sapere ciò che pensano gli altri.

[20]See the letters by Chelini to Bellavitis dated November 2nd, 1863, and June 23rd, 1863, and Bellavitis's letter to Chelini dated July 21st, 1871.

[21]On July 23rd, 1863, Chelini wrote to Bellavitis: "Let me introduce to you our mutual Friend Mr. E. Beltrami, a remarkable young man, outstanding both for his goodness of temper and liveliness of mind; I believe that he possesses all of the conditions necessary to become one of the most important mathematicians of our century."

[22]See Chelini (1863).

[23] Ho letto qui, non è molto, una Memoria sopra i diversi sistemi (che io chiamo semplici) di coordinate, e sopra la discussione dell'equazione generale di 2° grado espressa prima in coordinate triangolari, e poscia in coordinate tetraedriche, sia che la linea e la superficie rappresentata si voglia considerare come luogo di punti, sia che si voglia considerare come inviluppo di rette o di piani. Non mi è noto che una simile discussione metodica e completa sia stata ancor fatta. Il Prof. G. Battaglini ha toccato, è vero (e con molta eleganza com'Ella ha giustamente notato nella sua Rivista) questo argomento, ma solo in parte. La via che io tengo è al tutto diversa, e parmi assai più diritta, chiara e spaziosa, non avendo bisogno né di punti circolari, né di rette, né di piani all'infinito. Quando sarà stampata, Ella ne giudicherà seppure avrà tempo di fermarvi sopra alquanto l'attenzione; potrei però essermi ingannato.

[24] Se da gran tempo non le ho scritto, come avrei dovuto, spero che vorrà perdonarmi ove rifletta alle peripezie le quali ho dovuto soffrire senza avervi dato la minima cagione, e che mi hanno non poco afflitto e tribolato.

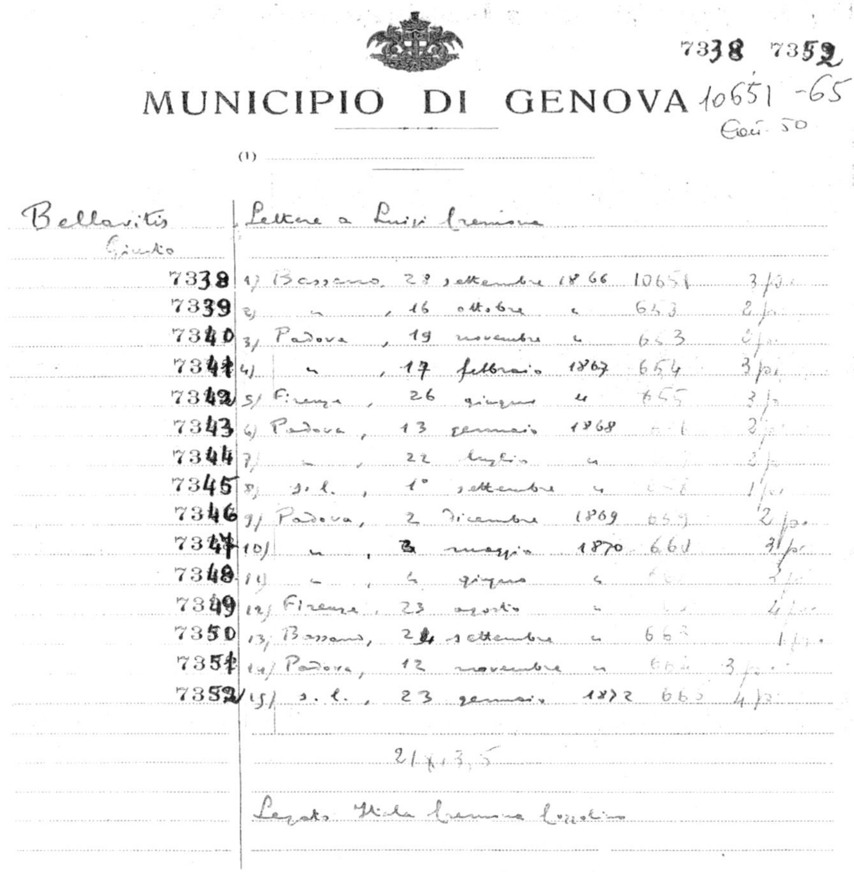
MUNICIPIO DI GENOVA

7338 7352 10651-65

(1)

Bellavitis Giusto — Lettere a Luigi Cremona

7338	1) Bassano,	28 settembre	1866	10651	3 p.
7339	2) " ,	16 ottobre	"	653	2 p.
7340	3) Padova,	19 novembre	"	653	2 p.
7341	4) " ,	17 febbraio	1867	654	3 p.
7342	5) Firenze,	26 giugno	"	655	3 p.
7343	6) Padova,	13 gennaio	1868	[illegible]	2 p.
7344	7) " ,	22 luglio	"	[illegible]	2 p.
7345	8) s. l.,	1° settembre	"	[illegible]	1 p.
7346	9) Padova,	2 dicembre	1869	659	2 p.
7347	10) " ,	3 maggio	1870	660	3 p.
7348	11) " ,	4 giugno	"	[illegible]	[illegible]
7349	12) Firenze,	23 agosto	"	[illegible]	4 p.
7350	13) Bassano,	24 settembre	"	663	1 p.
7351	14) Padova,	12 novembre	"	664	3 p.
7352	15) s. l.,	23 gennaio	1872	665	4 p.

21 x 13,5

Legato Itala Cremona Cozzolino

Fig. 2 List of the Bellavitis-Cremona letters at the Museo Mazziniano in Genoa

4 Luigi Cremona and Bellavitis

Now, let us consider some points of correspondence with Luigi Cremona. The Legato Itala Cremona Cozzolino in the Museo Mazziniano of Genoa houses 60 letters sent by Bellavitis to Luigi Cremona and one by Ernesto, Giusto Bellavitis's son. They are divided into four well-preserved envelopes: some of them are not dated or only partially. The period they refer to is from June 22nd, 1860, to January 31st, 1880 (Fig. 2).

Luigi Cremona was born on December 7th, 1830, in Pavia, and died on June 10th, 1903, in Rome. He took part in the defence of Venice, during the first Italian war of Independence in 1848–1849 (Pepe 2012). He graduated with a degree in mathematics in 1854, and then became a teacher in a secondary school. In 1860, Cremona was nominated Professor of Advanced Geometry in Bologna, where he developed the theory of birational transformations, later known as the Cremona transformations. In 1863, he published his first works on this topic, becoming famous and greatly esteemed in Europe as a result and winning the Steiner prize in 1866, one of the most prestigious awards of the time. The same year, he became Professor of Geometry at the Polytechnic of Milan, having been recommended by Brioschi (Lacaita 2012). In 1873, Cremona was offered a political post as General Secretary of the new Italian Government. Political pressures led him to serve the new Italian State. On March 16th, 1879, he was appointed senator. In 1898, he became Minister of Public Education (though only for a month) and ended his political career as Vice-President of the Senate.

Here are some of Cremona's most faithful readers: Cayley, Clifford, Rosanes, and Noether, who followed his research in Geometric Algebra, together with other distinguished mathematicians. Cremona's interest was not limited to research alone, as he was also involved in the problems concerning the teaching of mathematics in secondary schools, as can be seen from the efforts he made to bring about legislative innovation in these studies.

The main topics in the letters deal with teaching, mathematical research, the Chelini question, the Società dei XL, Risorgimento political events, the publication of scientific articles, the Rivista di Giornali, and personal family matters.

The letters that Bellavitis wrote to Cremona[25] definitely point out the age difference between the two correspondents. However, the initial paternal attitude expressed by Bellavitis gradually faded away as Cremona gained more scientific relevance and political power.

On June 22nd, 1860, Bellavitis wrote to the young Cremona, Professor of Advanced Geometry in Bologna:

> One day before your letter arrived, I received another from our mutual friend, who informed me of your assignment, which I am glad for and I envy you. I have never enjoyed teaching, except when, some years ago, I gave free lessons in superior Geometry; you will do very well, only remember that, as you are still young, success is not gained by the many we teach but by the few who learn well.[26]

In the earliest letters, it is clear that Bellavitis hoped that the young geometrician could use his capabilities to achieve creation of the equipollence method, since geometry of the plane had lay dormant for 20 years. At this point, we would like to mention some abstracts of undated letters in which Bellavitis wrote:

> … things concerning my studies are not so successful and my equipollence method remains static.[27]

---

[25] See note 2.

[26] Un giorno prima della vostra ne ebbi una dal comune amico, in cui mi dava la notizia della vostra nomina, ne sono lieto e v'invidio, non ho mai sentito piacere ad insegnare se non quando alcuni anni sono feci alcune lezioni libere di Geometria superiore; voi farete benissimo, soltanto, perché giovine, bisognerà che vi ricordiate che il profitto è misurato non dal molto insegnato ma dal poco che è bene appreso.

[27] le cose relative ai miei studii non sono molto prospere ed il mio metodo delle equipollenze rimane arenato.

And then:

> Do you think you will continue studying to achieve the equipollence method? Mind that you do not discredit the fixed principles, because science must go forward and never backward.[28]

Moreover:

> You are dealing with the equipollence method. I am afraid that it will take too long to achieve it, and I am an old man. The French should be in favour—now that Saint-Venant and Cauchy are working on it, but it seems that they do not appreciate Italian methods so much. What do you think about it?[29]

Cremona did not fulfil Bellavitis's aspirations and considered his colleague's positions on non-Euclidean geometry and imaginary numbers to be outdated (see some letters in as an example of this attitude). On September 15th, 1862, Bellavitis wrote about it:

> However, I do not agree with you on imaginary numbers, nor can I understand how you can admit imaginary numbers and then deny that a circle can become an infinite radius. I will keep on fighting against imaginary numbers, I have been doing it for about forty years; you see, time is long; I do not know if you understood that I associate $\surd$ with any other imaginary number. We will, however, always be dear friends, and this flatters me greatly.[30]

The correspondence points out some contrasts in their views on questions concerning Public Education, the publication of magazines and scientific reviews, and the Società dei XL, and also testifies to a period in the 1870s of clear conflict, later resolved, at a personal level.

On December 2nd, 1869, Bellavitis wrote:

> ....since it seems that you are complaining about me, let me propose something to you. We both greatly esteem Brioschi's moral integrity and feel honoured by his friendship, and considering his justifiable appreciation of and friendly deference towards you, I am sure that it is not inappropriate to propose him as an arbiter; if he thinks that I offended you in

[28] Hai intenzione di riprendere lo studio del metodo delle equipollenze per portarlo a compimento? Bada peraltro di non mettere in dubbio i principi già stabiliti, perché le scienze deggiono progredire e non mai indietreggiare.

[29] Vi occupate voi altri del metodo delle equipollenze. Temo che ci voglia troppo per portarlo a compimento, ed io son vecchio. I Francesi dovrebbero essergli favorevoli—ora che dopo Saint-Venant e Cauchy si cominciano ad occuparsene—ma parmi che abbiano poca simpatia pei metodi italiani. Che ve ne pare?

[30] Ma sugli immaginari non andiamo d'accordo ne' so capacitarmi come tu ammetta gli immaginari, e poi non volessi passarmi per buono un circolo che diventava di raggio infinito. Io già continuerò sempre a combattere contro gli immaginarii, la è una mia idea fissa da forse quaranta anni; tu vedi che il tempo è ben lungo; non so se tu abbia capito che al $\surd$ io associo qualche altro immaginario. Basta comunque sia noi resteremo sempre teneri amici, il che sommamente mi lusinga.

any way, I will be glad to apologize, since I really want to show myself as I am: a sincere and affectionate Friend.[31]

This did not, however, discourage the development of a strong affection between the two mathematicians, and they continued to write letters that, from the beginning, clearly showed their mutual friendship and their appreciation for Domenico Chelini.

In the letter to Cremona of December 7th, 1862, in favour of Chelini becoming a supporting member of the Società dei XL, Bellavitis wrote:

> Once again, it has not been possible to designate our excellent Chelini as a member; I already knew that Brioschi preferred Gaspar[r]ini, and I complied with the Mathematician [Brioschi], even though I do not usually put much faith in the Naturalists' reputation; moreover, I would prefer a predominance of mathematicians to continue in the Italian Society; I do not know the exact result of the voting yet, but there are great hopes that Chelini will follow the recently deceased Bizio.[32]

And subsequently, in the letter of January 29th, 1864:

> In the Gazzetta of December19th, I noticed a royal ordinance concerning the dismissal of two Professors in Bologna; could one of them be our good friend?
>
> To be honest, I think that the time has come for an oath to be required from everyone, but to be fair, it should be taken by all of the salaried staff; an oath should not offend anyone, it is just about loyalty[?]. I kindly ask you to present him with my best compliments and happy wishes. I won't write to him, unless I get information about his situation, and please keep in touch, because I am deeply concerned about it.[33]

---

[31] [...] siccome mi pare che tu creda d'avere qualche motivo di lagnarti di me, così permettimi di farti una proposta. Noi tutti e due stimiamo altamente la rettitudine del Brioschi, e quantunque egli mi onori della sua amicizia, prove per la giustissima stima ed amichevole deferenza che egli ha per te sono certo che non trovarsi sconveniente che te lo proponga ad arbitro; se egli giudicherà che in alcun modo io ti abbia offeso, sarò pronto a presentartene le mie scuse, poiché tengo sopra ogni cosa di mostrarmi quale veramente ti sono sincero ed affez. Amico

[32] Anche questa volta non riuscì la nomina dell'ottimo nostro Chelini; avevo già saputo che Brioschi preferiva Gaspa[r]rini io mi attenni al Matematico, tanto più che in generale non ho molta fiducia della fama di alcuni Naturalisti; poi desidererei che nella Soc. Ital. si mantenesse una preponderanza di Matematici; non conosco ancora esattamente l'esito della votazione, la quale dà fondata speranza che il Chelini succeda al Bizio testè mancato.

[33] Veggo nella Gazzetta che il giorno 19/12 fu sottoposto alla firma reale un decreto per la cancellazione di due Professori di Bologna; uno di questi sarebbe il nostro ottimo amico?
Per vero dire parrebbe che fosse giunto il tempo di richiedere da tutti il giuramento, ma perchè la cosa fosse giusta bisognerebbe che ciò si fosse fatto verso tutti gli stipendiati; la forma del giuramento parmi che non dovrebbe offendere alcuna coscienza, si tratta soltanto di obbedienza. Ti prego di presentargli per me i più cordiali complimenti ed augurio di felicità, io non gli scrivo, finchè non sappia al giusto come sia la cosa a suo riguardo, e ti prego di darmene notizie che molto mi interessano.

Throughout the correspondence, Bellavitis's concern about keeping and protecting the main features and aims of the Società dei XL is always present, as is shown, for example, in the same letter:

> I believe it is appropriate to be in favour of the oldest, in order to let all of the distinguished Mathematicians be a part of the Society that has such a beautiful name; and I hope it will be kept unchanged, if only in respect to the ancient wish that Lorgna tried to attain.[34]

As for the interest, mentioned in the letters, concerning the political events of the Risorgimento,[35] we propose some passages of Bellavitis's letter to Cremona dated July 19th, 1866, before the Cormons Armistice (12-8-1866), but after the Italian military operations started (June 23, 1866):

> Today, Tuesday, July 18th, I have found your letter of the 13th, which was also sent from Switzerland. You are getting closer and closer, since yesterday morning, together with two Paduans, I had the honour of presenting to His Majesty, our beloved King, the regards of the town of Padua; he welcomed us with his usual kindness and, I dare say, familiarity; and even in his words, I could find a touch of foreign wickedness.[. . .]
>
> In the night between Tuesday and Wednesday (the 10th and the 11th), the last Austrians and pro-Austrians, who considered themselves to be too compromised, left Padua. On Thursday, a Captain and a Sergeant of the Italian Army arrived and the entire town was decked with flags.[36]

[34] Credo opportuno di dar la preferenza ai più vecchi, acciocchè se è possibile tutti i Matematici più distinti facciano a lor volta parte della Società che porta un si bel nome; e che desidero rimanga invariata, se non fosse altro come memoria dell'antico desiderio a cui il Lorgna cercò di soddisfare nel modo per lui possibile.

[35] In order to get a sense of the political situation of this period, we point out some events concerning the Third Italian War of Independence (June 19th, 1866–October 3rd, 1866). Victor Emanuel the Second of Savoy was to be crowned King of Italy on March 17th, 1861, but he could not control Veneto and Lazio. On April 8th, 1866, the Italian government signed a military alliance with Prussia, through the mediation of Napoleon the Third of France. Prussia began hostilities on June 16th, 1866. Three days later, Italy declared war on Austria, starting military operations on June 23rd. The cessation of hostilities was marked by the Armistice of Cormons, signed on August 12th, followed by the Treaty of Vienna of October 3rd, 1866.
The terms of the Treaty included the cession of Veneto (with Mantua and western Friuli) to France (which had earlier ceded it to Italy) and of the Iron Crown (worn by the old Lombard Kings of Italy and by the Holy Roman Emperors). The diplomatic role of Napoleon the Third was very important.

[36] Oggi 18 luglio Mardì ritrovo la tua del 13 giunta questa pure per la via della Svizzera. Tu ti fai vicinissimo, perché ieri mattina insieme con due Padovani, ebbi l'onore di presentare a S.M. il nostro benamato Re l'omaggio della Città di Padova; egli ci accolse con la sua solita bontà e direi quasi famigliarità; anche dalle sue parole ebbi ragione di scorgere la nequizia straniera [...]
Nella notte dal Martedì nel mercoledì (10-11) partirono da Padova gli ultimi Austriaci, e gli Austriacanti che si credevano troppo compromessi. Giovedì giunse un Capitano ed un Sergente dell'esercito Italiano, e la città fu tutta imbandierata.

Furthermore, after the Cormons Armistice, Bellavitis wrote to Cremona in his letter of August14th, 1866:

> What anxious days! But they are not ended, if we consider Austria's disloyalty, the reasons for disagreement between Prussia and France, the affection on the part of Napoleon's followers towards Austria, so I won't feel secure as long as all of the fortresses are evacuated. To A., even disgraces can be profitable. I think that the Archduke was wrong when he agreed to fight in Custoza; if the Italians had had a general at the front , they would have won; and even if they had had no generals, every Brigadier could have disguised himself as a king, and no one would have written to Cialdini that the army had been defeated; without striking a blow, he would have seized Rovigo and all of the Polesine up to the Adige. The battle of Padua taught the Austrians the best strategy: to leave strong armies in the fortress and to draw back the whole army to Trentino and Isonzo (they knew there was no danger from the navy); the Italians could do nothing but occupy the abandoned provinces, and they were in the worst position: in front of them, there were two armies that had never been defeated and 5 or 6 fortresses at the back. What the A. had to do spontaneously was necessary to defend themselves: Napoleon's intermediation, the combat truce, the Prussia-Austria armistice all hastened the events in Italy, which had to beg Nap [oleon's] intervention for an indispensable armistice [...][37];

and later, he made a critical remark:

> Of course, the Commissioner had to follow the advice of those who approached him, having served the Italian case well. Until now, his behaviour has not met with disapproval in Padua; it has been so in Florence and Turin, according to some people who believe that all Austrian administration automatically becomes an Italian administration just through the signing of all of the administrative acts by Vitt.[orio] Eman.[uele], instead of Fran.[cesco] G.[iuseppe] and by ending every document by shouting "Long live the King" as loud as one's Austrian faith recommends. A curious way of ruling.[38]

[37] Quanti giorni ansiosi! né questi sono terminati, perché vista la mala fede dell'Austria i motivi di dissidio tra Prussia e Francia, e l'affetto che i Napoleonidi ebbero sempre per l'Austria io non sarò tranquillo nemmeno dopo fissata la pace, finchè non veggo sgombrate tutte le fortezze. All'A. sono proficue anche le disgrazie. Io credo che l'Arciduca aveva sbagliato nell'accettare una battaglia campale a Custoza, se gl'Italiani avevano un generale in campo riportavano una gran vittoria: ed anche dopo avuto il di sotto, se allora gli Italiani avevano la fortuna di non avere nessun generale in campo sicchè ogni Brigadiere si acconciasse da re, e niuno scrivesse a Cialdini che l'esercito era disfatto; questi senza colpo ferire occupava Rovigo e tutto il Polesine fino all'Adige. La battaglia di Padova insegnò agli Austriaci la migliore strategia, lasciare forti corpi nella fortezza ritirare l'esercito intero restante nel Trentino e all'Isonzo (già sapevano che poco avevano da temere dalla flotta); gli Italiani non potevano astenersi dall'occupare le province abbandonate, e si ponevano nella peggiore delle posizioni, con in faccia due eserciti non mai battuti in forti posizioni e con 5 o 6 fortezze dietro alle spalle. Ciò che gli A. dovevano fare spontaneamente, lo fecero per necessità di difesa: la mediazione di Napoleone la sospensiva d'armi l'armistizio della Prussia precipitarono le cose d'Italia, che dovette mendicare da Nap. l'armistizio ormai necessarissimo.

[38] Naturalmente che il Commissario non potè che attenersi ai consigli di quelli che lo avvicinarono, e che avevano già ben meritato della causa Italiana. Finora a Padova la sua condotta non è disapprovata; lo fu a Firenze e Torino da quelli che credono che tutta una amministrazione Austriaca divenga una amministr. italiana quando si è ottenuto che essa intesti i suoi atti con Vitt. Eman. anziché con Fran. G. e termini ogni carta con Viva il Re gridando tanto più forte quanto si fu più Austriacante. Curiosa maniera di governare.

In the letter of September 24th, 1870, after Napoleon the Third's capture at the Battle of Sedan (September 1st, 1870), and after September 20th, 1870, the date of the breach of Porta Pia and the seizure of Rome, Bellavitis wrote:

> Our best friend has miserably fallen and France is in such bad condition! Meanwhile, lucky Italy has been achieved. Long live Italy![39]

5 Angelo Genocchi and Bellavitis

Finally, a hint about Bellavitis's friendship with Angelo Genocchi, as reported in the correspondence containing:

100 letters from Bellavitis to Genocchi, Biblioteca Passerini Landi, Piacenza;
28 letters from Genocchi to Bellavitis, Fondo Istituto Veneto, Venezia; since this is one of the most complete correspondences, with letters from both of them, it provides an exhaustive reconstruction of many topics, ranging from scientific, cultural, social and political aspects to personal opinions.

Genocchi had been one of Peano's teachers at the University of Turin. The calculus of equipollence greatly influenced the realization of Peano's volume, *Applicazioni geometriche del calcolo infinitesimale* (1887).

Angelo Genocchi was born on March 5th, 1817 in Piacenza (Giacardi 2000). He earned a law degree in 1838 and began working as a lawyer. In 1845, he was appointed Substitute Professor at the law faculty in University of Parma, and the following year, he became full Professor in civil institutions. In 1848, he joined the revolts. While the Austrian troops were withdrawing to Piacenza, he fled to Stradella and then to Turin, where he dedicated himself to the study of mathematics with Giovanni Plana and Felice Chiò, starting with the numbers theory and publishing his first work in 1851. After obtaining a post at the university in 1857, he began to teach algebra and complementary geometry, becoming a full Professor in 1859. Then, he taught superior analysis, and subsequently complementary algebra and analytic geometry; during his Turin university years from 1861 to 1862, he taught advanced analysis (Garibaldi 1991). In 1865, he followed Plana in the teaching of infinitesimal analysis, which lasted until his death. In 1858, he was on the editorial board of the Annali. In 1886, he was appointed as a Senator. He died on March 17th, 1889 (Carbone et al. 2001).

His relationship with Bellavitis began between 1853 and 1854, spurred by the fact that Bellavitis knew of Genocchi's works, published by Tortolini in the Annali. In their letters, they wrote about university and academic matters and, though they did delve into matters of political organization, the two correspondents dealt especially with the teaching of algebra by debating the priorities of different subjects.

[39] Il nostro migliore amico è miserabilmente caduto, e la Francia è in così tristi condizioni! Intanto l'Italia sempre fortunata si è compiuta Viva l'Italia !

Although Genocchi was very reserved, he was one of the most influential Italian mathematicians in the mid-nineteenth century (Roero 1999, 2012), most likely thanks to his international relationships (Luciano and Roero 2012). In the letters, we can find discussions on the basis and language of mathematics, topics that raised very interesting questions for both correspondents; the details of the meaning of the mathematical terms at the basis of accomplished theories exist within mind maps that required time to develop. We can find whole pages about the definitions of 'integral' and 'derivative,' about the nature of complex numbers (imaginary numbers), as well as the reality of the entities of non-Euclidean geometries. More precisely, during the 1850s, equipollence and the imaginary numbers prevailed, in the 1860s, it was algebra and teaching, and in the 1870s, the Società dei XL and non-Euclidean geometry; other topics are cited throughout the correspondence.

As far as the imaginary numbers are concerned, here are some abstracts from a long letter by Genocchi dated October 23rd, 1868, that illustrate the approach to the debate very well:

> Kummer's ideal numbers are not those you suppose. If ρ is the imaginary root of $\rho^n = 1$ and a_0, a_1, a_2 are real numbers, the polynomial $a_0 + a_1\rho + a_2\rho^2 + \ldots + a_{n-1}\rho^{n-1}$ is an effective complex number (wirklick) and not an ideal number. Now, as it is asserted that an entire x function can be divided by a second degree real factor $(x - a)^2 + b^2$, we can also state that it can be divided by two first degree imaginary factors $x - a - bi$, $x - a + bi$; Then, effective complex numbers that cannot be divided into other effective complex factors are defined as numbers that can be divided into prime complex ideal factors.[40]

And further on:

> It seems to me (be astonished by my audacity) that, with no absurdity, the theory of imaginary numbers in algebra can be established as follows.
>
> I consider i to be an indeterminate quantity (that is real) introduced in the calculus by the convention, according to which every time we have the square i^2, the result must be transformed by considering −1; and I define as imaginary any expression containing that same indeterminate quantity i. By that assumed convention, the results will not be values of the function we are dealing with, but, on the contrary, transformations of it; they will not be equal, but equipollent to it itself. Then, if we have to multiply 1 + i by 2 + i, we will get first $2 + 3i + i^2$, which, by the above-mentioned convention, will become 1 + 3i, and this result will not be equal to any values of i to the product (1 + i) (2 + i), but it will be a transformation of the same product and it will be defined as equipollent to it. If you think about these ideas just for a moment (although I doubt you will), you will realize that they are not absurd in any way, and that all of the operations shown are simply explained with

[40] I numeri ideali di Kummer non sono quelli che supponi. Se ρ è radice immaginaria della $\rho^n = 1$, e a_0, a_1, a_2... sono numeri reali il polinomio $a_0 + a_1\rho + a_2\rho^2 + \ldots + a_{n-1}\rho^{n-1}$ è un numero complesso effettivo (wirklick) e non ideale. Ora allo stesso modo che quando una funzione intera di *x*è divisibile per un fattor reale di secondo grado $(x - a)^2 + b^2$, si dice che è divisibile per due fattori immaginarj di primo grado $x - a - bi$, $x - a + bi$; cosi i numeri complessi effettivi non decomponibili in altri fattori complessi effettivi si dicono decomponibili in fattori primi complessi ideali.

no opposition to any dictate by reason. Furthermore, I believe that these ideas can be given a convenient generality by the theorem establishing that any x function can be represented by a convergent series of the form $A_0 + A_1(x - a) + A_2(x - a)^2 + \cdots$.[41]

Bellavitis's answer is dated November 28th, 1868:

It is evident that, between the two opinions proposed by Gauss and by me, the former had the higher probability of being adopted; then, ramuno (or the root of -1) as a sign of perpendicularity is connected to a different theory, and "complex" was a meaningless and compromising word, and since maybe even Mathematicians felt uneasy in dealing with imaginary beings, they profited off the complex numbers' disguise..... Maybe this is the reason why Cauchy thought that this disguise was too transparent and he abandoned this theory of justification in favour of another one.[42]

Another subject that thrilled Bellavitis and many other interlocutors was non-Euclidean geometry: he manifested himself as an opponent and conservator, who was not so keen on new ideas, but he definitely contributed to a stronger strictness in the studies of his correspondents and readers, and to a skimming of all of the theories, destroying the old in favour of the new.

Their mutual friend and correspondent Eugenio Beltrami, with his work *Saggio di interpretazione della geometria non-euclidea* (Beltrami 1868), stimulated many debates in the letters. We can easily understand Bellavitis's difficulty in imagining all of those objects that mathematicians had to build little by little (lett. 21-9-76):

Since a piece of a flexible sphere can take another form, we can imagine that some portions of the sphere itself gather all around it, and so on, extending indefinitely; it seems possible

[41] A me sembra (stupisci della mia audacia) che senza assurdi la teoria degl'immaginari nell'algebra si possa stabilire come segue. Rappresento con i una quantità (reale) indeterminata che s'introduce nel calcolo con questa convenzione, che ogni qualvolta si presenta il quadrato i^2 si debba trasformare il risultato ponendo invece -1 ; e chiamo immaginaria ogni espressione che contenga quella stessa quantità indeterminata i. Coll'ammessa convenzione i risultati non saranno valori della funzione sulla quale si opera ma trasformazioni di essa, non saranno eguali ma equipollenti alla medesima. Così se devesi moltiplicare $1 + i$ per $2 + i$, si otterrà dapprima $2 + 3i + i^2$, che poi per la convenzione indicata si ridurrà ad $1 + 3i$, e questo risultato non sarà eguale per alcun valore di i al prodotto $(1 + i)(2 + i)$ ma sarà una trasformazione dello stesso prodotto e si dirà equipollente allo stesso. Se vorrai riflettere un momento (del che a dir vero dubito molto) su queste idee vedrai che non contengono alcun assurdo e che tutte le operazioni prescritte si spiegano semplicemente e senza contravvenire ad alcun dettame di ragione. Credo di più che a tali idee si possa dare la conveniente generalità profittando del teorema oggidì stabilito per cui ogni funzione di x si può rappresentare con una serie convergente della forma $A_0 + A_1(x - a) + A_2(x - a)^2 +$ ecc.

[42] E' evidente che tra due parole proposte da Gauss e da me, la prima aveva una immansa probabilità maggiore d'essere adottata; poscia ramuno come segno di perpendicolarità si collega con una teoria differente, e complesso era una parola insignificante e era compromettente, e siccome forse anche i Matematici sentivano un po' di rossore a occuparsi di esseri immaginari, così profittarono della maschera di complessi [...] Forse fu perché al Cauchy parve che questa fosse una maschera troppo trasparente che egli abbandonò tal maniera di giustificazione per tentarne un'altra.

to avoid being included in its area, maybe given that the area cuts itself. On the sphere, all of the geodesics starting from one point gather at a second one; will this happen in the second area too? If the geodesics meet at 2 points, they will meet again at infinite points periodically. Might this possibly happen in the second area? Is it possible that an entire spherical lune between two maximum semi-circles cannot take another shape (without cutting itself)? I think this is possible, but maybe not for an entire piece made by half a sphere; what is the limit between this possibility and its impossibility? Then, what is the maximum angle of 2 geodesics in order that, in the second area, it is cut into a second point? You understand that I cannot give a complete meaning to my ideas, since they have not been suggested by the examination of material things. If this is so difficult for the surface 2nd curvature of the sphere, which is so well-known, it will be worse for the third surface deriving from the askew curve (pseudo-sphere).[43]

Genocchi had written to Bellavitis on June 24th, 1873:

I have informed De Tilly about his mistake on the geodesics of the askew curve by mentioning you as befits; I have also protested against his and Hoüel's demonstration about the impossibility of proving Euclid's postulate. I do not know if the Accademia will print my letter, but at any rate, I could print it (and perhaps I will do so) in Boncompagni's Bulletin.[44]

Even if Bellavitis had prepared his announcement of death many years earlier, his sudden demise left a remarkable gap among his friends and correspondents, especially in Genocchi.

6 Conclusions

Man is what has been and what he dreams to become: one of the most sophisticated ways to know a man living in the 1800s is to study the thoughts and feelings that, without any inhibitions, he confessed in the letters to his friends and correspondents.

[43] Dal momento che un pezzo di sfera flessibile può prendere un'altra forma possiamo immaginare che intorno ad essa si attacchino porzioni della stessa sfera, e così in seguito in modo da estendersi indefinitamente; sembra che si possa schivare di rientrare nella superficie stessa, forse peraltro sarà indispensabile che la superficie tagli se stessa. Sulla sfera tutte le geodetiche che partono da un punto si riuniscono in un secondo, avverrà questo anche nella superficie seconda? Se le geodetiche s'incontrano in 2 punti torneranno ad incontrarsi periodicamente in infiniti punti. Sarebbe mai possibile che nella superficie seconda ciò non avvenisse? Forse che un intero fuso sferico compreso tra due semicircoli massimi non può (senza spezzarsi) prendere un'altra forma? Mi pare che ciò possa farsi, ma forse non lo può farsi per un pezzo intero costituito da mezza sfera, qual è il limite tra questa possibilità e l'impossibilità? Così qual è l'angolo massimo di 2 geodetiche perché sulla superf. seconda essa torna a tagliarsi in un secondo punto? Tu capisci che non so dare un completo significato alle mie idee quando non mi furono suggerite dall'esame di esseri materiali. Se ciò mi è tanto difficile per la sup. 2ª piegatura della sfera, così bene conosciuta; peggio sarà per la superficie terza derivante dal trattoide.

[44] ....ho avvertito il De Tilly sul suo errore rispetto alle geodetiche del trattoide citando te come si conveniva; ho anche esposto alcune obiezioni contro la dimostrazione sua e di Hoüel sull'impossibilità di provare il postulato d'Euclide. Non so se l'Accademia vorrà stampare la mia lettera ma in ogni caso potrei (e forse lo farò) stamparla nel Bollettino del Boncompagni.

This is valid for Giusto Bellavitis too. In the decade 1830–1840, he lived within the cultural background of Veneto, which surrounded him to the full, and at the same time, he collected and updated all of the scientific material of that period, reaching many fields of knowledge of his time. Since he loved epistolary correspondence, he encouraged others to write to him. Among his correspondents, we can find doctors, physicians, pharmacists, astronomers, engineers, historians, etc. Thanks to his work at the Istituto Veneto in 1840, as a retired member, and at the University of Padua in 1845, his epistolary relationships spread all over Italy and abroad. In the oldest part of the material that has been analysed, Bellavitis's interests dealt with a wide range of topics, but subsequently, mathematics became more relevant. The principles of geometric calculus linked to the interpretation of imaginary numbers are fundamental. In total autonomy, as a self-taught man, he processed the principles of the method of equipollence, with strong analogies in the contemporary works by H. G. Grassmann, confirming an element of modernity in his studies that was not readily received or accepted, or, at times, even understood by his readers-correspondents. Other themes treated in the letters are infinitesimal calculus, algebra, the classification of curves, and descriptive and projective geometry, all of which were widely developed and discussed.

In conclusion, we can affirm that Bellavitis's correspondence is remarkably interesting for historians. As for the documents that have not yet been copied down, in some cases, reading them has been difficult, because of the authors' handwriting. Unfortunately, some letters have been damaged by the passing of time and neglect, and some parts are missing. However, there are some projects whose preservation and reproduction should allow everyone to enjoy these documents in the future.

References

Alasia, C. 1906. Giusto Bellavitis: sa correspondance scientifique. *Enseignement Mathématique*: 98–117.

Baldassarri, F. 2008. L'evoluzione della matematica a Padova dal 1800 alla stagione d'oro, *I Matematici nell'Università di Padova dal suo nascere al XX secolo*, Padova, Università degli studi di Padova, Dipartimento di Matematica, Esedra editrice.

Belcastro, A., G. Canepa, G. Fenaroli, and M. Modonesi. 2003. Alcuni manoscritti relativi all'insegnamento del calcolo delle probabilità presenti nelle carte di Giusto Bellavitis (1803–1880), *Atti Istituto Veneto di Scienze, Lettere ed Arti. Classe di Scienze Fisiche, Matematiche e naturali* 161(II): 331–370. ISSN: 0392-6680.

Bellavitis, G. 1832. Sopra alcuni teoremi di geometria. *Annali delle Scienze del Regno Lombardo-Veneto* 2 (1832): 241–249.

———. 1837. Metodo delle equipollenze. *Annali delle Scienze del Regno Lombardo-Veneto* 7 (1837): 243–261.

———. 1838. Metodo delle equipollenze. *Annali delle Scienze del Regno Lombardo-Veneto* 8: 17–37, 85–121.

———. 1847. Dimostrazione col metodo delle equipollenze di alcuni teoremi in parte considerati dai signori Bellati e Ridolfi, *Atti dell'Istituto Veneto di Scienze. Lettere ed Arti* 6: 53–59.

———. 1851. Saggio sull'algebra degli immaginarii. *Memorie (IVSLA)* 4 (1852): 243–344.

Beltrami, E. 1868. Saggio di interpretazione della geometria non-euclidea. *Giornale di Mathematiche* VI: 285–315.

Borgato, M.T. 2012. On the historiography of mathematics in Italy. In *The Circulation of Science and Technology: Proceedings of the 4th International Conference of the ESHS, Barcelona*, 18–20 November 2010, ed. A. Roca-Rosell, 40–45. Barcelona: SCHCT-IEC.

Bottazzini, U. 1989. I matematici nell'Italia del Risorgimento: osservazioni sull'emergere di una comunità scientifica. In *La scienza accademica nell'Italia post-unitaria: discipline scientifiche e ricerca universitaria*, ed. V. Ancarani, 37–52. Milano: Franco Angeli.

———. 1994. *Va' pensiero: immagini della matematica nell'Italia dell'Ottocento*. Bologna: il Mulino.

———. 2003. *Il Flauto di Hilbert: storia della matematica*. Torino: UTET.

Canepa, G. 1994. Le carte Bellavitis. *Le Scienze Matematiche nel Veneto dell'Ottocento*, 49–59. Venezia: Istituto Veneto di Scienze, Lettere ed Arti.

———. 2010. Tematiche affini nelle opere di G. Bellavitis e G. Peano. *Giuseppe Peano e la sua Scuola fra matematica, logica e interlingua*, ed. C.S. Roero, 531–543. Atti del Congresso internazionale di studi, Torino, 6–7 ottobre 2008, Deputazione Subalpina di Storia Patria.

———. 2012. L'Istituto Veneto di scienze, lettere e arti e i suoi matematici alle soglie dell'Unità. In *Europa matematica e Risorgimento Italiano*, ed. L. Pepe, 365–375. Bologna: CLUEB.

Canepa, G., and P. Freguglia. 1991. Alcuni aspetti della corrispondenza tra Giusto Bellavitis e Angelo Genocchi. In *Angelo Genocchi e i suoi interlocutori scientifici*, ed. A. Conte and L. Giacardi, 211–219. Torino: Deputazione subalpina di Storia patria.

———. 2009. Il Carteggio Bellavitis-Tardy (1852–1880). In *Materiali per la ricostruzione delle biografie di Matematici italiani dell'Unità*, ed. A. Brigaglia and P. Testi. Milano: Mimesis.

Canepa, G., G. Fenaroli, and P. Freguglia. 2012a. Giusto Bellavitis e le matematiche nel Veneto. In *Europa matematica e Risorgimento Italiano*, ed. L. Pepe, 349–364. Bologna: CLUEB.

Canepa, G., G. Fenaroli, and I. Gambaro. 2012b. The "Rivista di Giornali" (1859–1880) and the circulation of the European mathematical culture in XIX century Italy: A case study. In *The Circulation of Science and Technology: Proceedings of the 4th International Conference of the ESHS*, Barcelona, 18–20 November 2010, ed. A. Roca-Rosell, 593–601. Barcelona: SCHCT-IEC.

———. 2014. The "Rivista di Giornali" (1859–1879) within the Italian and European Mathematical Culture of the 19th century. In *L'émergence de la presse mathématique en Europe au 19ème siècle. Formes éditoriales et études de cas [France, Espagne, Italie et Portugal]*, ed. C. Gerini and N. Verdier, 129–154. Milton Keynes: College Publications.

Caparrini, S. 2003. Early theories of vectors. In *Essays on the History of Mechanics: in Memory of Clifford Ambrose Truesdell and Edoardo Benvenuto*, ed. M. Corradi, A. Becchi, F. Foce, and O. Pedemonte, 179–198. Basel: Birkhäuser.

Carbone, L., R. Gatto, and F. Palladino. 2001. L'Epistolario Cremona-Genocchi (1860–1886). In *La costituzione di una nuova figura di matematico nell'Italia unificata*, Archivio della Corrispondenza degli Scienziati Italiani, vol. 15. Firenze: Leo S. Olschki.

Carnot, L. 1803. Géométrie de Position, Paris, de l'Imprimerie de Crapelet.

Cerroni, C., and G. Fenaroli. 2007. Il carteggio Cremona-Tardy (1860–1886). In *Materiali per la ricostruzione delle biografie di Matematici italiani dell'Unità*, ed. A. Brigaglia and P. Testi. Milano: Mimesis.

Chasles, M. 1830. Mémoire de géométrie pure, sur les systèmes de forces et les systèmes d'aires; et sur les polygones, les polyèdres, et les centres des moyennes distances. *Correspondance Mathématique et Physique* 6 (1830): 92–120.

Chelini, D. 1838. Saggio di geometria analitica, trattata con nuovo metodo. *Giornale Arcadico di Scienze Lettere ed Arti* 75: 80–130, 279–308; 76 (1839), 3–65, 257–286 (Reprint Tipografia delle Belle Arti, Roma).

———. 1849. Sull'uso sistematico de' principi relativi al metodo delle coordinate rettilinee, *Raccolta Scientifica di Palomba* 5: 227–263, 333–374.

———. 1860. *Elementi di meccanica razionale con appendice sui principi fondamentali delle matematiche*. Bologna: Legnani.

———. 1863. Sulla teoria de' sistemi semplici di coordinate e sulla discussione dell'equazione generale di secondo grado in coordinate triangolari e tetraedriche. In *Memorie dell'Accademia delle Scienze dell'Istituto di Bologna*, Serie II, Tomo III, Bologna, Tipografia Gamberoni e Parmeggiani, 3–81.

Dell'Aglio, L. 2008. Dal calcolo geometrico alle forme differenziali. In *Giuseppe Peano e la sua Scuola fra matematica, logica e interlingua*. Atti del Congresso internazionale di studi, Torino, 6–7 ottobre 2008, ed. C. S. Roero, 475–492. Deputazione Subalpina di Storia Patria.

Enea, M.R. 2009. Il carteggio Beltrami-Chelini (1863–1873). In *Materiali per la ricostruzione delle biografie di Matematici italiani dell'Unità*, ed. A. Brigaglia and P. Testi. Milano: Mimesis.

Enea, M.R., and R. Gatto. 2009. Le carte di Domenico Chelini dell'Archivio generale delle Scuole Pie e la corrispondenza Chelini-Cremona (1863–1878). In *Materiali per la ricostruzione delle biografie di Matematici italiani dell'Unità*, ed. A. Brigaglia and P. Testi, 2009. Milano: Mimesis.

Freguglia, P. 1992. *Dalle equipollenze ai sistemi lineari*. Urbino: QuattroVenti.

———. 1994. Il calcolo delle equipollenze di Giusto Bellavitis. In *Le Scienze Matematiche nel Veneto dell'Ottocento*, 11–48. Istituto Veneto di Scienze, Lettere e Arti, Venezia.

———. 1998. Bellavitis's equipollences calculus and his theory of complex numbers. In *Around Caspar Wessel and the Geometric Representation of Complex Numbers*. The Royal Danish Academy of Sciences and Letters, Copenhagen, August 11–15 (C.A. Reitzels Forlag, Copenhagen, 2001, 181–203).

Garibaldi, A.C. 1991. Sui rapporti tra Angelo Genocchi e Placido Tardy. In *Angelo Genocchi e i suoi interlocutori scientifici*, ed. A. Conte and L. Giacardi, 281–292. Torino: Deputazione subalpina di Storia patria.

Giacardi, L. 2000. DBI, Angelo Genocchi, 53.

———. 2012. Pel lustro della scienza italiana e pel progresso dell'alto insegnamento. L'impegno dei matematici risorgimentali. In *Le Università e l'Unità d'Italia (1848–1870)*, ed. A. Ferraresi and E. Signori, 233–254. Bologna: CLUEB.

Grassmann, H.G. 1844. *Die Lineale Ausdehnungslehre ein neuer Zweig der Mathematik*. Leipzig: Wiegand.

Guerraggio, A., and P. Nastasi. 2010. *L'Italia degli scienziati. 150 anni di storia nazionale*. Milano: Bruno Mondadori.

Hamilton, W.R. 1853. *Lectures on Quaternions*. Dublin: Hodges and Smith.

Lacaita, C.G. 2012. Scienza e modernità nelle riviste milanesi dell'800: Il Politecnico e gli Annali di fisica chimica e matematiche. In *Europa matematica e Risorgimento Italiano*, ed. L. Pepe, 267–281. Bologna: CLUEB.

Legnazzi, E.N. 1881. *Commemorazione del Conte Giusto Bellavitis letta il 6 Dicembre 1880 trigesimo della sua morte nell'Aula Magna della R*[a]. Università di Padova, Padova, Prosperini.

Luciano, E., and C.S. Roero. 2012. From Turin to Göttingen: dialogues and correspondence (1879–1923). *Bollettino di Storia delle Scienze Matematiche* 31: 2012.

Möbius, A. 1827. Der barycentrische Calcul. Leipzig: J. A. Barth.

Penso, G. 1978. *Scienziati italiani e Unità d'Italia. Storia dell'Accademia Nazionale delle Scienze detta dei XL*. Roma: Bardi.

Pepe, L. 2012. Esperienze internazionali di matematici e fisici italiani prima dell'Unità. In *Le Università e l'Unità d'Italia (1848–1870)*, ed. A. Ferraresi and E. Signori, 321–332. Bologna: CLUEB.

Roero, C.S. 1999. La Facoltà di Scienze Matematiche Fisiche Naturali di Torino 1848–1998. In *Ricerca, Insegnamento Collezioni scientifiche*, vol. 1, 283–314. Torino: Deputazione Subalpina di Storia Patria.

———. 2012. Politica e istruzione scientifica a Torino nell'età del Risorgimento. In *Europa matematica e Risorgimento Italiano*, ed. L. Pepe, 219–242. Bologna: CLUEB.

Volterra, V. 1902. Betti, Brioschi, Casorati, trois analystes italiens et trois manières d'envisager les questions d'analyse. In *Compte rendu du deuxième Congrès international des mathématiciens tenu à Paris du 6 au 12 août 1900*, 43–57. Paris: Gauthier-Villars

Poincaré-Week in Göttingen, in Light of the Hilbert-Poincaré Correspondence of 1908–1909

Scott A. Walter

Abstract

The two greatest mathematicians of the early twentieth century, David Hilbert and Henri Poincaré transformed the mathematics of their time. Their personal interaction was infrequent, until Hilbert invited Poincaré to deliver the first Wolfskehl Lectures in Göttingen in the spring of 1909. A correspondence ensued, which fixed the content and timing of the lecture series. A close reading of the exchange throws light on what Hilbert wanted Poincaré to talk about, and on what Poincaré wanted to present to Hilbert and his colleagues. To answer the latter question, reference is made to the published version of Poincaré's six talks, with a focus on two of them, concerning the propagation of Hertzian waves, and the theory of relativity.

1 Introduction

At the turn of the twentieth century, two mathematicians, David Hilbert and Henri Poincaré were at the top of their field. As an indicator of their standing, recall that the first two Bolyai Prizes were awarded in sequence to Poincaré and Hilbert in 1905 and 1910, the latter on Poincaré's recommendation (Poincaré 1911). If Hilbert and Poincaré were both bright stars by any measure, their ideas about the nature of mathematics, and in particular, about the relation between mathematics and the phenomenal world, were quite dissimilar. Hilbert's pursuit of a formalist program, launched on the heels of the success of his axiomatization of Euclidean geometry (Hilbert 1899), could find no place in Poincaré's conventionalist worldview. On this basis, Mehrtens (1990) considered Hilbert and Poincaré to be polar opposites, as

S. A. Walter (✉)
University of Nantes, Centre François Viète, Nantes, France
e-mail: scott.walter@univ-nantes.fr

M. T. Borgato et al. (eds.), *Mathematical Correspondences and Critical Editions*,
Trends in the History of Science, https://doi.org/10.1007/978-3-319-73577-1_15

far as mathematical modernism is concerned. A more detailed examination of their respective contributions, however, reveals that both mathematicians contributed significantly to what Gray (2008) calls the "modernist transformation of mathematics".

Historically there are few instances in which Hilbert and Poincaré came into close personal contact. Similarly, there are few instances where their research interests overlapped, in contrast to the case presented by Hilbert's senior colleague in Göttingen, Felix Klein.[1] In the early 1880s, Klein and Poincaré engaged in a race of sorts to map out the new domain of automorphic functions (Gray 2000). The surviving correspondence between Klein and Poincaré is significant, counting twenty-six letters over a span of 15 months ending in September, 1882. The surviving correspondence between Hilbert and Poincaré is less voluminous, with a total of ten letters, seven of which concern the arrangements for Poincaré's delivery of the first Wolfskehl lectures in April, 1909. This private exchange, which has been published on the website Henri Poincaré Papers, and is transcribed here in Sect. 5, offers insight into the Hilbert-Poincaré relationship, and into Poincaré's approach to questions of theoretical physics.

2 Hilbert's Invitation

The summer of 1908 was a dark time for David Hilbert, as he was overcome by depression, and required several months of rest in a sanitorium in the Harz mountains. The cure was effective enough for him to resume his lectures in the fall, and to work out the first proof of Waring's Theorem (Reid 1986, 112). In the fall of 1908, his friend and colleague Hermann Minkowski was working on a theory of electrons, compatible with his successful spacetime theory of the electrodynamics of moving media. Minkowski's four-dimensional spacetime was inspired in part by Poincaré's Lorentz-covariant theory of gravitation, and yet there were significant differences in their conceptualizations of fundamental elements of their theories, having to do with the notion of time. And while Minkowski and Poincaré should have had an occasion to discuss the theory of relativity with Lorentz during the ICM in Rome in the spring (Walter 1999), Poincaré fell ill during the meeting, and was unable to deliver his plenary lecture.

It may be imagined that both Minkowski and Hilbert viewed the Rome ICM as a missed occasion. Whatever the case may be, Hilbert wrote to Poincaré in the fall of 1908, inviting him to deliver a series of lectures under the auspices of the Wolfskehl-Stiftung, which provided a substantial honorarium of 2500 Marks (Hilbert 1910). Poincaré accepted the invitation, but neither Hilbert's letter of invitation nor Poincaré's response has been located. Hilbert thanked Poincaré for accepting his invitation, and suggested that he begin his lecture series in late February or late April (Sect. 5.1). Poincaré replied that he was about to be inducted into the Académie Française, but he did not know precisely when this would take

[1]On Klein, Hilbert, and mathematics in Göttingen, see Rowe (1992).

place (Sect. 5.2). On November 19, Hilbert interpreted Poincaré's response to mean that he preferred to lecture in late April, and expressed his "great interest" in receiving his lecture program.

3 Poincaré's Lecture Program

Up to this point, the Hilbert-Poincaré exchange was perfunctory, but now the terms of the encounter between Poincaré and the Göttingen mathematicians were to be fixed. Sometime after the 19th of November, Poincaré wrote to Hilbert to advise him of his lecture program. This program included two topics: applications of Fredholm's method, and the reduction of Abelian integrals. Poincaré expressed his wish to retain the power to modify his program, "if need be" (Sect. 5.4).

Hilbert may have taken awhile to respond to Poincaré's program. On 12 January, 1909, his good friend Minkowski, who was "a thousand times more a brother" to him, died suddenly from a ruptured appendix (Sect. 5.5). This "bolt from the blue" *(Schlag aus dem heitersten Himmel)* effectively mooted any ideas Hilbert might have had about getting Minkowski and Poincaré together.

Undoubtedly, others in and about Göttingen were still interested in discussing Poincaré's electron theory with Poincaré, among other topics. The only problem was that Poincaré had not offered to lecture on this topic, or others in theoretical physics. Ostensibly in the interest of enticing physicists, astronomers, and logicians to attend what Hilbert dubbed "Poincaré-Week", Hilbert asked Poincaré to add two topics to those he had already proposed. One might be on theoretical physics or astronomy, the other with a "logico-philosophical coloration".

Poincaré's response must have come as a surprise to Hilbert, as Poincaré claimed that his original proposition concerning Fredholm's equation included topics from both theoretical physics (i.e., Hertzian waves), and astronomy (i.e., the theory of tides). He agreed to add a lecture on a logico-philosophical topic, stemming from a forthcoming paper on Richard's paradox (Poincaré 1909d).

Hilbert's response to Poincaré's genial proposition has not been located, but he must have been in agreement, because Poincaré wrote back (Sect. 5.7) with a list of five lectures:

1. On the reduction of Abelian integrals
2. On applications of Fredholm's method
3. The theory of tides and Fredholm's equation
4. Hertzian waves and Fredholm's equation
5. On the notion of transfinite cardinal numbers

Whether Hilbert was pleased with Poincaré's program or not is difficult to know, although he had successfully negotiated the inclusion of topics from theoretical physics and mathematical logic. Poincaré's focus on Fredholm's equation was surely welcomed by Hilbert, who by 1904 had seen therein the possibility of developing a new framework for the study of boundary value problems, which would lead eventually to the concept of a Hilbert space (Archibald and Tazzioli 2014).

The correspondence between Hilbert and Poincaré tells us much we didn't know about Poincaré's lecture series, but it also raises a few new questions. In particular, one wonders why Poincaré added a sixth lecture to the five he announced to Hilbert, and why this sixth lecture was not on Fredholm's equation or mathematical logic, but on the theory of relativity? Perhaps Poincaré took to heart Hilbert's suggestion to add a topic on theoretical physics or astronomy, as this sixth lecture included both.

4 Historical Upshot of the Hilbert-Poincaré Exchange in 1908–1909

Hilbert's invitation to Poincaré on behalf of the Wolfskehl-Stiftung had multiple objectives. It was meant to reinforce ties between French and German mathematicians, as Hilbert's opening speech made clear from the outset.[2] Beyond this explicit and laudable goal, Poincaré's visit was designed to stimulate research by the members of the Göttingen Mathematical Society, and among German mathematicians in general. Hilbert may have understood the invitation as a way of encouraging Poincaré to take an interest in the ongoing research of GMS members, himself included, via informal exchanges.

On all three of these counts, Poincaré-Week must be rated at least a relative success. The credit here belongs in part to Poincaré, who took the risk of presenting work-in-progress. His lecture on the diffraction of Hertzian waves is one example; the lecture on the new mechanics is another. These two lectures will be discussed in what follows; for an overview of all six lectures, see Gray (2013, 416).

The topic of the propagation of Hertzian waves was one that was well-chosen for Göttingen. Poincaré's interest in Hertzian waves may be dated from his correspondence with Hertz in 1890 (Walter et al. 2007, § 2-30-1); he lectured on Hertzian waves at the Sorbonne (Poincaré 1894), and was particularly interested in explaining wave propagation over great distances (Poincaré 1903). In Göttingen, Max Abraham was a leading expert on Hertzian waves, as was another member of Poincaré's audience, the former assistant to Felix Klein, and since 1906, professor of theoretical physics in Munich, Arnold Sommerfeld. Poincaré may have noticed Sommerfeld's long paper on the propagation of Hertzian waves (Sommerfeld 1909), published in the leading German physics journal *Annalen der Physik* in the first week of February, 1909.

Poincaré had promised Hilbert a lecture on Hertzian waves as an application of Fredholm's method, and the lecture that he delivered in Göttingen fit the bill precisely. His lecture did not stray far from the content of a triplet of notes Poincaré published in the *Comptes rendus* of the Paris Academy of Science on 22 February, 29 March, and 13 April, 1909 (Poincaré 1909b,e,f).

[2]Nachlass Hilbert 579, Handschriftenabteilung, Niedersächsiche Staats- und Universitätsbibliothek.

Upon his return to Paris, Poincaré continued to work on the problem of wave propagation, and published again on 7 June (Poincaré 1909c). The published version of Poincaré's lecture thus represents a work-in-progress, up to and including the results contained in the notes published in the *Comptes rendus* on 13 April, 1 week before the beginning of his Wolfskehl lectures.

The question then arises of the effect, if any, that Hilbert's invitation had on Poincaré's engagement in 1909 with the problem of wireless wave propagation over a curved surface. It is plausible that Sommerfeld's publication renewed Poincaré's interest in the problem, inasmuch as he found it to be a good candidate for the application of Fredholm's method. Hilbert's interest in applying and extending Fredholm's method may have been a consideration in Poincaré's topic choice, as well, but Poincaré does not appear to have been conversant with Hilbert's results in this area.

For example, in Poincaré's first note of the year 1909, entitled "On some applications of Fredholm's method", he acknowledged his neglect of one of Hilbert's results:

> I take this opportunity to make amends for an involuntary omission that Mr. Picard pointed out to me.
> In a recent Note, I pointed out a series of results relative, respectively, to the cases in which the kernel of Fredholm's equation becomes infinite of order $< \frac{1}{2}, < \frac{2}{3}, < \frac{3}{4}, \ldots$; the first of these results had already been obtained via a different method by Mr. Hilbert.[3] (Poincaré 1909g)

Poincaré referred here to his note of 21 December (Poincaré 1908), and to Hilbert's first communication to the Göttingen Academy of Science on "basic features of a general theory of linear integral equations" (Hilbert 1904), reedited in Hilbert (1912).

As for the physical question of Hertzian wave propagation, Poincaré concluded his Wolfskehl lecture on this topic with the observation that intercontinental wireless telegraphy was not ruled out by his mathematical analysis. This was surely a welcome result, given that Marconi had succeeded in sending a signal by wireless from Poldhu to St. John's, Newfoundland in 1901. However, Poincaré soon realized that his analysis was faulty; he corrected himself in a subsequent paper, finished on 15 October, 1909 (Poincaré 1910d), and according to the corrected calculation, long-distance telegraphy was no longer possible! He did not correct the error in the proofs of his Wolfskehl lecture, published the next year (Poincaré 1910a), but appended a short note in French, alerting the reader to his error.

Sommerfeld took note of Poincaré's contribution, and set his student H. W. March the task of using his own approach to solve the same problem, i.e., applying an integral expansion in the case of Hertzian waves propagating over a spherical

[3]"Je profite de l'occasion pour réparer un oubli involontaire qui m'a été signalé par M. Picard. Dans une Note récente, j'ai signalé une série de résultats relatifs respectivement aux cas où le noyau de l'équation de Fredholm devient infini d'ordre $< \frac{1}{2}, < \frac{2}{3}, < \frac{3}{4}, \ldots$; le premier de ces résultats avait déjà été obtenu par une autre voie par M. Hilbert."

conductor. March's result disagreed with that of Poincaré, and in the month of March 2012, Sommerfeld wrote to Poincaré to see if he could find the reason for the divergence. Poincaré wrote back to inform Sommerfeld that he had located the point of divergence: March's integration of Hankel's function was incorrect, due to a defective asymptotic expansion; see Poincaré to Sommerfeld, in Walter et al. (2007), § 2-54-1. In the note Poincaré communicated to the Paris Academy on this topic, he observed that, once March's error was corrected, the result of March's analysis was identical to his own, such that his earlier result had been "confirmed" by Sommerfeld's student (Poincaré 1912). He remarked further that the latest measurements by Louis Austin off the coast of Virginia of the power of electric-arc-generated wireless waves pointed to a serious disagreement with his theory, such that there was "something here to discover".[4]

It appears that Poincaré-Week facilitated an exchange between Sommerfeld and Poincaré on the topic of wireless wave propagation, which was mutually beneficial. Much the same may be said of the sixth and final lecture of Poincaré-Week, on the new mechanics of relativity.

Poincaré's decision to add a sixth lecture to the program he had announced to Hilbert circa March, 1909, is not easily understood. While Hilbert had specifically requested that Poincaré treat a topic on mathematical physics or astronomy (Sect. 5.5), his lecture on Hertzian waves surely satisfied this desideratum. Why then did Poincaré choose to speak on the new mechanics in Göttingen?

A tentative answer to this question may be formulated by recalling the dual contexts of relativity theory in Paris and Göttingen.[5] In both places, in the early years of the twentieth century, theorists turned to theories of the electron in order to address the experimental results of electron beam deflection by crossed magnetic fields, and in magneto-optics, electrodynamics of moving bodies, and black-body radiation. In 1905, Poincaré proposed a modification of Lorentz's electron theory, in which the laws of all physical interactions are governed by covariance with respect to what he called the "Lorentz group". The law of gravitation was clearly a potential spoiler for his theory, but Poincaré showed that as long as the propagation velocity of gravitation is no greater than that of light, this law, too, could be Lorentz-covariant (Walter 2007).

In order to prove the latter result, Poincaré introduced a four-dimensional space with one imaginary dimension, which he used to form Lorentz-invariant quantities. Minkowski noticed this novel method of Poincaré's, and realized that it could be generalized into a four-dimensional vector formalism. Furthermore, the geometry of phenomenal space could be taken to be the geometry of these four-dimensional vectors. He announced this new "spacetime" with great pomp at the meeting of the German Association in Cologne, in September, 1908 (Walter 2008, 2010).

---

[4]Poincaré was right about this; for the subsequent history of long-distance wireless-wave propagation, see Yeang (2013).

[5]For an overview of research on relativity in Paris, see Walter (2011), and for research in Göttingen, see Pyenson (1979), Walter (1999), and Corry (2004).

Although Minkowski had earlier acknowledged Poincaré's fundamental contribution to relativity theory, he failed to do so in his Cologne lecture, prompting a worried reaction from Poincaré's allies (Walter 1999). As mentioned above, Minkowski had no further occasion to characterize Poincaré's contribution to relativity theory, as he died in January, 1909. The printed version of his Cologne lecture appeared 3 weeks after his death (Minkowski 1909), and it is not unlikely that Poincaré had the occasion to read it before delivering his Wolfskehl lectures.

Poincaré's lecture on the new mechanics bears no explicit reference to the work of Einstein or Minkowski. The order of arguments resembles that of the plenary lecture, longer and more detailed than the Wolfskehl talk, that Poincaré delivered on 3 August to the French Association for the Advancement of Science in Lille (Poincaré 1909a). The latter circumstance does not explain fully why Poincaré broached the topic of relativity in Göttingen in April, 1909.

I'd like to suggest that in the early months of 1909, Poincaré realized the potential sway of the Einstein-Minkowski theory of relativity, and sought to defend his own theory in the city that was just beginning to be identified with the relativity revolution. As an indicator of this identification, recall that the *second* Wolfskehl lecture series was assigned to another founder of relativity theory, H.-A. Lorentz, who prefaced his remarks on relativity with the following words:

> It is a particularly welcome task for me to discuss the Einsteinian principle of relativity here in Göttingen, where Minkowski worked.[6] (Lorentz 1913, 74)

Lorentz went on to mention two other Göttingen scientists who had contributed powerfully to the construction of the "mathematical side" of relativity theory: Max Abraham and Arnold Sommerfeld, both of whom attended Poincaré's lectures in 1909.

In addition to the above considerations of place and time, the content of Poincaré's Wolfskehl lecture on new mechanics features two conceptual novelties, which may be linked to these considerations. Prior to Poincaré-Week, Poincaré had promoted a view of relativity in which clocks are always at rest with respect to the ether. This view stands in contrast to the theories backed by Einstein and Minkowski, in which clocks in uniform motion are not distinguished from clocks at rest. The "proper time" *(Eigenzeit)* of a particle in motion, in Minkowski's spacetime theory, is just the time read by a comoving ideal clock, and this time will differ in general from the time read by non-comoving ideal clocks.

Once in Göttingen, Poincaré decided to allow clocks to move. In order to preserve the principle of relativity, this meant that the time read by clocks in motion is deformed with respect to the time read by clocks at rest with respect to the ether. To drive home the idea of time deformation for his audience, Poincaré introduced two observers *A* and *B* in relative motion, equipped not just with timekeepers, but with wireless transmitters and receivers. In keeping with his third lecture on wireless

[6]"Die Einsteinsche Relativitätsprinzip hier in Göttingen zu besprechen, wo Minkowski gewirkt hat, erscheint mir eine besonders willkommene Aufgabe."

wave propagation, Poincaré equipped his observers with the means of transmitting time-stamped position data on the fly. In spite of this high-technology equipment, Poincaré's comoving observers were still unable to detect their absolute motion:

> *A* can believe he is at rest, and *B*'s apparent speed will be 400000 km/s. If *A* knows the new mechanics he will say to himself: "*B* has a speed that he cannot attain, so it must be that I, too, am in motion." It seems that he could determine his absolute situation. But he would have to be able to observe *B*'s motion. To make this observation, *A* and *B* begin by setting their watches, then *B* sends telegrams to *A* indicating his successive positions; putting these signals together, *A* can give an account of *B*'s motion, and trace its curve. Well, the signals propagate at the speed of light; the watches marking apparent time vary at every instant and it all will go down as if *B*'s watch were fast.[7] (Poincaré, 1910b, 54–55)

The tabulation of telemetric data would, in principle if not yet in practice, show that the watches of the two observers in relative motion did not run at the same rate.

In fact, in the circumstances described by Poincaré, relativity requires that *B*'s watch retard with respect to that of *A*. The sign error notwithstanding, Poincaré's Wolfskehl lecture on the new mechanics was his first-ever invocation of the deformation of time due to translation (Walter 2014).

In summary, Poincaré's Wolfskehl lectures on the new mechanics and on Hertzian wave propagation reflect a possible awareness on his part of recent advances in these areas by Minkowski and Sommerfeld, respectively. In light of the subsequent history of these two topics, both Hilbert and Poincaré had reason to be satisfied with the lecture series. Their epistolary exchange in 1908–1909 gives us a better idea of Hilbert's motivation in inviting Poincaré to Göttingen, and of Poincaré's intentions in accepting the invitation.

5 Annex: The Hilbert-Poincaré Correspondence, 1908–1909

5.1 Hilbert to Poincaré

Göttingen d. 6.11.08

Sehr geehrter Herr Kollege

Ihre Zusage hat uns alle hoch erfreut und auch in der mathematischen Gesellschaft, in der ich gestern Ihren Brief mitteilte, wurde allgemein Freude ausgedrückt.

[7]"*A* peut se croire au repos et la vitesse apparente de *B* sera, pour lui, 400000 kilomètres. Si *A* connaît la mécanique nouvelle il se dira: *B* a une vitesse qu'il ne peut atteindre, c'est donc que moi aussi je suis en mouvement. Il semble qu'il pourrait décider de sa situation absolue. Mais il faudrait qu'il puisse observer le mouvement de *B* lui-même; pour faire cette observation *A* et *B* commencent par régler leurs montres, puis *B* envoie à *A* des télégrammes pour lui indiquer ses positions successives; en les réunissant, *A* peut se rendre compte du mouvement de *B* et tracer la courbe de ce mouvement. Or les signaux se propagent avec la vitesse de la lumière; les montres qui marquent le temps apparent varient à chaque instant et tout se passera comme si la montre de *B* avançait."

Was nun die Zeit Ihres Herkommens betrifft, so möchte wir als das Optimum bezeichnen, wenn Sie Ihre Vorträge innerhalb der Zeitraumes

27 Febr. bis 10 März

verlegen könnten; allenfalls liesse sich dieser Spielraum noch um einige Tage am Anfange und Ende erweitern. Sollte Ihnen diese Zeit nicht möglich sein, so müssten wir die letzte Aprilwoche (Anfang des Sommersemesters) in Aussicht nehmen.

Vorbereitungen unsererseits bedarf es ja nicht; aber, da wir die Zeit, sowie die Gegenstände Ihrer Vorträge gern zeitig genug in den *Jahresberichten der Deutschen Mathematikervereinigung* bekannt machen und auch unseren auswärtigen Freunden und Kollegen mitteilen möchten, so bitte ich Sie um Mitteilung Ihrer Entschlüsse, sobald Ihnen dies möglich ist.

Mit den besten Grüssen

Hochachtungsvoll und ergebenst

Hilbert

ALS 2p. Private collection, Paris 75017.

5.2 Poincaré to Hilbert

[Between 6 and 18.11.1908]

Mon cher Collègue,

Je suis très flatté de votre proposition et je suis très disposé à l'accepter. Seulement il y a un obstacle. Je ne sais si je serai libre à l'époque que vous fixez.[8] L'Académie française n'a encore choisi ni le jour de ma réception, ni celui des élections. Mais tout fait prévoir que ce sera à la fin de février ou au commencement de mars.[9]

Pourriez-vous me dire entre quelles limites on pourrait faire varier la date de mon voyage à Göttingen ; si au besoin on pourrait le remettre au semestre d'été, et à quel moment il convient que je vous donne une réponse définitive.

Veuillez agréer, mon cher Collègue, l'assurance de mes sentiments affectueux et de mon admiration pour votre talent. Seriez-vous assez bon pour me rappeler au souvenir de M. Klein.

Votre bien dévoué Collègue,

Poincaré

ALS 2p. Cod. Ms. D. Hilbert 312, Niedersächsische Staats- und Universitätsbibliothek, Handschriftenabteilung.

---

[8]Hilbert a suggéré par lettre à Poincaré du 25.02.1909 que la série de conférences ait lieu entre le 27.02 et le 10.03.1909 (Sect. 5.1).

[9]Poincaré souhaitait éviter un conflit d'emploi du temps avec sa réception à l'Académie française. Sa réception a eu lieu le 28.01.1909, alors que la série de conférences Wolfskehl a eu lieu du 22 au 28 avril, 1909 (Poincaré 1910c).

5.3 Hilbert to Poincaré

Göttingen den 19 Nov. 08.

Sehr geehrter Herr Professor.

Wir rechnen nun darauf, dass Sie Ihre Vorträge in die Woche vom 22–28sten April nächsten Jahres verlegen, da diese Tage für uns wegen des Beginnes der Sommersemester die beste Zeit sind.[10] Ich sehe mit grossem Interesse der Mitteilung Ihrer Programmes entgegen.

Mit ergebensten Grüssen

Ihr

Hilbert

ALS 1p. Cod. Ms. D. Hilbert 312, Niedersächsische Staats- und Universitätsbibliothek, Handschriftenabteilung.

5.4 Poincaré to Hilbert

[After 19.11.1908]

Mon cher Collègue,

Voici les titres des sujets que je me propose de traiter.

Sur quelques applications de la méthode de Fredholm.

Sur la réduction des intégrales abéliennes.

Je suppose que je reste libre de modifier ce programme s'il y a lieu.

Je serai très heureux d'avoir l'occasion de vous voir.

Veuillez transmettre mes compliments à M. Klein et croire à ma sincère amitié et à mon entier dévouement,

Poincaré

ALS 2p. Cod. Ms. D. Hilbert 312, Niedersächsische Staats- und Universitätsbibliothek, Handschriftenabteilung.

5.5 Hilbert to Poincaré

Göttingen den 25.2.09

Hochgeehrter Herr Kollege,

Wie ich Ihnen schon mitzuteilen mir erlaubte, beabsichtigen wir zu der Göttinger ‚Poincaré-Woche' 22–28 April, auch einige Nicht-Göttinger Mathematiker heranzuziehen. Würde es Ihnen vielleicht möglich sein, auch ein Thema aus der mathematischen Physik oder der Astronomie und ein solcher Logisch-philosophischer

[10]In his previous letter to Hilbert (Sect. 5.2), Poincaré informed his German colleague that his upcoming reception at the *Académie Française* conflicted with the period Hilbert had proposed for the lecture series.

Färbung zu behandeln? Wir könnten in diesem Falle auch die betreffenden Göttinger Fachkollegen zu Ihren Vorträgen einladen.

Auch beabsichtigen wir an einem oder anderen Abend jener Woche eine Sitzung der hierigen mathematischen Gesellschaft abzuhalten, wo wir dann unsererseits nach unseren Kräften etwas zum Besten geben könnten.

Endlich ist für den 30sten April, dem Geburtstage von Gauss, in dem benachbarten Dransfeld auf dem „hohen Hagen" (der einen Ecke des Gaussischen geradlinigen Dreieckes, für welches er die Winkelsumme π beobachtet hat) die Einweihung einer Gaussturmes projektiert. Ihre Anwesenheit dabei wäre dringend wünschenswert.[11]

Leider sind wir – ganz besonders aber ich – durch den vor kurzem erfolgten Tod Minkowski's in tiefe Trauer versetzt. Ich habe an ihm meinen liebsten und treuesten Jugendfreund, der mir tausendmal mehr wie ein Bruder war, ganz plötzlich und jäh (durch Blinddarm-Entzündung) verloren. Es war ein Schlag aus dem heitersten Himmel.[12]

Mit den besten Grüssen
Hochachtungsvoll
Hilbert

ALS 3p. Collection particulière, Paris 75017.

5.6 Poincaré to Hilbert

[Après le 25.02.1909]

Mon cher Collègue,

Mon programme sur les applications de l'équation de Fredholm comprend des applications à la Physique Mathématique et à l'Astronomie, en particulier à l'étude

[11]The cornerstone-laying ceremony of the Gauss monument was scheduled to take place on the anniversary of Gauss's birthday, on April 30, 1909 (see the notice in the *Jahresbericht der deutschen Mathematiker-Vereinigung* 17, 1908, 121).

According to a story popular in Göttingen at the time, Carl Friedrich Gauss (1777–1855), director of the Göttingen Observatory, and professor of mathematics at the University of Göttingen, tested the Euclidicity of space in the 1820s, by employing his heliotrope to measure the angle sum of a triangle formed by the mountaintops of Brocken, Inselsberg and Hohenhagen (Scholz 2004). In 1908, Felix Klein solicited donations from astronomical and mathematical societies around the world, in order to build a tower on the Hohenhagen commemorating the work of the eminent Göttingen geometer.

Poincaré accepted Hilbert's invitation to attend the cornerstone-laying ceremony (Sect. 5.6). His presence at the inauguration of the Gauss monument was poignant, in light of what Study (1914, 117) later called the polemic between Poincaré and the writings of Gauss, Riemann and Helmholtz. For these authors the geometry of space was in some sense empirically determined, a position contested by Poincaré.

[12]Hermann Minkowski (1864–1909) died in Göttingen on 12.01.1909.

des marées et à celle des ondes hertziennes.[13] Je pourrais aussi, si vous le désirez, prendre comme sujet relatif aux ensembles, une note qui va prochainement paraître dans les *Acta Mathematica*.[14]

Je pourrai assister à l'inauguration de la tour de Gauss.

Je suppose que je puis faire mes conférences en français; s'il en était autrement, je pourrais m'en tirer, mais je vous prierais de m'en avertir un certain temps d'avance.

Votre bien dévoué Collègue,

Poincaré

ALS 2p. Cod. Ms. D. Hilbert 312, Niedersächsische Staats- und Universitätsbibliothek, Handschriftenabteilung.

5.7 Poincaré to Hilbert

[Ca. 03.1909]

Mon cher Collègue,

Merci de votre lettre. Nous pourrions alors prendre pour titres des diverses communications.

Sur la Réduction des Intégrales Abéliennes.

Sur quelques applications analytiques de la méthode de Fredholm.

La théorie des Marées et l'équation de Fredholm.

Les ondes hertziennes et l'équation de Fredholm.

Sur la notion de nombre cardinal transfini.[15]

Maintenant il y a un point sur lequel je désire attirer votre attention. Je suis encore sous le coup de l'accident qui m'a frappé l'année dernière à Rome et je suis impérieusement obligé à certaines précautions. Je ne puis boire ni vin, ni bière, mais seulement de l'eau. Je ne puis assister à un banquet, ni à un repas prolongé.[16]

Cette circonstance m'avait fait hésiter à accepter votre invitation, mais j'ai pensé que vous sauriez arranger les choses en conséquence.

Je pense qu'il y a moyen de voir nos collègues dans d'autres circonstances que dans des banquets et j'espère dans ces conditions, avoir le plaisir de faire leur connaissance. Je serai enchanté en particulier d'avoir l'occasion de vous voir.

Votre bien dévoué Collègue,

Poincaré

ALS 2p. Cod. Ms. D. Hilbert 312, Niedersächsische Staats- und Universitätsbibliothek, Handschriftenabteilung.

---

[13]Poincaré répond au souhait exprimé par Hilbert (Sect. 5.5) qu'il augmente son programme, en traitant une question de physique mathématique ou d'astronomie, et et en abordant un sujet logico-philosophique.

[14]Il s'agit des remarques sur le paradoxe de Jules Richard; voir Poincaré (1909d).

[15]Poincaré a prononcé des conférénces sur les cinq sujets annoncés ici, ainsi que sur un sixième sujet, intitulé "La mécanique nouvelle." Uniquement cette sixième conférence sera publiée en français, les autres paraîtront en allemand; voir Poincaré (1910c).

[16]Lors du Congrès international des mathematicians tenu à Rome en avril 1908, Poincaré a eu une malaise, liée alors par les médicins à une hypertrophie du prostate; voir Darboux (1916, LXVI).

References

Archibald, T., and R. Tazzioli. 2014. Integral equations between theory and practice: the cases of Italy and France to 1920. *Archive for History of Exact Science* 68: 547–597.

Corry, L. 2004. *David Hilbert and the Axiomatization of Physics (1898–1918): From Grundlagen der Geometrie to Grundlagen der Physik.* Dordrecht: Kluwer.

Darboux, G., N. E. Nörlund, and E. Lebon, eds. 1916. *Œuvres d'Henri Poincaré, Volume 2.* Paris: Gauthier-Villars.

Gray, J. 2000. *Linear Differential Equations and Group Theory from Riemann to Poincaré*, 2nd ed. Boston: Birkhäuser.

—. 2008. *Plato's Ghost: The Modernist Transformation of Mathematics.* Princeton: Princeton University Press.

—. 2013. *Henri Poincaré: A Scientific Biography.* Princeton: Princeton University Press.

Hilbert, D. 1899. *Die Grundlagen der Geometrie.* Leipzig: Teubner.

—. 1904. Grundzüge einer allgemeinen Theorie der linearen Integralgleichungen (erste Mitteilung). *Nachrichten von der Königlichen Gesellschaft der Wissenschaften zu Göttingen, mathematisch-physikalische Klasse*: 49–91.

—. 1910. Wolfskehl-Stiftung. *Nachrichten von der Königlichen Gesellschaft der Wissenschaften zu Göttingen; Geschäftliche Mitteilungen*: 13.

—. 1912. *Grundzüge einer allgemeinen Theorie der linearen Integralgleichungen.* Leipzig/Berlin: Teubner.

Lorentz, H.A. 1913 Das Relativitätsprinzip und seine Anwendung auf einige besondere physikalische Erscheinungen. In *Das Relativitätsprinzip; Eine Sammlung von Abhandlungen.* ed. O. Blumenthal. Fortschritte der mathematischen Wissenschaften in Monographien, Vol. 2, 74–89. Leipzig: Teubner.

Mehrtens, H. 1990. *Moderne–Sprache–Mathematik.* Frankfurt am Main: Suhrkamp.

Minkowski, H. 1909. Raum und Zeit. *Physikalische Zeitschrift* 10 (3): 104–111.

Poincaré, H. 1894. *Les oscillations électriques.* Paris: Carré et Naud.

—. 1903. Sur la diffraction des ondes électriques : à propos d'un article de M. Macdonald. *Proceedings of the Royal Society of London* 72 : 42–52.

—. 1908. Remarques sur l'équation de Fredholm. *Comptes rendus hebdomadaires de l'Académie des sciences de Paris* 147 (25): 1367–1371.

—. 1909a. La mécanique nouvelle. *Revue scientifique* 12: 170–177.

—. 1909b. Les ondes hertziennes et l'équation de Fredholm. *Comptes rendus hebdomadaires de l'Académie des sciences de Paris* 148 (8): 449–453.

—. 1909c. Les ondes hertziennes et l'équation de Fredholm. *Comptes rendus hebdomadaires de l'Académie des sciences de Paris* 148 (23): 1488–1490.

—. 1909d. Réflexions sur les deux notes précédentes. *Acta mathematica* 32: 195–200.

—. 1909e. Sur la diffraction des ondes hertziennes. *Comptes rendus hebdomadaires de l'Académie des sciences de Paris* 148 (13): 812–817.

—. 1909f. Sur la diffraction des ondes hertziennes. *Comptes rendus hebdomadaires de l'Académie des sciences de Paris* 148 (15): 966–968.

—. 1909g. Sur quelques applications de la méthode de M. Fredholm. *Comptes rendus hebdomadaires de l'Académie des sciences de Paris* 148 (3): 125–126.

—. 1910a. Anwendung der Integralgleichungen auf Hertzsche Wellen. In *Sechs Vorträge über ausgewählte Gegenstände aus der reinen Mathematik und mathematischen Physik*, 23–31. Leipzig/Berlin: Teubner.

—. 1910b. La mécanique nouvelle. In *Sechs Vorträge über ausgewählte Gegenstände aus der reinen Mathematik und mathematischen Physik*, 51–58. Leipzig/Berlin: Teubner.

—. 1910c. *Sechs Vorträge über ausgewählte Gegenstände aus der reinen Mathematik und mathematischen Physik.* Leipzig/Berlin: Teubner.

—. 1910d. Sur la diffraction des ondes hertziennes. *Rendiconti del Circolo Matematico di Palermo* 29: 169–259.

—. 1911. Rapport sur le prix Bolyai (décerné à M. David Hilbert). *Rendiconti del Circolo Matematico di Palermo* 31: 109–132.
—. 1912. Sur la diffraction des ondes hertziennes. *Comptes rendus hebdomadaires de l'Académie des sciences de Paris* 154: 795–797.
Pyenson, L. Physics in the shadow of mathematics: the Göttingen electron-theory seminar of 1905. *Archive for History of Exact Sciences* 21(1) (1979): 55–89.
Reid, C. 1986. *Hilbert-Courant*. Berlin: Springer.
Rowe, D.E. 1992. *Felix Klein, David Hilbert, and the Göttingen mathematical tradition*. Ph.D. dissertation, City University of New York, New York.
Scholz, E. 2004. C. F. Gauß' Präzisionsmessungen terrestrischer Dreiecke und seine Überlegungen zur empirischen Fundierung der Geometrie in den 1820er Jahren. In *Form, Zahl, Ordnung: Studien zur Wissenschafts- und Technikgeschichte, Festschrift für Ivo Schneider zum 65. Geburtstag*. ed. by R. Seising, M. Folkerts, and U. Hashagen, 355–380. Boethius 48. Wiesbaden: Franz Steiner Verlag.
Sommerfeld, A. 1909. Über die Ausbreitung der Wellen in der drahtlosen Telegraphie. *Annalen der Physik* 333 (4): 665–736.
Study, E. 1914. *Die realistische Weltansicht und die Lehre vom Raume: Geometrie, Anschauung und Erfahrung*. Braunschweig: Vieweg.
Walter, S.A. 1999. Minkowski, mathematicians, and the mathematical theory of relativity. In *The Expanding Worlds of General Relativity*, ed. H. Goenner, J. Renn, T. Sauer, and J. Ritter. Einstein Studies, Vol. 7, 45–86. Boston/Basel: Birkhäuser.
—. 2007. Breaking in the 4-vectors: the four-dimensional movement in gravitation, 1905–1910. In *The Genesis of General Relativity, Volume 3*, ed. J. Renn, and M. Schemmel, 193–252. Berlin: Springer.
—. 2008. Hermann Minkowski's approach to physics. *Mathematische Semesterberichte* 55(2): 213–235.
—. 2010. Minkowski's modern world. In *Minkowski Spacetime: A Hundred Years Later*, ed. V. Petkov, Fundamental Theories of Physics, Vol. 165, 43–61. Berlin: Springer.
—. 2011. Henri Poincaré, theoretical physics and relativity theory in Paris. In *Mathematics Meets Physics: A Contribution to Their Interaction in the 19th and the First Half of the 20th Century*. ed. K.-H. Schlote, and M. Schneider, 213–239. Frankfurt am Main: Harri Deutsch.
—. 2014. Poincaré on clocks in motion. *Studies in History and Philosophy of Modern Physics* 47 (1): 131–141.
Walter, S.A., Bolmont, E., and Coret, A., eds. 2007. *La correspondance d'Henri Poincaré, Volume 2: La correspondance entre Henri Poincaré et les physiciens, chimistes et ingénieurs*. Basel: Birkhäuser.
Yeang, C.-P. 2013. *Probing the Sky with Radio Waves: From Wireless Technology to the Development of Atmospheric Science*. Chicago: University of Chicago Press.

The Correspondence of Bartel Leendert van der Waerden (1903–1996)

Erwin Neuenschwander

Abstract

The aim of this paper is to present an overview of the extensive correspondence of B. L. van der Waerden housed at the Library of the Eidgenössische Technische Hochschule (ETH) in Zurich, and to provide some first results and conclusions of our ongoing partial edition of these papers. In particular, we present a short biography of van der Waerden based on the numerous books and articles about him that have recently been published, discuss some of the conclusions about his work that can be drawn from the letters archived in Zurich, and give a list of the most extensive correspondences in that Zurich collection.

1 Introduction

van der Waerden was one of the most influential mathematicians of the twentieth century. For decades, his textbook *Moderne Algebra* (van der Waerden 1930, 1931) was considered a standard work that introduced many generations of students to modern algebra as it had arisen from the works of R. Dedekind, H. Weber, D. Hilbert, E. Steinitz, E. Artin, and Emmy Noether. Moreover, van der Waerden was a polymath. He made contributions to practically all branches of mathematics, ranging from algebraic geometry and abstract algebra to number theory, topology, axiomatic geometry, analysis and probability theory right through to applied mathematics and quantum mechanics. Equally impressive is his work in the field of the history of science, to which he contributed seven books and nearly 200 articles over a period of fifty years.

E. Neuenschwander (✉)
Institute of Mathematics, University of Zurich, Zurich, Switzerland
e-mail: neuenschwander@math.uzh.ch

M. T. Borgato et al. (eds.), *Mathematical Correspondences and Critical Editions*, Trends in the History of Science, https://doi.org/10.1007/978-3-319-73577-1_16

On the occasion of van der Waerden's 90th birthday, and likewise at his death, a relatively large number of mostly smaller articles appeared, along with a bibliography (Top and Walling 1994), but they were often flawed and incomplete. It took more than ten years before two major monographs, by Alexander Soifer and Rüdiger Thiele, were published that analyzed van der Waerden's life and work in greater depth. Soifer undertook, in his *The Mathematical Coloring Book* (2009), a critical analysis of van der Waerden's early years up to the beginning of his Zurich period, based on extensive archival research. He expanded his work into a recently published biography entitled *The Scholar and the State: In Search of van der Waerden* (2015). Thiele (2009) published a richly illustrated assessment of van der Waerden's Leipzig years, with an appendix listing his Leipzig lectures and publications, the dissertations supervised by van der Waerden at Leipzig University, but also including excerpts from the files of the Ministry of Education in Dresden concerning van der Waerden. In autumn 2011, the well-researched PhD thesis by Martina R. Schneider also appeared, subsequently revised for publication, entitled *Zwischen zwei Disziplinen. B.L. van der Waerden und die Entwicklung der Quantenmechanik.*

However, the majority of these publications do not or only rarely make use of the huge correspondence left behind by van der Waerden, which is now housed at the Library of the Eidgenössische Technische Hochschule (ETH) in Zurich. This correspondence consists of around 15,000 letters, stretching from 1943 until his death in 1996; van der Waerden's earlier papers were destroyed during a bombing raid in Leipzig on December 4, 1943. In order to make this important collection known to a wider public, I am currently preparing a partial edition of van der Waerden's correspondence, together with a critical assessment of his scientific and administrative activities during his Zurich years. The present paper is based on my talk at the 6th International Conference of the European Society for the History of Science in Lisbon in September 2014, and gives a brief survey of ongoing work on this project since 2010. It first presents a short account of van der Waerden's life and work, and then provides an overview of his correspondence.

2 A Short Biography of B.L. van der Waerden

van der Waerden was born on February 2, 1903, in Amsterdam, as the first child of Dr. Theodorus van der Waerden and his wife Dorothea Adriana, née Endt. His father was trained as a civil engineer, yet earned his living as a teacher at the *Hogere Burgerschool* in Amsterdam. From 1919 onwards, van der Waerden studied mathematics, physics, and chemistry at the University of Amsterdam. There, he came into contact with L.E.J. Brouwer and R. Weitzenböck, but was influenced more by Gerrit Mannoury and Hendrik de Vries. It was under the direction of de Vries that van der Waerden received his doctorate with a thesis on algebraic geometry.[1] In 1924, after taking his first degree in Amsterdam, van der Waerden went with a letter of introduction by Brouwer for about a year to the University of Göttingen, where, as he later stated, a new world opened up to him. In Göttingen, he established especially close contacts with Emmy Noether, Hellmuth Kneser, and Richard Courant. In 1925, van der Waerden returned to the Netherlands, where he wrote his doctoral dissertation and carried out his military service at the marine base in Den Helder. On March 24, 1926, he successfully defended his doctoral thesis *De algebraiese grondslagen der meetkunde van het aantal* (The algebraic foundations of enumerative geometry) in the grand hall of Amsterdam University. In the meantime, he had been awarded a Rockefeller fellowship for seven months and, following the semester in Göttingen with Emmy Noether, he went to Hamburg to study algebra with Emil Artin, Wilhelm Blaschke, and Otto Schreier. It was in the summer term of 1926 that van der Waerden attended Artin's lectures on algebra. These formed the basis for his best-known book, *Moderne Algebra* (van der Waerden 1930, 1931), through which he achieved international fame in that field.[2] At the beginning of 1927, he returned to Göttingen, where, on February 26, he gained the venia legendi (habilitation) for mathematics under the supervision of Richard Courant. In May 1927, he became a scientific assistant to Richard Courant.

In 1928, when he was only 25 years old, van der Waerden was made full professor at the University of Groningen in the Netherlands. On his return to Göttingen during the summer of 1929, as a visiting professor, he met Camilla Rellich (1905–1998), sister of the mathematician Franz Rellich (1906–1955). They were married in September of the same year. Two years later, van der Waerden was appointed to a full professorship at the University of Leipzig, where he remained until shortly

[1] Soifer (2015, pp. 15 ff.) has unearthed an impressive amount of new material from van der Waerden's family that sheds light on his early life before he went to Germany in 1924.

[2] Cf. Soifer (2015, pp. 39–66). Soifer states that, according to van der Waerden's correspondence with Courant (New York University, Archives, Courant Papers), Artin and van der Waerden were initially supposed to write the 'Algebra Book' together. Quoting from letters written by van der Waerden, Soifer (2009, p. 381 f., 2015, pp. 40–43) writes: "So, Artin has given his course [...], Artin is making his material 'more precise in details through conversations,' but 'Artin himself writes very little'." Soifer then continues: "Clearly, Artin refused to write the book with Van der Waerden, and thus 'astonished' Courant. He was obviously offended by Van der Waerden, but how? [...] But never mind the Master, the Student has gotten everything he needs, and can now publish The Book by himself, with the blessing of his mentor and 'Yellow Series' Founder and Editor Richard Courant." Soifer's implied accusation that van der Waerden had "stolen" his famous algebra book from Artin—as Günther M. Ziegler puts it in his review of Soifer (2009) in *Jahresbericht DMV* 116, 2014, p. 267—seems to me much too harsh. Why would Artin have invited van der Waerden on January 20, 1948, to deliver a lecture at Princeton, and even offered to accommodate him in his home, if the relationship between the two men was damaged in such a way as that suggested by Soifer? (For the invitation, see the handwritten letter in German from Artin to van der Waerden, in uncatalogued papers of van der Waerden, cardboard folder with the caption "Selected Letters, Baltimore + Laren 1946–1949"). Furthermore, van der Waerden stated in the preface of his book that he had included so many other results that it would be difficult to recognize Artin's lectures in it. For further information, see Mechthild Koreuber (2015, p. 236), who presents a detailed history of the book (ibid., pp. 232–245) or van der Waerden (1975a).

before the end of the Second World War. In Leipzig, van der Waerden developed close ties to the physicists Werner Heisenberg and Friedrich Hund. This led to, among other things, his well-known book, *Die gruppentheoretische Methode in der Quantenmechanik* (The Group-Theoretical Method in Quantum Mechanics) (1932),[3] and stimulated his life-long interest in physics. At the same time, he began working on his series of articles, "Zur algebraischen Geometrie" (On Algebraic Geometry) (1933–1971), while his friendship with the Leipzig philosopher Hans-Georg Gadamer awakened his interest in Greek mathematics and Plato.

The fact that van der Waerden remained in Germany during the entire Nazi period has since been a source of criticism by many authors, and led to, among other things, the refusal, in 1946, by the Dutch Minister of Education to accept van der Waerden's appointment to a public Dutch university. Soifer has set forth in detail in his book that van der Waerden signed administrative letters partly with the mandatory "Heil Hitler!" (Soifer 2009, p. 405 f.), gave the Hitler oath (Soifer 2015, p. 105), declined an invitation to a guest semester in Princeton in 1933 (Soifer 2015, pp. 97–104), and, in 1944, after much hesitation, did the same with an invitation to a professorship at the University of Utrecht (Soifer 2015, pp. 171–179). On the other hand, it must be noted that van der Waerden was exposed to numerous hostilities by the Nazi regime in Leipzig (Eisenreich 1981, p. 240; Frei 1993, p. 6, 1994; Thiele 2009, pp. 29–36; Schneider 2011, pp. 158–177), and that he later repeatedly argued that, by staying in Germany, he was able to do more for persecuted scientists than he could have done from abroad. In fact, it emerges from personal files at the University of Leipzig recently studied by M. Schneider (2011, pp. 161–165), R. Siegmund-Schultze (2011, pp. 206–209), and A. Soifer (2015, pp. 113–139) that van der Waerden, in the spring of 1935, protested in the faculty against the dismissals of the four so-called Jewish "front soldiers", namely Benno Landsberger, Friedrich Wilhelm Levi, Joachim Wach, and Fritz Weigert. He was joined in these protests by Werner Heisenberg, Friedrich Hund, and Bernhard Schweitzer. It is also apparent from the same sources that he had earlier, in September 1934, spoken up for the "half-Jewish" mathematician Harald Bohr at the annual DMV Meeting at Bad Pyrmont against the founder of "German mathematics," Ludwig Bieberbach (Schneider 2011, pp. 165–167; Soifer 2015, pp. 168–170). This led to a reprimand by the president of the University of Leipzig (Soifer 2015, pp. 141–151). In addition, the head of the NS-Dozentenbund at the University of Munich, the astronomer Bruno Thüring, was opposed to the appointment of van der Waerden to that university in the summer of 1938 on the grounds that he was "extremely philo-Semitic and [considers] antisemitism superfluous." Furthermore, Thüring remarked of van der Waerden: "He belongs to that type of university lecturer we no longer want to see today" (Litten 1994, p. 156 f.).[4]

[3]For a detailed report about the history and content of van der Waerden's book, see Schneider (2011, esp. 179–189).

[4]In addition, Soifer (2015, pp. 148–151) presents a document with similar remarks from the head of the NS-Dozentenbund in Leipzig, the Austrian anatomist Max Clara, dated April 20, 1940.

While Soifer adopts, in his books on van der Waerden, a mostly hypercritical, accusatory attitude, Thiele (2009, pp. 29–35) sets the emphasis on his having adopted a morally upright stance by protesting against the dismissal of the Jewish "front soldiers," by criticizing Bieberbach at the DMV conference in Bad Pyrmont, and by having spoken up in the USA, unfortunately in vain, for his endangered colleague Otto Blumenthal.[5] Martina R. Schneider (2011, pp. 173–177) adopts a well-balanced position between these two opposing viewpoints. On the one hand, she tries to understand van der Waerden's motives for remaining in Germany and his argument that he felt obliged to defend European culture—particularly science, or, more precisely, the "nucleus of German mathematics and physics"—against the culture-destroying actions of the Nazis. On the other hand, she suggests that by remaining in Germany, van der Waerden was collaborating in a certain sense with the National Socialists, their forces having occupied his homeland since 1940.[6] Moreover, it should be pointed out that before the Second World War, a professorship at a first-rate German university was considered the high point in the career of any mathematician or physicist. This explains why Albert Einstein, Erwin Schrödinger, and Hermann Weyl, along with others, left Switzerland for Germany to take up appointments there, although they all subsequently regretted having done so. Understandably, van der Waerden left the Netherlands for the foremost research center in physics, then located in Leipzig. The pressure to emigrate from Nazi Germany was, for him, not as strong as in the case of Einstein and Weyl, who were in a much more tenuous position because of their Jewish backgrounds. Furthermore, he felt responsible for his family and his three children, this sense of responsibility undoubtedly playing a major role in his decisions. After the bombing of his apartment in Leipzig, on December 4, 1943, and the provisional removal of his family to Bischofswerda—more than 100 kilometers away from Leipzig—van der Waerden must have been in a desperate situation. This explains, at least in part, why he sought to improve his circumstances by looking for help from the Nazi-sympathizing mathematicians Helmut Hasse and Wilhelm Süss. In 1944, van der Waerden tried, without success, to move out of the heavily bombarded city of Leipzig and to solicit an academic offer from either one of the much less endangered

[5]For the case of Otto Blumenthal see Segal (2003, pp. 231–244).

[6]A slightly different standpoint is taken by Siegmund-Schultze, who regards van der Waerden's claim that he wanted to defend "European culture" against the Nazis as merely a protective assertion ["reine Schutzbehauptung", e-mail to E.N., Febr. 17, 2016]. As to the attitude and statements of Soifer, he expressed himself in his review of Soifer's book as follows (Siegmund-Schultze 2015, p. 924): "While the book is in many ways an admirable effort, it is also deeply flawed, exhibiting insufficient understanding of the historical and political era of van der Waerden and of the languages he spoke, and lacking proper attribution to other work on which the book depends. The topic clearly resonates with Soifer, both politically and emotionally, driving him to amass a tremendous amount of material and to endow the book with great passion. Unfortunately, that same passion has compromised his objectivity and judgment." Siegmund-Schultze's review provoked a rather harsh response from Soifer in *Geombinatorics* XXV (3), 2016, 123–132 with the title "Siegmund-Schultze Proposal of German Monopoly on the Third Reich History."

cities of Göttingen and Jena.[7] On the other hand, he hesitated to accept an offer from the University of Utrecht, which he had already received in 1943, after the retirement of Johan Antony Barrau, because he feared this would preclude his future return to Germany.[8] After the war, his brother-in-law Franz Rellich, director of the Institute of Mathematics in Göttingen since 1946, tried unsuccessfully, together with Richard Courant, to bring van der Waerden back to Göttingen to help with the post-war reconstruction of the mathematical institute there. But after 1948, van der Waerden had received several other offers and held some reservations against such a proposal. In a letter written during his one year stay in Baltimore, on February 21, 1948, he told Franz Rellich: "I'd love to go to Goettingen for the world, if it was possible at all. But I cannot take provisions of food and clothing for 5 or 7 years, which are the prerequisites you set. I am afraid it will not work out."[9]

After the war, van der Waerden was working at the Bataafse Petroleum Maatschappij (BPM; today: Royal Dutch Shell) for some time, whereupon, in September 1947, he began a six-month stay as visiting professor at Johns Hopkins University in Baltimore, USA. From there, he was appointed professor at the University of Amsterdam in 1948, and in 1951, he received his appointment to the University of Zurich.[10] In Zurich, he was successor to the late Rudolf Fueter and

[7]Soifer (2015, pp. 181–198) describes van der Waerden's appeals to Hasse and Süss to get him an offer from Göttingen in great detail, but he overlooks van der Waerden's similar efforts with respect to Jena (see van der Waerden's correspondence with Hellmut Bredereck and Helmuth Kulenkampff from Jena in ETH, Hs 652). In addition to Soifer's description, it is perhaps also worth mentioning that van der Waerden stated, in a post-war letter to Courant of November 20, 1945, that Hasse and Herglotz wanted to have him in Göttingen in 1944, but he was unacceptable ("untragbar") to the NS-Regime. Soifer (2009, p. 424, 2015, p. 218) states that he was unable to locate van der Waerden's crucial letters to Hopf of July 19 and 21, 1945. However, these two letters are among Hopf's papers at the ETH, which Soifer consulted in connection with Neugebauer's letters to Hopf. Cf. ETH, Hs 621: 1041, 1451 and 1452.

[8]Cf. van der Waerden to Hasse, March 6, 1944 in SUB Göttingen, Cod. Ms. H. Hasse 25:2, 110. For Barrau's succession at Utrecht see Soifer (2015, pp. 171–179).

[9]I would like here to add the full quotation of the German original of this important handwritten letter from van der Waerden: "Lieber Franzi, Es ist mir furchtbar leid, daß ich jetzt antworten soll ob ich nach Göttingen gehen würde oder nicht. Ich wüßte nichts in der Welt was ich lieber täte als nach Göttingen gehen, wenn es möglich wäre. Aber ich kann nicht für 5 oder 7 Jahre Vorräte an Essen und Kleidung mitbringen, was Du als unerläßliche Bedingung aufgestellt hast. Ich fürchte also, daß es nicht gehen wird. Für das kommende akademische Jahr 1948/1949 werde ich in den nächsten Wochen entweder Johns Hopkins oder Amsterdam mein Wort geben; da bin ich also schon vergeben. Was danach kommt, weiß Gott allein, und ich möchte jetzt nicht gern für eine so weite Zukunft ja oder nein sagen. Morgen früh fahren wir alle nach New York, und Mittwoch müssen wir für 3½ Monate Abschied nehmen. Es ist sehr traurig. Herzliche Grüße Dein Bartel". Handwritten draft in a cardboard folder with the caption "Selected Letters, Baltimore + Laren 1946–1949", which van der Waerden kept at his home until his death. Further information about the post-war offer from Göttingen, which van der Waerden did not accept, can be obtained from van der Waerden's correspondence with Courant in NYU Archives, or from Schneider (2011, p. 365).

[10]For more information about van der Waerden's "employment history" from 1945 to 1951, see Soifer (2015, pp. 283–369) and Schneider (2011, pp. 355–374).

acted as director of the Mathematical Institute from 1951 to 1970. He supervised nearly 50 Ph.D. theses and was second examiner of roughly ten further doctoral theses. This testifies to his immense commitment to and impact on the development of mathematics in Switzerland. He was a member of the editorial boards of several well-known series and journals, among them the *Grundlehren der Mathematischen Wissenschaften*, the *Mathematische Annalen* and the *Archive for History of Exact Sciences*. Alongside these tasks, he continued to write and publish unremittingly, so that we have today the impressive number of altogether more than 20 books and about 300 articles, which have been compiled in several (unfortunately not complete) lists by Gross (1973) and Eisenreich (1981), in his *Zur algebraischen Geometrie* (1983b), as well as by Top and Walling (1994).

3 The Correspondence of B.L. van der Waerden

van der Waerden's correspondence, together with his scientific estate, is housed at the Library of the Eidgenössische Technische Hochschule (ETH) in Zurich. The largest part of the Zurich collection was donated by van der Waerden himself to the ETH in 1982, after his resignation as head of the small Department for the History of Science within the Institute of Mathematics at the University of Zurich, when he had to vacate his office. Additional correspondence files followed in 1983 and 1984. All of these papers are fully catalogued, and are also accessible via a 206-page inventory on the Internet (Jakob 1985). Other parts, which came to the ETH through the author after van der Waerden's death in 1998 or which remained at the Institute of Mathematics together with a small part of his scientific papers, have not yet been completely catalogued, and are therefore only partly accessible to the general public.[11] van der Waerden very rarely archived his notes and materials for his scientific papers. During his work, he assembled the relevant papers in specific cardboard folders, on which he noted the title of the relevant research or publication project. After publication, he normally threw the contents away and used the cardboard folder for another project. Many of these cardboard folders carry informative lists of titles of up to ten former projects (cf. Fig. 1), but most of them were, in fact, empty at the time of his death. On the other hand, van der Waerden carefully archived his scientific correspondence. Not only was it usual for him to keep all of the letters from his correspondents, but he also saved the drafts or photocopies of his own letters. Therefore, the approximately 15,000 letters in Zurich allow for deep insights into the scientific development, work and personality of van der Waerden.

As van der Waerden's last assistant and longtime collaborator at our former Department for the History of Science at the Institute of Mathematics at the University of Zurich, I have long felt the need to preserve for posterity my memories

[11]Information according to the accession book of the ETH Library, based on two emails from Marion Wullschleger (ETH Library) from April 23, 2012, and January 22, 2016.

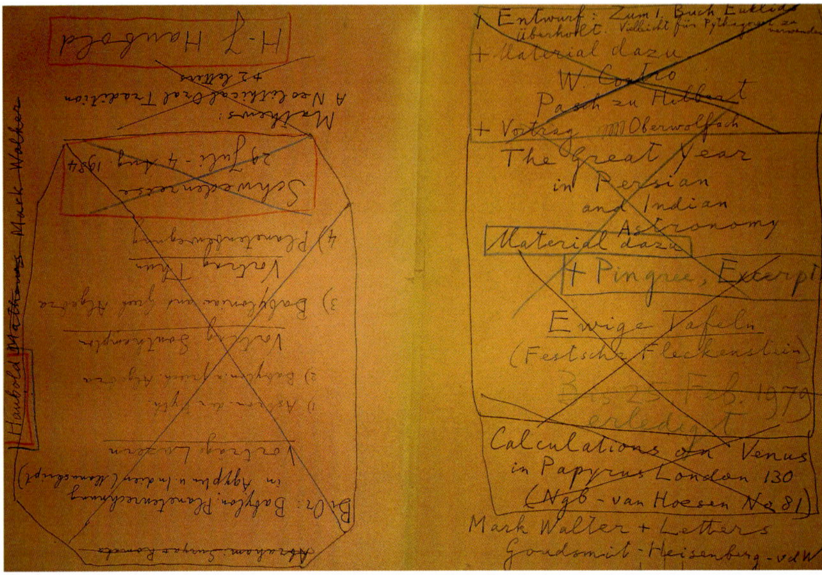

Fig. 1 Cardboard folder of B.L. van der Waerden, used for, among other things, his studies on the Great Year in Persian and Indian astronomy, his calculations on Venus in Papyrus London 130, his lectures in Lucerne, Southampton, and Thun, a trip to Sweden in 1984, and finally, for his correspondence with H.J. Haubold, in Uncatalogued Papers of B.L. van der Waerden

of him and the years we spent working together. Already in 1996, shortly after his death, I started writing down my memories in a short biography in German that was used for my English biographical entry about van der Waerden in the encyclopaedic work *Writing the History of Mathematics: Its Historical Development* (cf. Neuenschwander 2002). Unfortunately, my intention to enlarge my manuscript into a monographic van der Waerden biography was constantly delayed by my twenty years of working as an editor for all the natural sciences at the *Historical Dictionary of Switzerland* (2002–2014) (cf. http://www.hls-dhs-dss.ch/) and my large historical overview for the centenary of the Swiss Mathematical Society (Neuenschwander 2010, 2016). After the appearance of Soifer (2009) and Thiele (2009), I restarted my work on the van der Waerden project. As van der Waerden's early years have already been thoroughly researched by Schneider, Soifer, and Thiele, and as this period is hardly documented in the Zurich estate because of the irretrievable damage caused to van der Waerden's manuscripts by the bombing raid in Leipzig, I have decided to concentrate my work on van der Waerden's Zurich years. The objective of my efforts is to represent as comprehensively as possible his career in Zurich, roughly in the same way as Thiele has done for the Leipzig period.

In 2010, I began, for this reason, to examine the letters at the ETH in order to select and edit the most significant ones. Furthermore, I was able to compile a more accurate catalogue of van der Waerden's publications, which contains seventy new titles, because, after his death, I was entitled by his family to take over his personal collection of offprints of his own publications. Subsequently, I started to research his professional activity as the long-standing director of the Mathematical Institute in Zurich and his work as a historian of science. This amounted, in 2013, to a manuscript of more than three hundred pages. In July 2013, on the occasion of the 24th International Congress of the History of Science, Technology and Medicine in Manchester, I showed it, amongst others, to the editor of the Birkhäuser publishing house, which later published (Soifer 2015). In 2014, the same manuscript served as the basis for my lecture in Lisbon. Unfortunately, work on my van der Waerden book was again interrupted by the need to reissue all of the biographies and articles dealing with natural sciences that had appeared under my editorship in the thirteen-volume *Historical Dictionary of Switzerland* (2002–2014) in a new one-volume encyclopaedia. As long-time president of the Swiss National Committee of the International Union for the History and Philosophy of Science and Technology (IUHPST), I naturally did not want to forego the exceptional opportunity to make this *Historical Encyclopaedia of the Natural Sciences in Switzerland* a substantial up-to-date contribution to the bicentenary of the Swiss Academy of Sciences (SCNAT).

In order to provide the interested reader with an easily accessible overview of van der Waerden's huge correspondence, I have put all of the correspondents with whom he exchanged at least 25 letters together in a table (cf. Appendix A.1). In those cases in which the respective collection comprises at least fifty documents, the name of the correspondent appears in bold face type, while in the cases of a hundred documents or more, bold and italic type have been used. In the following, I shall briefly describe the four correspondences that contain more than a hundred letters. Furthermore, I set out a few noteworthy conclusions that can be drawn from van der Waerden's correspondence.

Hans Freudenthal, a German-born Jewish mathematician, was a good and almost lifelong friend of van der Waerden. He grew up in Luckenwalde, a small town near Berlin, and studied mathematics in Berlin from 1923 to 1930. It was through his interest in intuitionism that he was recruited, in 1930, by L.E.J. Brouwer as an assistant and moved to Amsterdam. From then on, he lived in Holland. He made substantial contributions to algebraic topology and also took an interest in history of science and teaching of mathematics. In 1940, when Germany invaded the Netherlands, Freudenthal was suspended from his duties at the University of Amsterdam by the Nazis, and later on sent to a labour camp in the Dutch village of Havelte. It is not known exactly when van der Waerden and Freudenthal first met. As Freudenthal remarked in an after-dinner speech on the occasion of van der Waerden's 80th birthday in 1983, they must have been introduced to each other by either Witold Hurewicz or David van Dantzig in the 1930s, when van der Waerden travelled from Leipzig to the Netherlands from time to time in order to attend the monthly assembly of the Dutch mathematical society, the Wiskundig Genootschap. They seem to have come into closer contact around 1940, when Freudenthal was working on Lie groups, a topic on which van der Waerden had published a number of articles around 1933.[12] However, when van der Waerden was repatriated to Holland in late June 1945, he needed to obtain employment as soon as possible. In this difficult situation, Freudenthal introduced him to the Bataafse Petroleum Maatschappij, and on October 1, 1945, van der Waerden got a job as an analyst for that company.

Their correspondence,[13] mainly from the 1950s and 1960s and largely conducted in Dutch, contains letters on various subjects, such as newly published scientific books or articles, questions concerning their own publications in the history of mathematics and philosophy, scientific competitions (held by the Wiskundig Genootschap te Amsterdam; Freudenthal was an editor of its journal, the Nieuw Archief voor Wiskunde), as well as invitations to lectures or plans to meet. One of the highlights of the correspondence concerns the information about van der Waerden's unpublished textbook *Introduction to Topology and Riemann Surfaces*. It is scarcely known that, apart from his famous textbook *Moderne Algebra*, van der Waerden also wrote a textbook on topology, which was based on a lecture course he had given in Baltimore in 1948. When he had finished the manuscript of 264 pages, in the early 1950s, he sent it to John Wiley & Sons, who had it reviewed by two eminent topologists. Both argued that the book had attractive features for student beginners: "a clear easy to read style, a minimum of notational complication, numerous figures," but that it was written in a somewhat outdated, old-fashioned style. So, van der Waerden sent the manuscript with the two expert reviews to his lifelong friend Freudenthal and asked him for his opinion (letter of December 18, 1952). Freudenthal analysed the manuscript in detail, but was not in favour of its publication either, and so it remained unpublished. Another highlight of the collection in the Noord-Hollands Archief is a letter of 45 pages from van der Waerden to Freudenthal, written in 1956, in which he describes in detail his views about synthetic a priori knowledge and its role in natural science. Nine years later, he published an article entitled "Synthetische Urteile a priori" in *Acta Philosophica Fennica* 18, 1965, 277–291. Most of the main thoughts that van der Waerden wrote down in his letter to Freudenthal can be found in this publication, although some show a slightly different view and some express the same view more concisely.

Edward S. Kennedy was a specialist in medieval Islamic astronomy, who, around 1956, came into contact with van der Waerden. It was about this time that Kennedy had published his magisterial work *A Survey of Islamic Astronomical Tables*, and

[12]The oldest letter of their correspondence that can be found in the archives was written by Freudenthal on August 2, 1940. It points to a couple of errors in van der Waerden's article "Die Klassifikation der einfachen Lieschen Gruppen," in *Mathematische Zeitschrift* 37, 1933, 446–462, and in his paper with Hendrik Casimir entitled "Algebraischer Beweis der vollständigen Reduzibilität der Darstellungen halbeinfacher Liescher Gruppen," in *Mathematische Annalen* 111, 1935, 1–12.

[13]Freudenthal's part of the correspondence is held by the Noord-Hollands Archief in Haarlem (NL) in the collection of Freudenthal's papers (inventory number: 615/89). There, one can find about one hundred and fifty letters (1940–1986), of which around forty are also held as drafts or originals in the ETH Archives.

that van der Waerden became interested in Islamic astronomy. So, he studied Kennedy's work and wrote a very detailed review of it in *Bibliotheca Orientalis* 14, 1957, 109 f. This led, in due course, to their joint paper "The World-Year of the Persians" in the *Journal of the American Oriental Society* 83, 1963, 315–327, and to a lifelong friendship. The correspondence contains intensive discussions and an extensive exchange of ideas in preparation for that paper. Especially treated are van der Waerden's ideas about the transmission of Babylonian and Hellenistic notions about the "Great Year" to the Persians, the difficult dating of the first two horoscopes of Māshā'allāh, and the criticism by David Pingree, which led to a revision of certain parts of the paper.[14]

Otto Neugebauer was one of van der Waerden's oldest friends. van der Waerden had been acquainted with him since his stay in Göttingen, and he also attended Neugebauer's lectures on the history of Greek geometry there. Neugebauer was one of the editors of the pioneering *Quellen und Studien zur Geschichte der Mathematik, Astronomie und Physik* (1929–1938), where van der Waerden's first larger historical article about Egyptian fractions, entitled *Die Entstehungsgeschichte der ägyptischen Bruchrechnung*, appeared in 1938. After Neugebauer's emigration to Copenhagen, in 1934, and later to the USA, their correspondence was interrupted until the end of the Second World War, when van der Waerden reopened the exchange of letters in the summer of 1945. The correspondence contains a wealth of interesting information and discussions on the research into the history of astronomy in ancient cultures. The extensive exchange of letters is also of particular interest for the biography of Neugebauer, because the latter did not keep his correspondence. The research methods of the two scholars differed considerably. Whereas van der Waerden based his historical reconstructions mainly on mathematical deduction and enjoyed developing far-reaching theories, Neugebauer focused his historical research on the careful analysis and editing of the fragmentary texts handed down to us, and felt more and more perturbed by the discovery of new manuscripts. Already on December 9, 1947, he wrote to van der Waerden: "I do not see how you can put the Hilprecht text into the interval from 1400 to 1900. The text is undoubtedly Old-Babylonian. I am extremely skeptical against any definite arrangement of a few scattered fragments which we have from a period of 1500 years. It is my experience that every new piece of information completely destroys all attempts to reach a consistent picture. Consistent history can be written only if you have no documents."[15] And about thirty years later, on April 28, 1979, he reiterated this

[14]For this revision, see above all the letter of Kennedy to George C. Miles, Associate Editor of the 'Journal of the American Oriental Society', dated July 3, 1963 [sic] in ETH, Hs 652: 4699.

[15]Typed, hand-signed letter in English, in the uncatalogued papers of van der Waerden, cardboard folder with the caption "Neugebauer + Sachs 1948". In the same folder, one also finds a draft for the extremely revealing reply from van der Waerden, dated December 17, 1947: "Dear Otto. No, I do not belong to the gifted authors who are able to write consistent history without documents. Das kommt weil ich 1. keine Phantasie habe, 2. kein Philologe bin und deswegen keine Texte amendieren [sic] kann, 3. nicht kritisch, sondern leichtgläubig bin, und nie ein Zeugnis verwerfe das nicht in meinen Kram passt, 4. keine vorgefassten Meinungen und überhaupt keine

view after the reception of van der Waerden's new book about the Pythagoreans (van der Waerden 1979): "Dear Bartel, it is very kind of you to have sent me your 'Pythagoreans,' although you know that I am a totally unphilosophical mind [as Pauli once put it: 'You will drown in the facts' (a pleasant manner of death, in my opinion)]. You have the capability to construe from scattered fragments a consistent development, while I only see more difficulties the better I know a field. It follows that historiography is essentially a literary endeavor and not an objective science."[16] van der Waerden was quite depressed by Neugebauer's reservations about his theories, as can be seen from his later correspondence with Derek Thomas Whiteside.[17] Besides Neugebauer and Kennedy, van der Waerden also corresponded with Richard A. Parker, David Edwin Pingree, William Kendrick Pritchett, and Abraham Sachs, who were all working on the same topics. With some of them, he published joint papers, with others—for example, Pingree—he argued about the correct interpretation of texts.

Clifford A. Truesdell was a leading authority on the history of mechanics and since 1960 had been the editor of two journals, the *Archive for Rational Mechanics and Analysis*, from 1957 onwards, and the *Archive for History of Exact Sciences*. His correspondence with van der Waerden includes a great deal of information about the foundation of the latter journal and the decade-long activity of van der Waerden as one of the co-editors of the *Archive for History of Exact Sciences*. During a visit to Zurich, Truesdell discussed the foundation of the *Archive* with van der Waerden, and in their subsequent correspondence, van der Waerden helped to perfect the relevant "Suggestions to Members of the Editorial Board." van der Waerden belonged to the Editorial Board from 1960 to 1993 and, as editor, submitted more than eighty papers to the *Archive*. In addition, van der Waerden was also one of the editors of the Springer book series *Grundlehren der Mathematischen Wissenschaften* (1934–1975) and the journal *Mathematische Annalen* (1934–1968). From van der Waerden's correspondences with Richard Courant, Heinrich Behnke,

originellen Ideen habe. Ich muß mir deswegen das lustige Märchenschreiben versagen, und finde eine Erklärung nur dann wenn sie die einzige ist die zu allen Texten passt. [...] I am quite perplexed by your statement concerning Hilprecht's text "The text is undoubtedly Old-Babylonian". In QS [Quellen und Studien] B 3 p. 277, you ascribed it to ± 12th century and so did Thureau-Dangin according to your own quotation. However, this does not matter much. Meinentwegen mag der Text altbabylonisch sein. You missed my point. [...]". For van der Waerden's "point" that Hilprecht's text does not prove the existence of a highly developed Old-Babylonian scientific astronomy, but is simply a mathematical exercise, see van der Waerden (1949, p. 6); for a photograph of the first page of van der Waerden's draft, see Fig. 2.

[16]"Lieber Bartel, es ist sehr nett von Dir, dass Du mir Deine 'Pythagoreer' geschickt hast obwohl Du weisst, dass ich ein völlig unphilosophischer Kopf bin (wie Pauli es einmal ausdrückte: 'Sie werden noch in den Tatsachen erdrinken [sic]' [eine angenehme Todesart, nach meiner Ansicht]). Du hast die Fähigkeit aus zerstreuten Fragmenten eine einheitliche Entwicklung zu konstruieren während ich immer nur mehr Schwierigkeiten sehe je besser ich ein Gebiet kenne. Woraus folgt, dass Geschichtsschreibung im Wesentlichen eine literarische Angelegenheit ist und keine objektive Wissenschaft." [Handwritten letter in German of April 28, 1979; ETH, Hs 652: 6492].

[17]Cf. handwritten letter (draft) from van der Waerden to Whiteside dated August 13, 1983, in: ETH, Hs 652: 12276.

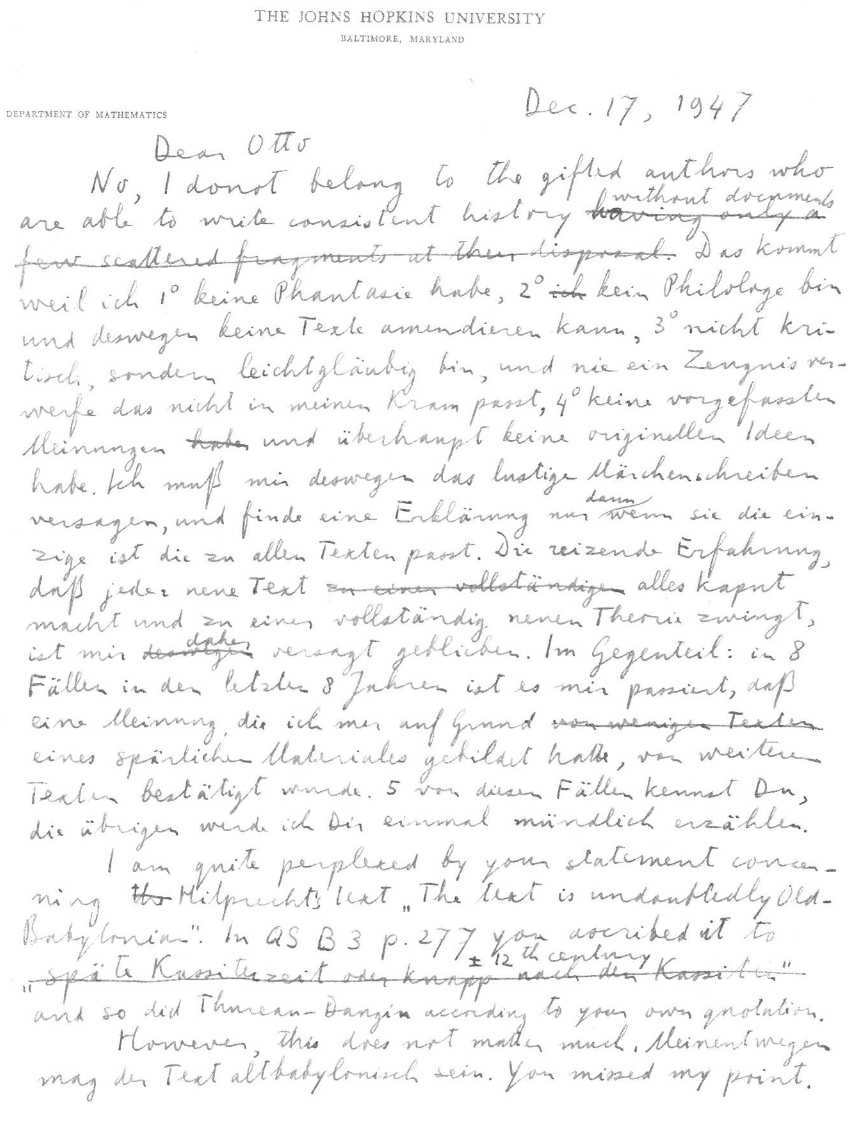

THE JOHNS HOPKINS UNIVERSITY
BALTIMORE, MARYLAND

DEPARTMENT OF MATHEMATICS

Dec. 17, 1947

Dear Otto

No, I donot belong to the gifted authors who
are able to write consistent history without documents ~~having only a~~
~~few scattered fragments at their disposal.~~ Das kommt
weil ich 1° keine Phantasie habe, 2° ~~ich~~ kein Philologe bin
und deswegen keine Texte amendieren kann, 3° nicht kri-
tisch, sondern leichtgläubig bin, und nie ein Zeugnis ver-
werfe das nicht in meinen Kram passt, 4° keine vorgefassten
Meinungen ~~habe~~ und überhaupt keine originellen Ideen
habe. Ich muß mir deswegen das lustige Märchenschreiben
versagen, und finde eine Erklärung nur dann ~~wenn~~ sie die ein-
zige ist die zu allen Texten passt. Die reizende Erfahrung,
daß jeder neue Text ~~zu einer vollständigen~~ alles kaputt
macht und zu einer vollständig neuen Theorie zwingt,
ist mir daher ~~deswegen~~ versagt geblieben. Im Gegenteil: in 8
Fällen in den letzten 8 Jahren ist es mir passiert, daß
eine Meinung, die ich mir auf Grund ~~von wenigen Texten~~
eines spärlichen Materiales gebildet hatte, von weiteren
Texten bestätigt wurde. 5 von diesen Fällen kennst Du,
die übrigen werde ich Dir einmal mündlich erzählen.

I am quite perplexed by your statement concer-
ning ~~the~~ Hilprecht's text „The text is undoubtedly Old-
Babylonian". In QS B 3 p. 277 you ascribed it to
"± 12 th century ~~„späte Kassitenzeit oder knapp nach den Kassiten"~~"
and so did Thureau-Dangin according to your own quotation.
However, this does not matter much. Meinentwegen
mag der Text altbabylonisch sein. You missed my point.

Fig. 2 First page of van der Waerden's draft of his letter to Otto Neugebauer from Dec. 17, 1947, in Uncatalogued Papers of B.L. van der Waerden, cardboard folder with the caption "Neugebauer + Sachs 1948"

Friedrich Karl Schmidt, etc., we learn a great deal about the problems of editing and publishing during the Nazi era and the difficult years after the war.

There are, of course, several other important correspondences in van der Waerden's papers that should be mentioned in this short survey. First of all, one should particularly highlight the vast correspondence with the well-known classicist Walter Burkert, who exchanged long and eminently learned letters with van der Waerden about the origin and spread of ancient Mediterranean cultures, and especially about that of the Pythagoreans. Whereas van der Waerden accepted nearly all existing attributions of scientific discoveries to the Pythagoreans by classical authors as being true, Burkert was much more sceptical about such testimonies and regarded them mostly as later interpolations. This led to a completely different assessment of the scientific achievements of the Pythagoreans, which, of course, is reflected in the relevant works of the two scholars.[18] Nevertheless, they always maintained a very friendly relationship, characterized by mutual respect. Burkert helped van der Waerden in the interpretation of linguistically difficult Greek passages, while van der Waerden exposed him to his latest historical theories. This open, mutually respective attitude is particularly well reflected in one of the last letters from Burkert, wherein he thanked van der Waerden for sending him his book about the Pythagoreans (van der Waerden 1979) with the following words:

> Dear Mr van der Waerden,
>
> Finally, we have your book on the Pythagoreans into which—if we calculate starting out from your essay on Zeno—a full ten plus a full twenty-eight years of work has gone. We have often and thoroughly chewed over details. What altogether makes a great impression is, on the one hand, the impartiality with which you comprehend and represent things—from outside the historians' and philologists' guild—, and the immensely broad horizon of your work up to the Sabians, to whom I had not yet paid any attention in this context. My cordial thanks and all best wishes.
>
> Yours, Walter Burkert.[19]

The friendly accommodating tone in the correspondence with Burkert stands in sharp contrast to the sometimes harsh statements that one can find in van der Waerden's correspondence with David Pingree. Nevertheless, the latter is extremely informative for the development of van der Waerden's and Pingree's ideas about the transmission of astronomical theories between the Near East and India. Starting from Kennedy's paper about the Sasanian astronomical handbook *Zīj-i Shāh* (Kennedy 1958) and the 16 horoscopes of Māshā'allāh recorded in Ibn Hibintā's astrology, van der Waerden and Burckhardt (1968) tried, in their paper

[18]Cf. Burkert (1972) and van der Waerden (1979).

[19]„Sehr verehrter Herr van der Waerden, Nun liegt also Ihr großes Pythagoreer-Buch vor, in das – von Ihrem Zeno-Aufsatz an gerechnet – eine vollkommene 10 plus eine vollkommene 28 an Arbeitsjahren eingegangen sind. Über Einzelheiten haben wir genug hin- und herdiskutiert. Was insgesamt großen Eindruck macht, ist einerseits die Unbefangenheit, mit der Sie, von außerhalb der Historiker- und Philologenzunft, die Dinge erfassen und darstellen können, und der ungeheuer weite Horizont, bis zu den Szabiern, auf die ich in diesem Zusammenhang noch gar nicht geachtet hatte. Herzlichen Dank und beste Wünsche! Ihr Walter Burkert." [undated original; ETH, Hs 652: 10583].

"The Astronomical System of the Persian Tables I," to determine the constants of the astronomical tables and the computational methods by which the horoscopes of Māshā'allāh were calculated in order to arrive at probable dates for the horoscopes themselves.

The second part of the paper about the Persian Tables appeared only after nearly twenty years as van der Waerden (1987b). The main reason for the delay was probably that Kennedy and Pingree had published, in 1971, a major monograph on the horoscopes of Māshā'allāh, in which they arrived at significantly different dates for the first two horoscopes (Kennedy and Pingree 1971). In the second paper, van der Waerden compares the lunar longitudes and ascendants in the horoscopes with those computed by the Midnight System of the great Indian astronomer Āryabhaṭa. He arrives among others at the following conclusions: "1. The horoscopes were indeed computed by means of the *Tables of the Shāh*, as Ibn Hibintā asserts. 2. These tables were based on the Midnight System of Āryabhaṭa, with small changes in the numerical elements. 3. In the *Tables of the Shāh*, a mean conjunction of all planets at or near 0° Aries was assumed to have taken place at midnight between February 16 and 17 in the year −3101. This date is just one day earlier than the date of a mean conjunction in the Midnight System of Āryabhaṭa. 4. The year numbers of the first two horoscopes are −3320 and −3300. The years proposed by Pingree, −3380 and −3360, are impossible" (van der Waerden 1987b, 198 f.).

Thereafter, van der Waerden presents some further arguments against Pingree's dating of the first two horoscopes and, at the end, summarizes the early history of the Persian tables in a few words. He believed that the Persian tables were ultimately based on the heliocentric system of Aristarchus of Samos, who had a follower in Babylon, namely Seleucus of Seleucia or Babylon. Seleucus was a near-contemporary of Hipparchus, so he had trigonometrical methods and accurate observations at his disposal. Therefore, van der Waerden assumed that Seleucus determined the constants in the heliocentric theory and developed methods for computing planetary positions. This enabled him—or one of his successors, for example, "Teukros the Babylonian"—to compute tables to meet the needs of astrologers. The work of Teukros was well known in Sassanid Persia and in the Islamic world. From there, these pre-Ptolemaic theories were passed down to Āryabhaṭa, to the *Tables of the Shāh*, and ultimately to Māshā'allāh, as van der Waerden described in several other publications.[20]

van der Waerden explained in a long letter to Pingree of January 8, 1962, how he arrived at his theory.[21] Pingree answered on January 14, 1962, that he found van der Waerden's account of the "development of the theory of a Persian origin for Indian yuga-astronomy somewhat disappointing," and he described, in a similarly long letter, his own theory.[22] Several other letters followed. On February 24, 1962, van der Waerden replied:

[20]Cf. van der Waerden (1980a, b, 1987a, b, 1988).

[21]van der Waerden, carbon copy of typed letter in English, ETH, Hs 652: 6996.

[22]Pingree, typed, hand-signed letter to van der Waerden, ETH, Hs 652: 6998.

> Dear Dr. Pingree,
>
> Allow me to pick out of your letter of Febr. 10 a single sentence: "You will of course agree that the references to the inhabitants of Babel and the Chaldeans are ridiculous; the conjunctions must have been mean and therefore have been computed on the basis of epicyclic theory."
>
> No, I don't agree. On the whole, my way of thinking is rather different from yours. I know that I am groping in the dark. I collect evidence from all sorts of sources, without discarding any testimony. I try to make sense of all the texts, to find hypotheses which explain them all, to find confirmations of my conclusions from other sources, etc. It's guesswork, but in many cases my guesses proved justified afterwards.
>
> Your method, in many cases, seems to be different. You start with a statement which you consider obvious, and then you reject everything that is not in accordance with your initial statement.
>
> [...]
>
> Your method is wrong, because your fundamental assumptions are based upon nothing. You think you know exactly what goes on in the mind of a Greek astrologer, of a Chaldean, of an orthodox Iranian, of Aryabhata. Aryabhata chose the shorter mahayuga simply because it's easier to work with an ahargana of that period rather than one a thousand times as great is another example of your pretension to know everything. By a lucky accident, I was able to refute one of your basic assumptions, but the others are just as doubtful. It's the method which is wrong, not just one particular instance.[23]

Unfortunately, there is not enough space here to undertake a closer examination of this fundamental dispute over methods between Pingree and van der Waerden.[24] There are several other interesting controversies in van der Waerden's correspondence that were, however, dealt with in a much friendlier way. As early as 1944, van der Waerden had presented, in long letters to Jakob Heinrich Anderhub, Eduard J. Dijksterhuis, and Albert Rehm,[25] his controversial thesis that Plato, in his astronomical passages in the *Timaeus* and at the end of the *Republic*, alluded to a primitive epicycle theory of the older Pythagoreans for the Sun, Venus and Mercury. Rehm's objections helped van der Waerden to perfect and complete his theory as he set it out, in 1951, in his great monograph *Die Astronomie der Pythagoreer*, as well as in several later papers.[26] Furthermore, in January 1966, van der Waerden discussed with Thomas S. Kuhn the question as to whether intellectual achievements were more strongly influenced by environmental factors or by genetic endowments.[27] Then, in 1981, he argued with Adolf Yushkevich about a connection between the spread of Indo-European languages and of pre-Babylonian mathematics, which van der Waerden tried to establish in his articles entitled "On Pre-Babylonian Mathematics I and II" (1980c) and in his book *Geometry and Algebra in Ancient Civilizations* (1983a). van der Waerden was convinced that great

[23]van der Waerden, handwritten draft, ETH, Hs 652: 7001.

[24]In subsequent years, the relationship between Pingree and van der Waerden got even worse. Pingree wrote negative reviews of van der Waerden's publications and van der Waerden published counterstatements. Cf., for example, Pingree (1976a, b), van der Waerden (1976, 1977/1978, 1980a).

[25]Cf. van der Waerden, correspondence with Anderhub and Rehm, in ETH, Hs 652: 11743, 652:11746–47 et passim.

[26]Cf., for example, van der Waerden (1974, 1978, 1988).

[27]van der Waerden, handwritten draft in ETH, Hs 652: 5217–5219.

inventions (with very few exceptions) came about only once.[28] And finally, there is the century-old discussion about "Geometric Algebra" on which van der Waerden (1975b/1976) set out his case in the paper "Defence of a 'Shocking' Point of View," which is still highly topical, as a recent article (Blåsjö 2016) proves. So, it is not a surprise that van der Waerden's wife expressed the view several times to me and to others[29] that it would have been much better if her husband had concentrated his work more on mathematics than on the history of science.

Despite such criticisms about his contributions to our understanding of the history of ancient science, there can be no doubt that van der Waerden's scientifically-based and occasionally speculative approach provided an important stimulus for ongoing historical debates and engendered much in the way of further research on the topic. In so doing, his work has led to a number of now widely accepted new theories on the emergence and spread of the exact sciences in Antiquity, to which a considerable number of books and article-length publications can testify. While it is certainly the case that van der Waerden was sometimes criticized for his speculative ideas, we should not lose sight of his profound insights into the mathematical sciences and ancient cultures that continue to impress today. van der Waerden remains an important historian of science, not the least because of the more science-oriented approach he adopted, in contrast to many other studies with a stronger anchoring in philological analysis and cultural studies.

A.1 Appendix: Most Extensive Correspondences of B.L. van der Waerden[30]

Correspondent	Period	Letter Nos.
Abraham, George (1919 – 2009) (Madras Christian College)	1976–1982	46–127
Baer, Reinhold (1902–1979)	1950–1976	376–411 etc.

(continued)

[28]Cf. van der Waerden, correspondence with Yushkevich in ETH, Hs 652: 11195–11203.

[29]Cf. Dold-Samplonius (1994, p. 145, English translation: 1997, p. 319).

[30]The present list is based on the inventory of van der Waerden's correspondence at the ETH Library (Jakob 1985). To allow for an easier overview, we have only included correspondents with at least twenty-five items in that inventory. Correspondences with at least fifty items appear in bold face type, and those with at least a hundred items appear in bold and italics. As Jakob (1985) also lists references to other correspondences and enclosures, but, however, does not include hitherto uncatalogued material that was given to the library in 1998 by the author, the numbers of items in that inventory do not necessarily include all letters from a certain correspondent. For easier identification of the correspondents, we have, whenever possible, added their life data. For PhD students of van der Waerden, we included the graduation year and the diploma awarding body, if it was not the University of Zurich. For little-known correspondents, we added at least some identifying features, which make it easy for the reader to get additional information from Jakob (1985), Wikipedia or the Internet.

Correspondent	Period	Letters Nos.
Bandler, Wyllis (1916–1995) (PhD Student v.d. Waerden 1962)	1958–1967	437–518 10426–10427
Batschelet, Eduard (1914–1979)	1955–1975	558–611 etc.
Becker, Oscar (1889–1964)	1951–1962	649–678
Behnke, Heinrich (1898–1979) and Wife	1944–1972	682–706 10438–10459 etc.
Billard, Roger (1922–2000)	1977–1980	870–928
Birkhäuser–Verlag Basel	1951–1981	953–1046 etc.
Böker, Robert (1885–1980)	1944–1962	1110–1112 10480–10533
Bruin, Frans (1922–2001)	1959–1982	1283–1331 3397, 10565 f.
Burckhardt, Johann Jakob (1903–2006)	1952–1983	1382–1393 10575–10582 etc.
Burkert, Walter (1931–2015)	1963–1982	1402–1465 10583–10590 etc.
Cherniss, Harold (1904–1987)	1951–1982	1567–1589 10636–10639
Courant, Richard (1888–1972)	1945–1966	1705–1720 (cf. also Courant Institute)
Derwidué, Léon (1914–1971) (Belgian mathematician)	1951–1957	1817–1838 10657–10665
Dijksterhuis, Eduard Jan (1892–1965)	1943–1956	1888–1891 10689–10710
Eichler, Martin (1912–1992)	1954–1971	1993–2026
Fellmann, Annemarie (1949–) (math. assistant v.d. Waerden)	1979–1986	2131–2167 10762–10805 12320–12432
Fellmann, Emil A. (1927–2012)	1958–1983	2168–2182 10806–10817
Field, Gerard (1921–2012) (math. Birmingham and Hamilton CA)	1963–1968	2208–2210 10824–10863
Fierz, Markus (1912–2006)	1952–1982	2211–2235 10864–10867 etc.
Fleckenstein, J. O. (1914–1980)	1956–1980	2283–2332 etc.
Flegg, Graham H. (1924–2015) (The Open University, Milton Keynes, UK)	1974–1983	2333–2379 10885–10889
Frei, Günther (1942–) (PhD Student v.d. Waerden 1969)	1966–1983	2458–2488 10908–10912

(continued)

Correspondent	Period	Letters Nos.
Freudenthal, Hans (1905–1990)	1945–1983	2511–2596 10913–10927 etc.
Fritz, Kurt von (1900–1985)	1950–1983	2632–2683 10930–10931
Gross, Herbert (1936–1989) (PhD Student v.d. Waerden 1960)	1959–1981	3127–3156 4435–4436 etc.
Hartner, Willy (1905–1981)	1957–1981	3355–3399, etc.
Hasse, Helmut (1898–1979)	1943–1961	3400–3409 etc. (cf. Nachlass Hasse)
Hawkins Jr., Thomas W. (1938–)	1968–1985	3431–3476 11054–11089
Hegde, Keshava S.V. (1922–) (Mysore, India) (PhD Student v.d. Waerden 1955)	1953–1963	3495–3524
Heisenberg, Werner (1901–1976)	1959–1973	3549–3574 etc.
Herk, C. G. G. van (1900–1982) (PhD Student v.d. Waerden Amsterdam 1951)	1950–1982	3637–3683 11099–11109 etc.
Herrmann, Manfred (1932–1997)	1959–1979	3705–3743
Huber, Peter J. (1934–)	1955–1974	3963–4001 etc.
Hund, Friedrich (1896–1997)	1953–1983	4004–4027 etc.
Ingólfsson, Ketill (1936–) (PhD Student v.d. Waerden 1967) (math. Reykjavík, Philadelphia)	1963–1982	4173–4233 11182–11185
Isis (Editorial Office)	1956–1974	4275–4315 8915–8918
Kappos, Demetrios A. (1904–1985)	1945–1982	4397–4421 11221–11222
Keller, Ott-Heinrich (1906–1990) and Wife	1944–1983	4475–4608 11228–11240
Kennedy, Edward S. (1912–2009)	1956–1983	4626–4748 6985–6988 11245–11246
Köthe, Gottfried (1905–1989)	1953–1966	4922–4945 etc.
Kohli, Karl (1939–) (PhD Student v.d. Waerden 1967)	1966–1970	4953–4987 6895–6896 etc.
Kubli, Fritz (1942–) (PhD Student v.d. Waerden 1970)	1969–1983	5141–5175 etc.

(continued)

Correspondent	Period	Letters Nos.
Kuhn, Thomas S. (1922–1996)	1959–1981	5187–5242 11323–11326 etc.
Leung, Kam–Tim (1932–) (PhD Student v.d. Waerden 1957)	1954–1973	5492–5519
Levi, Friedrich Wilhelm (1888–1966)	1950–1963	5522–5541 11374–11376 etc.
Locher–Ernst, Louis (1906–1962)	1951–1962	5560–5636
Lorenzen, Paul (1915–1994)	1954–1982	5656–5673 11393–11400
Mercier, Raymond P. (1934–) (University of Southampton/Cambridge)	1962–1982	5971–5997 etc.
Meretz, Wolfgang (1936–)	1962–1983	5998–6046 etc.
Neuenschwander, Erwin (1942–) (PhD Student v.d. Waerden 1972)	1968–1983	6296–6371 9691–9701 11492–11495 etc.
Neugebauer, Otto (1899–1990)	1945–1980	6372–6494 11496–11500
Neumann, Hermann (1875–1966) (Senior teacher, Munich)	1951–1966	6498–6524
Nevanlinna, Rolf (1895–1980)	1950–1975	6534–6559 1896–1899 10511–10513 etc.
Nievergelt, Erwin (1929–2018) (PhD Student v.d. Waerden 1957)	1953–1968	6589–6629 etc.
Noordhoff Publishers (Groningen, Leyden)	1946–1978	6656–6690 11530–11534
Northrop, Filmer S.C. (1893–1992)	1976–1983	6700–6714 11546–11555
Papke, Werner (1944–) (Historian of science)	1977–1981	6845–6868 11614–11617
Pingree, David (1933–2005)	1960–1968	6985–7013
Pritchett, W. Kendrick (1909–2007)	1958–1982	7132–7221 11660–11664
Rawlins, Dennis (1937–)	1976–1983	11689–11740
Rechenberg, Helmut (1937–2016)	1982–1983	7348–7379 etc.
Reich, Theodor (1920–2013)	1954–1964	7415–7439
Reidemeister, Kurt (1893–1971)	1944–1963	7469–7477 11756–11783
Reiter, Hans J. (1921–1992)	1950–1983	7484–7570 11784–11792 etc.

(continued)

Correspondent	Period	Letters Nos.
Richter, Hans (1912–1978) (PhD Student v.d. Waerden Leipzig 1936)	1950–1975	7610–7658 etc.
Rose, Alan (1929–1987) (University of Nottingham)	1951–1960	7764–7788 10468–10469
Schmidt, Friedrich Karl (1901–1977)	1944–1965	8033–8064 11867–11879 etc.
Schneider, Ivo (1938–)	1965–1983	8094–8134 9702–9709, etc.
Schramm, Matthias (1928–2005)	1962–1982	8167–8196 etc.
Scott, David Bernard (1915–1993)	1951–1972	7923–7953
Seidenberg, Abraham (1916–1988)	1961–1983	8296–8336 etc.
Severi, Francesco (1879–1961)	1950–1957	8378–8409 11956–11964
Speiser, David (1926–2016)	1973–1983	8495–8582
Springer Verlag	1944–1983	8624–8696 11990–11996
Strebel, Kurt (1921–2013)	1953–1982	8809–8833 12019–12020 etc.
Studer, Herbert (1940–) (PhD Student v.d. Waerden 1966)	1963–1983	8863–8885 12027–12030
Szabó, Árpád (1913–2001)	1961–1983	8924–9007 etc.
Thesleff, Berndt Holger (1924–)	1963–1979	9129–9155 etc.
Tóth, Imre (1921–2010)	1967–1980	9233–9254 12086–12088
Trost, Ernst (1911–1982)	1955–1974	9283–9304 1949–1950 etc.
Truesdell, Clifford A. (1919–2000)	1957–1983	9307–9371 6818–6823 12091–12123 etc.
van der Waerden (Family etc.)	1944–1977	9464–9567 etc.
Weidner, Ernst F. (1891–1976)	1944–1971	9868–9934 12244–12263
Weizsäcker, Carl Friedrich v. (1912–2007)	1944–1983	9959–9984 12269–12272
Wette, Eduard (1925–2009) (German mathematician)	1956–1978	10001–10037
Whiteside, Derek Thomas (1932–2008)	1979–1983	12275–12279 10050–10074
Wiley, John & Sons, Publishers	1951–1966	10100–10131
Wyler, Armand (1939–) (Swiss mathematician)	1968–1982	10279–10302 etc.
Ziegler, Konrat (1884–1974)	1950–1979	10323–10365 etc.

References

Blåsjö, V. 2016. In defence of geometrical algebra. *Archive for History of Exact Sciences* 70: 325–359.

Burkert, W. 1972. *Lore and Science in Ancient Pythagoreanism*. Cambridge: Harvard University Press.

Casimir, H.B.G. 1996. Gedenkworte für Bartel Leendert van der Waerden. *Orden Pour le mérite für Wissenschaften und Künste. Reden und Gedenkworte* 26: 23–29.

De Boer, J.H. 1994. Van der Waerden's foundations of algebraic geometry. *Nieuw Archief voor Wiskunde* (4) 12: 159–168.

Dold-Samplonius, Y. 1994. Bartel Leendert van der Waerden befragt von Yvonne Dold–Samplonius. *NTM* (2) 2: 129–147 (Italian translation: *Lettera Matematica P.RI.ST.EM* 15 (1995). English translation: *Notices of the AMS* 44 (1997), 313–320).

———. 1997. In Memoriam: Bartel Leendert van der Waerden (1903–1996). *Historia Mathematica* 24: 125–130.

Eisenreich, G. 1981. B.L. van der Waerdens Wirken von 1931 bis 1945 in Leipzig. In *100 Jahre Mathematisches Seminar der Karl–Marx–Universität Leipzig*, ed. H. Beckert, and H. Schumann, 218–244. Berlin: VEB Deutscher Verlag der Wissenschaften (with a list of publications up to 1946).

Frei, G. 1993. Bartel Leendert van der Waerden. Zum 90. Geburtstag. *Historia Mathematica* 20: 5–11. Revised English translation: *Nieuw Archief voor Wiskunde* (4) 12 (1994), 137–144 (q.v. further articles that were presented at the "Van der Waerden Day" on May 28, 1993 in Groningen).

Gross, H. 1973. Herr Professor B.L. van der Waerden feierte seinen siebzigsten Geburtstag. *Elemente der Mathematik* 28: 25–32 (with publication list up to 1971).

Hlawka, E. 1995/1996. Bartel Leendert Van der Waerden. *Österreichische Akademie der Wissenschaften. Almanach* 146: 399–405.

Hogendijk, J.P. 1994a. Bibliography of publications by B.L. van der Waerden since 1983 on ancient and medieval science. *Historia Mathematica* 21: 71–72.

———. 1994b. B.L. van der Waerden's detective work in ancient and medieval mathematical astronomy. *Nieuw Archief voor Wiskunde* (4) 12: 145–158.

Jakob, B. 1985. Die Briefsammlung B.L. van der Waerden. Diplomarbeit der Vereinigung Schweizerischer Bibliothekare. Wissenschaftshistorische Sammlungen der ETH–Bibliothek, Zürich. Online: http://e–collection.library.ethz.ch/view/eth:22078

Jones, A., C. Proust, and J.M. Steele, eds. 2016. *A Mathematician's Journeys. Otto Neugebauer and Modern Transformations of Ancient Science*. Cham: Springer.

Kennedy, E.S. 1956. A survey of Islamic astronomical tables. *Transactions of the American Philosophical Society* 46: 123–177.

———. 1958. The Sasanian astronomical handbook Zīj–i Shāh and the astrological doctrine of "Transit" (Mamarr). *Journal of the American Oriental Society* 78: 246–262.

Kennedy, E.S., and D.E. Pingree. 1971. *The Astrological History of Masha'allah*. Cambridge: Harvard University Press.

Koreuber, M. 2015. *Emmy Noether und die Noether–Schule und die moderne Algebra*. Berlin: Springer Spektrum.

Litten, F. 1994. Die Carathéodory–Nachfolge in München 1938–1944. *Centaurus* 37: 154–172.

Mac Lane, S. 1997. Van der Waerden's "Modern Algebra". *Notices of the AMS* 44: 321–322.

Neuenschwander, E. 2002. van der Waerden, Bartel Leendert. In *Writing the History of Mathematics: Its Historical Development*, ed. J.W. Dauben and C.J. Scriba, 547–551. Basel: Birkhäuser.

———. 2010. 100 Jahre Schweizerische Mathematische Gesellschaft. In *math.ch/100. Schweizerische Mathematische Gesellschaft, Société Mathématique Suisse, Swiss Mathematical Society*, 1910–2010, ed. B. Colbois, and V. Schroeder, 23–105. European Mathematical Society.

———. 2016. 100 Years: The centenary history of the Swiss Mathematical Society. *Archives Internationales d'Histoire des Sciences* 66 (176): 119–150 (Slightly revised, shortened English translation of [Neuenschwander 2010]).

Pingree, D. 1976a. Review of: B.L. van der Waerden, Science awakening II. The birth of astronomy. *Historia Mathematica* 3: 90–91.

———. 1976b. Review of: B.L. van der Waerden, Das heliozentrische System in der griechischen, persischen und indischen Astronomie. *Centaurus* 20: 258–260.

Renschuch, B., and W. Vogel. 1982. Über den Bezoutschen Satz seit den Untersuchungen von B.L. van der Waerden. *Beiträge zur Algebra und Geometrie* 13: 95–109.

Saltzer, W. 1970. Zum Problem der inneren Planeten in der vorptolemäischen Theorie. *Sudhoffs Archiv. Zeitschrift für Wissenschaftsgeschichte* 54: 141–172.

Schappacher, N. 2007. A historical sketch of B. L. van der Waerden's work in algebraic geometry: 1926–1946. In *Episodes in the History of Modern Algebra (1800–1950)*, ed. J.J. Gray and K. Hunger Parshall, 245–283. Providence: American Mathematical Society.

Schneider, M.R. 2011. *Zwischen zwei Disziplinen. B. L. van der Waerden und die Entwicklung der Quantenmechanik*. Berlin: Springer.

Scriba, C.J. 1996a. Bartel Leendert van der Waerden 1903–1996. *Mitteilungen der Mathematischen Gesellschaft in Hamburg* 15: 13–18.

———. 1996b. Bartel Leendert van der Waerden (2. Februar 1903 – 12. Januar 1996). *Berichte zur Wissenschaftsgeschichte* 19: 245–251.

Segal, S.L. 2003. *Mathematicians under the Nazis*. Princeton: Princeton University Press.

Siegmund-Schultze, R. 1998. *Mathematiker auf der Flucht vor Hitler, Quellen und Studien zur Emigration einer Wissenschaft*. Braunschweig: Vieweg (revised and enlarged English edition published at Princeton University Press in 2009).

———. Bartel Leendert van der Waerden (1903–1996) im Dritten Reich: Moderne Algebra im Dienste des Anti–Modernismus?, in: *"Fremde" Wissenschaftler im Dritten Reich. Die Debye–Affäre im Kontext*, ed. by D. Hoffmann and M. Walker. Wallstein, Göttingen, 2011, 200–229.

———. 2015. Van der Waerden in the Third Reich. *Notices of the AMS* 62: 924–929 (Review of Soifer 2015).

Soifer, A. 2004–2007. In Search of Van der Waerden, *Geombinatorics* XIV (1) (2004), 21–40; XIV (2) (2004), 72–102; XIV (3) (2005), 124–161; XVI (3) (2007): 305–342.

———. 2009. *The Mathematical Coloring Book. Mathematics of Coloring and the Colorful Life of its Creators*. New York: Springer.

———. 2015. *The Scholar and the State. In Search of Van der Waerden*. Basel: Birkhäuser.

Springer, T.A. 1997. Bartel Leendert van der Waerden (2 februari – 12 januari 1996). *Koninklijke Nederlandse Akademie van Wetenschappen. Levensberichten en herdenkingen*, 45–50.

Thiele, R. 2007. Van der Waerden und Artin — Der Weg zur modernen Algebra. In *Emil Artin (1898–1962)*, ed. K. Reich and A. Kreuzer, 137–167. Augsburg: Rauner Verlag.

———. 2009. *Van der Waerden in Leipzig*. Leipzig: Edition am Gutenbergplatz.

Top, J., and L. Walling. 1994. Bibliography of B. L. van der Waerden. *Nieuw Archief voor Wiskunde* (4) 12: 179–193.

van Dalen, D. 1999–2005. *Mystic, Geometer, and Intuitionist. The life of L.E.J. Brouwer*, 2 vols. Oxford: Clarendon Press.

van der Waerden, B.L. 1930, 1931. *Moderne Algebra I und II*. Berlin: Springer (from the 4th edition entitled "Algebra", 7th/5th edition1966/67. Translations into English: 1949/50, 1970, 1991, 2003, etc.; Portuguese 1954; Chinese 1964; Russian 1976, ^{2}1979, reprint 1999).

———. 1949. Babylonian Astronomy II (The Thirty–Six Stars). *Journal of Near Eastern Studies* 8: 6–26.

———. 1950. *Ontwakende wetenschap*. Groningen: Noordhoff (English translation: *Science Awakening I*, Noordhoff, Groningen 1954, ^{2}1961, ^{4}1975, ^{5}1988. German translation: *Erwachende Wissenschaft I. Ägyptische, babylonische und griechische Wissenschaft*, Birkhäuser, Basel/Stuttgart 1956, ^{2}1966. Translations into Portuguese 1954, Russian 1959, Hungarian 1977, Greek 2000).

———. 1951. Die Astronomie der Pythagoreer. *Verhandelingen der Koninklijke Nederlandse Akademie van Wetenschappen, Afdeling Natuurkunde*, 1. Reeks 20 (1): 80 p.
———. 1965a/1968/1980. *Erwachende Wissenschaft II: Die Anfänge der Astronomie*. Groningen/Basel: Noordhoff/Birkhäuser (English translation: *Science Awakening II*, Noordhoff, Leyden 1974. Russian translation: Nauka, Moskau 1991).
———. 1965b. Wie der Beweis der Vermutung von Baudet gefunden wurde, *Abhandlungen aus dem Mathematischen Seminar der Universität Hamburg* 28: 6–15 (Reprint: *Elemente der Mathematik* 53 (1998), 139–148).
van derWaerden, B.L., and J.J. Burckhardt. 1968. Das astronomische System der Persischen Tafeln I. *Centaurus* 13: 1–28.
van der Waerden, B.L. 1971. The foundation of algebraic geometry from Severi to André Weil. *Archive for History of Exact Sciences* 7: 171–180.
———. 1974. The earliest form of the epicycle theory. *Journal for the History of Astronomy* 5: 175–185.
———. 1975a. On the sources of my book "Moderne Algebra". *Historia Mathematica* 2: 31–40.
———. 1975b/1976. Defense of a "Shocking" point of view. *Archive for History of Exact Sciences* 15: 199–210.
———. 1976. D. Pingree's review of my book "Science Awakening II". *Historia Mathematica* 3: 209–211.
———. 1977/1978. The Great Year in Greek, Persian and Hindu astronomy. *Archive for History of Exact Sciences* 18: 359–383.
———. 1978. On the motion of the planets according to Heraclides of Pontus. *Archives Internationales d'Histoire des Sciences* 28: 167–182.
———. 1979. *Die Pythagoreer. Religiöse Bruderschaft und Schule der Wissenschaft*. Zürich: Artemis.
———. 1980a. Two treatises on Indian astronomy. *Journal for the History of Astronomy* 11: 50–58.
———. 1980b. The conjunction of 3102 B.C. *Centaurus* 24: 117–131.
———. 1980c. On pre-Babylonian mathematics I and II. *Archive for History of Exact Sciences* 23: 1–46.
———. 1983a. *Geometry and Algebra in Ancient Civilizations*. Berlin: Springer.
———. 1983b. *Zur algebraischen Geometrie, selected papers*. Berlin: Springer.
———. 1987a. The heliocentric system in Greek, Persian and Hindu astronomy, in: *From Deferent to Equant*. A Volume of Studies in the History of Science in the Ancient and Medieval Near East in Honor of E.S. Kennedy. *Annals of the New York Academy of Sciences* 500: 525–545.
———. 1987b. The astronomical system of the Persian Tables II. *Centaurus* 30: 197–211.
———. 1988. *Die Astronomie der Griechen. Eine Einführung*. Darmstadt: Wissenschaftliche Buchgesellschaft.
———. 1996. *Bartel Leendert van der Waerden, 2. Februar 1903 – 12. Januar 1996*. Texts from the funeral ceremony at the Nordheim crematorium in Zurich.
———. 1997. Meine Göttinger Lehrjahre (Guest Lecture on the occasion of the Algebra lecture course, held on 26 January 1979 in Heidelberg; with an epilogue by P. Roquette). *Mitteilungen der Deutschen Mathematiker–Vereinigung* 5 (2): 20–27.

Index

M. T. Borgato et al. (eds.), *Mathematical Correspondences and Critical Editions*,
Trends in the History of Science, https://doi.org/10.1007/978-3-319-73577-1

Kolibri 360 GmbH
Ferdinand-Jühlke-Straße 7
99095 Erfurt, Deutschland
produktsicherheit@kolibri360.de